FULLERENES

Nanochemistry,
Nanomagnetism,
Nanomedicine,
Nanophotonics

Elena Sheka

CRC Press
Taylor & Francis Group
Boca Raton London New York

CRC Press is an imprint of the
Taylor & Francis Group, an **informa** business

CRC Press
Taylor & Francis Group
6000 Broken Sound Parkway NW, Suite 300
Boca Raton, FL 33487-2742

First issued in paperback 2016

ISBN 13: 978-1-138-19905-7 (pbk)
ISBN 13: 978-1-4398-0642-5 (hbk)

Library of Congress Cataloging-in-Publication Data

Sheka, E. F. (Elena Fedorovna)
Fullerenes : nanochemistry, nanomagnetism, nanomedicine, nanophotonics / Elena Sheka.
p. cm.
Summary: "Using an approach based on the author's original concepts, this book addresses the unique features of fullerenes that make them the keystones of carboneous nanoscience. Using tables, graphs, and equations to make comparisons, the book explores the similarities and differences between fullerenes, carbon nanotubes, and graphene. These visual aids help readers to better understand the possibilities of computational nanotechnology"-- Provided by publisher.
Includes bibliographical references and index.
ISBN 978-1-4398-0642-5 (hardback)
1. Fullerenes. I. Title.

QD181.C1S44 2010
620'.5--dc22 2010038202

Visit the Taylor & Francis Web site at
http://www.taylorandfrancis.com

and the CRC Press Web site at
http://www.crcpress.com

Dedication

To Daniil and Nathaly, my beloved grandchildren
and to the memory of my husband, Vladimir Broude.

Contents

Preface

> Calculations do not substitute and prevent the experiments but are the subject of informatic experiment in the virtual world that becomes more and more close to the real world.

Fullerenes came to the scientific world at a time when scientific methodology was definitely grown up. High skill and large facilities of a great number of different techniques not only provide their own field of interest among so far discriminated science branches, but also tightly interconnect them, giving an obvious synergetic relish to the modern science. This circumstance has provided a large front of simultaneous investigation of the objects from the very beginning, which in turn has allowed for constructing a perfectly designed building of fullerene nanoscience in a comparably short period. Fullerenes are the first objects for which the synergetic nature of modern science has been revealed most brightly. At the interface between chemistry and biology, chemistry and physics, and physics and biology the most prominent achievements have been obtained, addressed to both fullerene fundamental science and their applications. This avenue is now repeating with carbon nanotubes and graphene.

The present book is, to some extent, the first attempt to exhibit the synergism of fullerene nanoscience from the computational viewpoint. Obviously, it is practically impossible at the moment to consider all aspects comprehensively in a book of limited size. Therefore, I have decided to focus on a unified computational platform that embraces two distinguishing fullerene features related to the odd electron character of its electron system and exclusive donor–acceptor abilities. This choice has affected both presenting the theory and selecting the experimental material. In my exposé I tried to follow a pattern for which theory and experiment would be most closely intertwined and would complement one another, thus demonstrating the unique opportunities offered by the suggested approach in investigating some general problems, such as nanochemistry, nanomedicine, nanophotonics, and nanomagnetism. Finally, I proceeded from the fact that the nanoscience of fullerenes is an inseparable part of the nanoscience of sp^2 nanocarbons and tried to show an intimate similarity in the behaviors of fullerene, carbon nanotubes, and graphene.

E. F. Sheka
Chernogolovka

About the author

Elena F. Sheka graduated from Kiev State University in 1957 with degrees in physics and spectroscopy. Her postgraduate studies were done at the Institute of Physics of the Academy of Sciences of the Ukranian Republik in Kiev. Her PhD thesis, "Exciton Spectra of Naphthalene Crystals," was defended in Lebedev's Physical Institute of the Academy of Sciences of the USSR in Moscow in 1962. Her doctor of science dissertation, "Exciton Spectra of Molecular Crystals," was defended in Lebedev's Physical Institute of the Academy of Sciences of the USSR in Moscow in 1971. She has been an assistant professor of solid-state physics spectroscopy since 1978 and a full professor since 1985.

Sheka's affiliations include the Institute of Physics of the Academy of Sciences of the Ukranian SSR (Kiev), 1958–1966; the Institute of Solid State Physics of the Academy of Sciences of the USSR (Chernogolovka, Moscow district), 1966–1986; the Moscow Physical-Technical Institute (professorship), 1978–1986; the Peoples' Friendship University of the Russian Federation (Moscow), from 1986 to the present.

Sheka is currently a professor, principal scientist, and scientific curator of the Laboratory of Computational Nanotechnology of the General Physics Department of the Peoples' Friendship University of the Russian Federation.

Sheka has authored more than 280 papers and 3 monographs. She is a member of the editorial board of *Molecular Crystals and Liquid Crystals* (Taylor & Francis Publishing), an associate editor of the *Journal of Nanoparticle Research* (Springer Publishing), a full member of the European Society of Computational Methods in Science and Engineering (ESCMSE), a member of the American Chemical Society, and a grantee of the Russian Foundation of Basic Researches.

Sheka's fields of interests include excitonics of molecular crystals; phonon spectra of molecular crystals (inelastic neutron scattering, calculation); exciton–phonon interaction and vibronic spectra of molecular crystals; phase transformation in molecular solids with liquid crystal behavior (vibrational spectroscopy and neutron diffraction); vibrational spectroscopy of nanoparticles; quantum chemical simulations of nano-objects toward computational nanotechnology; quantum fullerenics; and simulations of carbon nanotubes and graphene.

Acknowledgments

The book could not have been possible without assistance of my colleague and friend Valentin Zayetz (deceased), an outstanding quantum chemist, whose ideas implemented in intelligent and heuristic computational programs made it possible to obtain results that form the groundwork of the book. My many thanks to Boris Razbirin, Dmitrii Nelson, Leonid Chernozatonskii, Masakuzo Aono, Tomonobu Nakayama, Ivgenny Katz, Erkki Brändes, David Tomanek, and David Avnir, with whom fruitful discussions were highly stimulating. My gratitude to Rostislav Andrievski for encouragement and to my daughter Maria Broude for her patience, understanding, challenging, and financial support. I am grateful to the authors of all cited papers for the copresence.

The author alone is responsible for any faults that may be found in the book.

chapter one

Concepts and grounds

One important area of research in modern material nanoscience concerns carbon-based materials, among which fullerenes take one of the first places. Nanochemistry of the twenty-first century, new light nanomagnets, new optical devices, new drugs—these and other excitedly sounded promises imply fullerene-based material can be heard throughout the world. What makes fullerene so exclusive for the properties of the materials to be outstanding? We shall try to answer this question from the viewpoint of computational nanoscience of fullerene. This consideration is based on the electronic properties of the molecule. Two cornerstones lay the foundation of the electronic fullerenics building. The first concerns the odd electron nature of the fullerene atomic system, which is intramolecular by nature. The second is the extremely high donor and acceptor characteristics of the molecules, and is intimately connected with intermolecular interaction that is smoothly transformed into the peculiar intramolecular property of fullerene derivatives and composites. It turns out that the two concepts make an allowance for interconnecting the fullerenics building constructions via tight interrelations between nanochemistry and nanomagnetism, nanomedicine and nanophotonics. In turn, the intrinsic concordance of the above building departments forms the ground for a unified theoretical approach. Before entering the building, let us look at the basic building blocks that form the construction fundamentals.

1.1 Why odd electrons and not π electrons?

The term *odd electrons* has come from organic chemistry, where it was introduced to describe the electronic structure of diradicals.[1,2] It naturally covers π electrons, magnetic electrons,[3] and open-shell molecule electrons.[4] With respect to benzenoid nanocarbons, where the latter term marks a hexagon motive of carbon atoms, such as fullerenes, carbon nanotubes, graphite, and graphene, it manifests that the number of valence electrons of each carbon atom is larger by one than that of the interatomic bonds it forms.

Aromaticity is the first and oldest concept applied to the odd electron problem, which started in 1825, when Michael Faraday discovered benzene. The 185-year development of the concept followed and reflected in the best way the buildup of molecular physics and chemistry in general and organic chemistry in particular.[5–9] The term π *electrons* was suggested by Hückel in 1931 for planar {4*n* + 2} aromatic molecules.[10] Since that time π electrons have become a peculiar qualitative characteristic of a molecule possessing two distinctive features: (1) odd electron delocalization over molecule atoms and (2) a complete covalent bonding of the odd electrons so that the corresponding π orbitals are occupied by the electrons in pairs. These characteristics of the electrons are well coherent with four aromaticity criteria, often referred to as classic criteria, which have originated in the due course of the long-time development of the aromaticity concept.[7] Those are concerned with the following molecular characteristics:

1. Chemical behavior—electrophilic aromatic substitution
2. Structural feature—bond length equalization due to cyclic delocalization
3. Energetic conditions—enhanced stability (large resonance energy)
4. Magnetic behavior—ring current effects:
   - Anomalous chemical shift
   - Large magnetic anisotropies
   - Diamagnetic susceptibility exaltation

Initially, the criteria were suggested quite empirically. The development of the electronic theory of molecules allowed for obtaining an electronic conceptual view of the phenomenon (started by Hückel) and for proving the criteria computationally. By the time the fullerene was discovered in 1985,[11] the main theoretical concept of aromaticity had been constructed. According to the latter, π electrons can be considered in the first approximation as an electron gas encasing the σ framework in a double skin. The wave functions of this electron gas are characterized by the angular momentum quantum numbers $l = 0, 1, 2, 3, \ldots$, corresponding to *s*, *p*, *d*, and *f* π shells; π electrons are subordinated to the Pauli principle and are accommodated on the orbitals by pairs (see Bühl and Hirsh[12] and references therein), which has given rise to the *restricted* formalism of the electron consideration.

When fullerene C_{60} was discovered, Kroto suggested that it could be "the first example of a spherical aromatic molecule."[11] This suggestion was supported by the discussion of its extremely stable aromatic character.[13,14] However, the hypothesis was soon rejected in the light of numerous chemical reactions undergone by fullerenes.[14–17] Nowadays, it is generally assumed that fullerenes have an ambiguous aromatic character,[18–21] with some properties that support the aromaticity of these systems in view of the above criteria, while others do not. Thus, there is broad evidence that fullerenes experience substantial ring currents (see Haddon[22] and Steiner et al.[23] and references therein). This fact, together with rather considerable stability, seems to stress their aromatic character.[20] In contrast, evidence from chemical reactivity is against aromaticity, since fullerenes are very reactive and easily undergo a large variety of chemical transformations that, unlike most aromatic compounds, are in most cases addition reactions to the conjugated π system.[15–17,24] The enthalpy of formation of fullerenes does not support the energetic criterion of the aromatic character either.[25] In addition, the existence of two types of C-C bonds in C_{60} ([6,6] and [5,6] bonds)[26,27] violates criterion 2 on the bond length equalization and indicates that π electrons of the fullerene are partially localized. Therefore, only magnetic and nuclear magnetic resonance (NMR) properties seem to evidence delocalization by ring currents (see Bühl and Hirsh[12] and references therein). Based on these findings, it was concluded that the scope of aromaticity is much broader for Hückel type aromatic molecules than for fullerenes, where it is essentially limited to criterion 4. At the same time, the 185-year history of aromaticity has convincingly highlighted that aromatic compounds cannot be characterized by a single property,[28,29] so that the concept has a multidimensional character. This stimulated looking for additional criteria that might prove fullerene aromaticity. Since no new empirical yardstick for fullerene aromaticity could be offered, researchers have concentrated on computational chemistry, relying on its sharp progress. A restricted formalism of the electrons approach has been kept unchanged.

Two main streams of modern computational chemistry are based on either the Hartree-Fock (HF) approach or density functional theory (DFT). Both computational schemes were applied to prove classic aromaticity criteria for fullerenes (see reviews[12,30,31]). A few computational characteristics that are outside the above classic criteria have been suggested:

1. The topology of DFT electron density, as well as properties derived from the density, such as electron localization function and local ionization potential[30]
2. Weighted highest occupied (HOMO) and lowest unoccupied (LUMO) molecular orbitals' gap as an index of kinetic stability (T),[32] as well as absolute and relative hardness (η)[30]
3. Delocalization index ($\delta(A,B)$)[33]

In addition to ring current criterion 4, the issues tend to assist in the description of the delocalization-aromatic behavior of fullerenes in the framework of restricted computational schemes. However, they are too "yang" to get a status similar to that of classic criteria.

Since odd electrons exist not only in benzenoid-structured compounds, and since their manifestation in various compounds has become clear for many researchers, the aromaticity approach has been expanded over these species, both two- and three-dimensional, among which there are conjugated heterocycles, [n]trannulenes, Möbius and metallacycles, transition metal (half) sandwiches such as ferrocenes, carbon-free sandwich complexes like $[(P_5)_2Ti\}^{2-}$, boron-based clusters, bimetallic and metallic clusters, and many others (see Schleyer[9] as well as a brief review in Chen and King[31] and references therein). However, as it turns out, the approach was able to catch only some traces of the odd electrons delocalization, while the electronic behavior of the species was not fit in the framework based on the classic aromaticity criteria. An unavoidable requirement to either significantly develop the aromaticity concept or suggest an alternative one has been widely anticipated.

An alternative approach to the odd electron problem has come from molecular magnetism of molecular complexes, in which transition metal ions form a net of spins caused by *magnetic electrons* that interact weakly.[3,34] The scheme in Figure 1.1 explains main points of the odd electron problem. Initially, doubly degenerated atomic levels Ψ_A and Ψ_B are split due to electron interaction with the energy difference ΔE. Two spins of the relevant electrons can be distributed over the levels in five different ways. Configurations I to IV are related to a singlet state, while only configuration V describes the triplet one. As a result, the triplet state is spin-pure at any ΔE, while the singlet state is either completely covalent, that is, in consonance with the aromaticity concept (configuration I) and, consequently, spin-pure at large ΔE, or is a mixture of configurations I to IV when the interaction is weak. The energy difference, ΔE, turns out to be the main criterion for attributing the species to either covalently bound or diradical species. A deep understanding of the underlying physics, which followed from the work of Löwdin, Nesbet, Anderson, Hay and coworkers, and others (see Noodleman[3] and references therein), has resulted in the necessity of the configuration interaction (CI) to be applied in the case. Moreover, the *unrestricted* formalism with respect to wave function construction, which oppositely to restricted aromaticity, takes the electron's spin into account, should be taken as a starting point for any further CI consideration.

Thus, if electron pairs are weakly coupled, the CI expansion will be slowly convergent and high-order excitation from the reference configuration (see ψ_2, ψ_3, and ψ_4 in Figure 1.1) will contribute significantly to the wave function.[35] Benard found that the unrestricted Hartree-Fock (UHF) energy

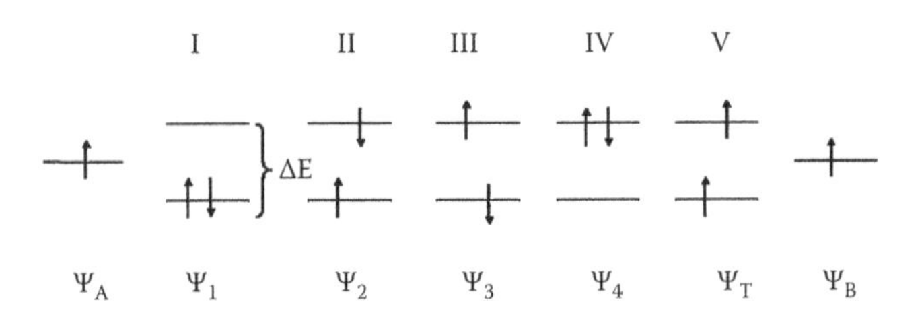

Figure 1.1 Scheme of the distribution of two electrons over split orbitals.

is much lower than the restricted one (RHF). Therefore, coming back to species, the aromatic character of which is under question, one can suggest a decisive computational criterion by misalignment of their energies,

$$\Delta E^{RU} = E^{R} - E^{U} \tag{1.1}$$

that are calculated following conceptually identical restricted (E^R) and unrestricted (E^U) computational schemes. Obviously, the aromaticity concept in its classical formulation should be applied to the species if only $\Delta E^{RU} = 0$.

The first attempt of a direct CI approach based on unrestricted wave functions and applied to the odd electron interaction was performed for metal dimers.[36,37] However, until now the application of the CI computational scheme was restrained by a severe complication of the computational procedure. Thus, the most popular complete active space self-consistent field (CASSCF) methods, which deal correctly with diradicals and some dinuclear magnetic complexes (see comprehensive review in Davidson and Clark[38]), cannot handle systems with a large number of odd electrons due to a huge number of configurations generated in the active space of the system. To imagine arising technical difficulties, it is worthwhile to cite Davidson and Clark, p. 1886:[38] "If there are *'m'* singly occupied orbitals on each of *'n'* identical centers, then 2^{mn} Slater determinants can be formed by assigning spins up or down each of the *'nm'* orbitals." Thus, no CASSCF type approach, including a simplified embedded-cluster CASSCF supplemented by the complete active space second-order perturbation theory (CASPT2),[39] seems feasible for many odd electron systems such as fullerenes. That is why addressing single-determinant approaches turns out to be the only alternative.

The open-shell *unrestricted broken symmetry* (UBS) approach suggested by Noodleman[3] is the most widely known among the latter. Since the early remark of Slater[40] and the symmetry dilemma later formulated by Löwdin,[41] the term *broken symmetry* has implied that the energetically best HF solution is not necessarily symmetry adapted. Noodleman showed that when the magnetic orbitals (singly occupied orbitals) of two dimer subunits were allowed to interact by overlapping in a self-consistent field (SCF) procedure, a state of mixed spin symmetry and lowered space symmetry was obtained. This spin-mixed or broken symmetry state is considered within the framework of the UBS approach.

A general view of the broken symmetry wave function, $|\psi_B\rangle$, for n pairs of magnetic electrons is the following:[3]

$$\begin{aligned} |\psi_B\rangle &= (N!)^{-1} M^{-1/2} \det\left[(a_1 + cb_1)\alpha, a_2\alpha \ldots a_n\alpha \middle| (b_1 + ca_1)\beta, b_2\beta \ldots b_n\beta\right] \\ &= M^{-1/2}\left(\phi_1 + c\phi_2 + c\phi_3\right) \end{aligned} \tag{1.2}$$

The function describes the singlet state; doubly occupied canonical molecular orbitals that describe paired electrons are omitted. Here a_i represents singly occupied magnetic orbitals centered on atom A, all magnetic electrons of which have spin α, while b_i does the same for atom B, with spin β attributed to its magnetic electrons. Mixing parameter c implies a slight nonorthogonality of orbitals: $\bar{a}_1 = (a_1 + cb_1)$ and $\bar{b}_1 = (b_1 + ca_1)$. The principal determinant, ϕ_1, describes pure covalent coupling of n odd electrons, while small amounts of the charge transfer determinants, ϕ_2 and ϕ_3, corresponding to $A^- - B^+$ and $A^+ - B^-$ configurations (see II and III in Figure 1.1), are mixed in due to nonorthogonality of atomic orbitals $\bar{a}_1$ and $\bar{b}_1$. Configuration IV is usually omitted since the relevant state energy is high. The open-shell manner for the function $|\psi_B\rangle$ is just appropriate to distinguish electron spins of atoms A and B. The weaker the interaction between the electrons, the bigger is the contribution of ϕ_2 and ϕ_3 constituents to $|\psi_B\rangle$ and, consequently, the bigger the difference in the energy of restricted (ϕ_1) and unrestricted ($|\psi_B\rangle$) configurations. The aromaticity concept corresponds to the first term in Equation 1.1 and should be attributed to strong interaction between odd electrons.

The UBS approach is not reduced to routine UHF or unrestricted DFT (UDFT) calculations, but consists in determining the energy of pure spin states on the basis of calculations using $|\psi_B\rangle$ within either UHF or UDFT computational schemes. As shown,[3] this determination is possible at the level of the second-order perturbation theory with respect to parameter c supplemented by variationally determined $S_{\bar{a}\bar{b}}$, that is, meeting the requirements

$$\frac{\partial E_B}{\partial S_{\bar{a}\bar{b}}} = 0 \text{ and } \frac{\partial E(S)}{\partial S_{\bar{a}\bar{b}}} = 0 \tag{1.3}$$

for the spin-mixed and pure spin state, respectively. Here $S_{\bar{a}\bar{b}}$ presents the magnetic orbitals' overlap. This level of theory is equivalent to that for explicit CI calculations.

Anyone can easily convince oneself that a complete coincidence of the energies E^R and E^U takes place for the benzene molecule (the energies slightly deviate for naphthalene), but the difference, ΔE^{RU}, grows when the number of benzenoid cycles increases. As for fullerenes, first attempts to compare the application of RHF and UHF computational schemes to fullerene C_{60} were undertaken by the author seven years ago.[42–44] It turned out that ΔE^{RU} constituted about 2% from the total energy of the molecule and is about two tens kcal/mol by absolute value. Later on it was found that analogous things happened with fullerene C_{70},[45–47] carbon nanotubes,[48,49] and graphene.[50] This finding highlights a considerable weakening of odd electron interaction in the species in comparison with benzene. It became obvious that difficulties with fitting fullerenes in the aromaticity concept were connected with this finding. On the other hand, the application of the UBS approach to fullerenes has opened new possibilities to quantitatively describe new features of the molecules, making fullerenics quite understandable and predictable. The concept of the UBS approach, and its implications within the framework of the Hartree-Fock and DFT computational scheme, is a guideline for the sp^2 carboneous species presentation in this book. Thus, the approach basic ground is presented in Chapter 2. Chapter 3 describes the basic electronic characteristics of the fullerene molecules in view of the UBS HF approach. The chemical activity of fullerenes and computational synthesis of fullerene derivatives of different kinds, subordinated to well-defined algorithms, are discussed in Chapters 4 to 8. Logically connected with the preceding chapters, Chapters 11 and 12 exhibit a new view on chemical reactivity of carbon nanotubes and graphene. All these topics are considered on the same platform on which UBS HF calculations are based, which allows for tracing a common origin laying the foundation of the properties discussed.

1.2 Donor–acceptor ability as a leitmotiv of intermolecular interaction of fullerenes

For more than a half of century, peculiarities of intermolecular interaction (IMI) in molecular donor–acceptor (D–A) complexes have mainly been associated with a long wavelength absorption band, known as a *charge transfer band*.[51–53] A widely used expression for the band position has the form

$$h\nu_{CT} = I_D - \varepsilon_A - \left(E_{ext} - E_N\right) \tag{1.4}$$

where I_D and ε_A are the molecular donor ionization potential (IP) and the molecular acceptor electron affinity (EA), respectively, while E_{ext} and E_N are the energies of complexes $[A^+B^-]$ and $[A^0B^0]$. The band recording has always been considered a qualitative indication of the D–A character of a binary molecular complex. But actually, the band is only a consequence of the IMI complex by nature.[54] As has been shown by Evans and Polanyi,[55] a qualitative description of the interaction can be suggested when comparing IMI potentials, the IMI terms below, of complexes $[A^+B^-]$ and $[A^0B^0]$. Since the IMI term of the complex $[A^0B^0]$ comes to zero at infinity, the difference in the terms' asymptotes is equal to $I_D - \varepsilon_A$. According to Equation 1.4, the difference is always positive. On the other hand, the $[A^+B^-]$ term is below that of the $[A^0B^0]$ complex within the main range of small intermolecular distances, so that there is always a space region where these terms intersect. However, the breakdown of adiabaticity between electrons and nuclei in the region causes replacing the terms' intersection by their splitting,[54,56] so that a region of an *avoided crossing* arises where a number of diabatic processes, such as photosynthesis, photodissociation, spin exchange reactions, and others, among which charges transfer, occur. This viewpoint has had much development in the study of charge transfers at colliding alkali metal atoms with halides and other molecules with high EA.[57,58] The suggested model of *chemoionization* is of a general nature and can be fruitfully applied to the consideration of features of the D–A interaction. The idea to apply the approach to molecular complexes involving fullerene came to the author in 2004.[59] Time has shown its high efficacy, which allowed for considering at the same platform different phenomena related to nanochemistry, such as features of chemical reactions in molecular complexes—dyads involving fullerene (Chapter 7), dimerization or oligomerization of fullerene (Chapter 8), photodynamic nanomedicine (Chapter 9) and nanophotonics (Chapter 10) of fullerene solutions, and the formation of composites based on fullerene–carbon nanotube and carbon nanotube–graphene dyads (Chapters 8 and 12). Let us consider the basic grounds of the approach in detail.

1.2.1 *Energy terms of ground and excited states of a binary complex with intense D–A interaction*

Intermolecular interaction is difficult for theoretical description. Repeatedly the question has risen to divide the total interaction into constituents, each of clear physical meaning, for the further determination of those to be possible. Thus, the concepts of electrostatic (E_{es}), inductive (E_{pl}), exchange (E_{ex}), dispersive (E_{disp}), and charge transfer (E_{ct}) interactions have appeared.[54] Concurrent with individual consideration of the above terms, there have been attempts to tackle them jointly within the framework of a unique computational scheme. That suggested by Morokuma et al.[60,61] should be accepted as the most successful. The total IMI is presented as a sum:

$$E_{\text{int}} = E_{es} + E_{pl} + E_{ex} + E_{ct} + E_{mix} \tag{1.5}$$

each term of which is determined within one session of calculations. The term E_{mix} completes contributions by interactions unable to be determined by SCF calculations, by the dispersion interaction in particular. Morokuma et al.'s analysis has usually been performed for a fixed geometry of the complex described by sets of internal $\{r\}$ and external $\{R\}$ coordinates. But actually, $E_{\text{int}}(r,R)$ is a complex function of the coordinates.

Analysis of interaction of alkali metal atoms with various molecules showed[55,57] that both complex stabilization and complex structure depend

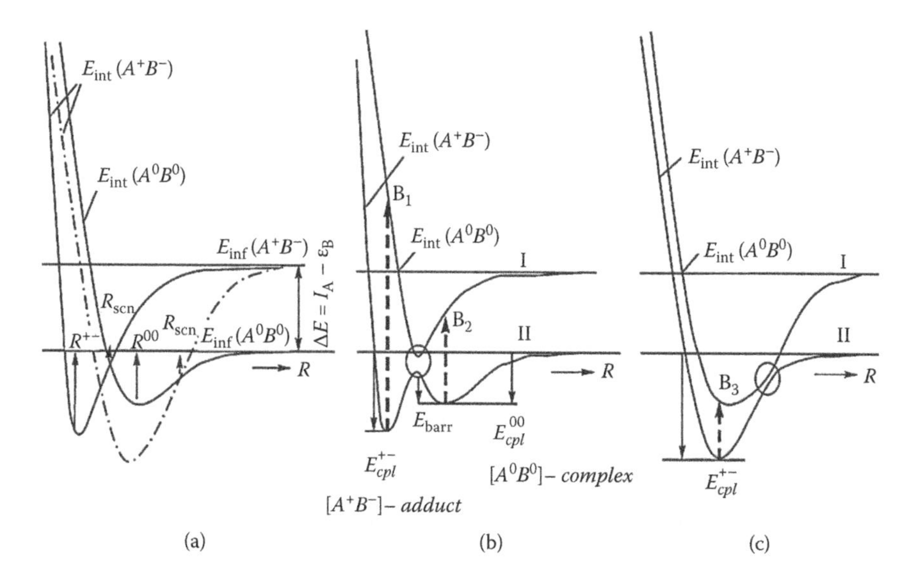

Figure 1.2 Scheme of the formation of two branches of intermolecular interaction terms, $|E^{+-}_{cpl}| > |E^{00}_{cpl}|$. (a) IMI terms for neutral molecules and ions; R_{scn} matches two points of the term intersection. (b and c) IMI terms of D–A dyads of types 1 and 2, respectively; B_1, B_2, and B_3 mark phototransitions from the corresponding minima of the ground-state terms.

on a composite IMI term of the ground state consisting of $E_{int}(A^0B^0)$ and $E_{int}(A^+B^-)$ terms that describe the ions' and neutral molecules' interaction, respectively. This results in the ion coupling at R^{+-} and neutral molecules at R^{00}. If $I_A - \varepsilon_B > 0$, the terms are split to avoid their crossing, thus forming two branches of composite terms. Figure 1.2 presents a sketch of a one-dimensional cross-section of the potential energy surface that demonstrates the formation of the above discussed two branches of the IMI term of complex $A + B$ for two different terms $E_{int}(A^+B^-)$ plotted in Figure 1.2(a).[59] Horizontal lines $E_{inf}(A^+B^-)$ (I) and $E_{inf}(A^0B^0)$ (II) mark asymptotic summary energies of molecular ions and neutral molecules at infinity, respectively. The latter is usually taken as the reference level. The lower branch is attributed to the ground state while the upper branch describes the excited one. Ovals mark the avoided crossing regions. Coupling energies E^{+-}_{cpl} and E^{00}_{cpl} as well as the chemoionization barrier energy, E_{barr}, are the main energetic parameters of the ground state.

As shown in Figure 1.2, the summary term of the ground state may be either a two- or one-well type, depending on the position of the intersection point R_{scn} on either the left- or right-hand side from the point R^{00}. In the first case (Figure 1.2(b)), classified as type 1, the complex is characterized by two states that correspond to coupling at the R^{+-} and R^{00} minima. As a result, two products are formed in due course by the IMI inside the

$A + B$ pair, namely, a chemically bound $[A^+B^-]$ adduct at the R^{+-} minimum and a weakly bound $[A^0B^0]$ complex at the R^{00} one. The former can be conditionally called ionic, implying that the structure and electronic properties of the relevant complex are mainly determined by the interaction of molecular ions. The latter state is obviously neutral since the interaction of neutral molecules is mainly responsible for the complex properties. In the second case (Figure 1.2(c)), classified as type 2, only the $[A^+B^-]$ adduct is formed. In both cases, $|E_{cpl}^{+-}| > |E_{cpl}^{00}|$ and the R^{+-} minimum plays the main role.

The IMI terms shown in Figure 1.3 correspond to the condition when $E_{cpl}^{+-} > 0$; $E_{cpl}^{00} < 0$, which means that in both cases presented in the figure, the R^{00} minimum is the deepest by energy. As seen in the figure, the ground-state term can also be either two- or one-well, depending on the availability of the term $E_{int}(A^+B^-)$ minimum and its depth. When the minimum exists (Figure 1.3(a)), generally both the $[A^+B^-]$ adduct and $[A^0B^0]$ complex may be produced (IMI potential of type 3). Therewith, if the latter is obviously thermodynamically preferable and easily formed, the $[A^+B^-]$ adduct can be obtained only kinetically, say, via B_2 band photoexcitation of the $[A^0B^0]$ complex followed by the produced ions' relaxation over the $E_{int}(A^+B^-)$ term curve toward the R^{+-} minimum. If the barrier between the two states is high enough to prevent reversing of the $[A^+B^-]$ adduct back to the $[A^0B^0]$ complex, the former can be obtained in addition to the $[A^0B^0]$ complex. This kind of photostimulated chemical reaction involving fullerene C_{60} will be considered in Chapter 7.

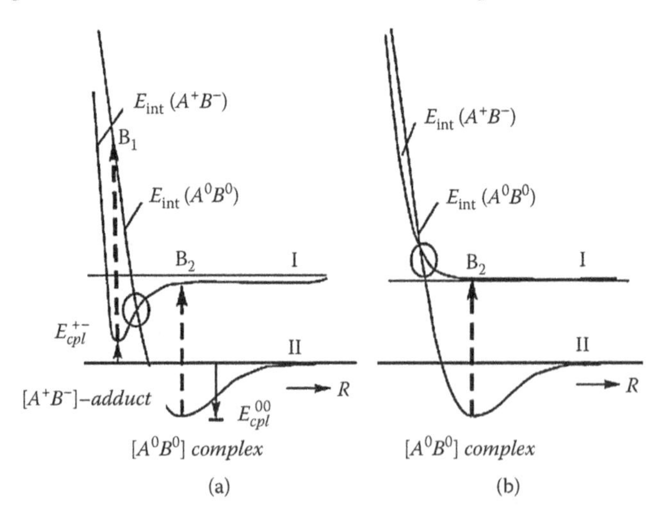

Figure 1.3 Scheme of the formation of two branches of intermolecular interaction terms, $E_{cpl}^{+-} > 0$; $E_{cpl}^{00} < 0$. (a and b) IMI terms of D–A dyads of types 3 and 4, respectively.

A limit case of the one-well IMI term (Figure 1.3(b)) is related to the term $E_{int}(A^+B^-)$ not having a minimum and showing the lack of a coupled ionic state (IMI potential of type 4). The only $[A^0B^0]$ complex formation can be expected for the relevant $A + B$ pair.

The four types of IMI potential considered above cover practically all classes of possible chemical reactions that might be expected for fullerene. Important to note is that in three of the four cases, one might expect a barrierless formation of $[A^0B^0]$ complexes that are charge transfer complexes in the energy region determined by the position of the B_2 band. The excluded case related to the IMI potential of type 2 concerns a rather peculiar and quite rare situation when the formation of a $[A^+B^-]$ adduct occurs at large $R^{+-} \sim R^{00}$, which greatly exceeds standard chemical bond lengths. This is typical for, say, interaction of fullerene with atoms of alkali metals.[59] As for $[A^+B^-]$ adducts, their formation is energetically profitable only for IMI potentials of type 1 and 2. However, whiles in the latter case the formation occurs without barrier, in the former case the barrier should be overcome, which makes this stage of reaction kinetically dependent. This circumstance will be discussed in greater detail in Chapter 8.

A particular discussion of $[A^0B^0]$ complexes is presented in Chapters 9 and 10 in regards to photodynamic therapy and nanophotonics of C_{60}-based fullerene dissolved in various molecular solutions. The charge transfer nature of the clusters causes them both to act as spin-flippers and nanosize amplifiers of light, thus providing spin-flip in the oxygen molecule, transforming its multiplicity to a singlet in the first case, and enhancing both the spectral and nonlinear optical properties of the solutions in the second.

1.2.2 *Ionic components of a D–A binary system*

If a $[A^0B^0]$ complex structure is generally clear, then one of the $[A^+B^-]$ adducts will depend on the state of ions that are produced at the intersection point.[57] This is caused by two circumstances. First, the diabatic transition molecule-to-ion is vertical, which means that the space configuration of the formed ions corresponds to the equilibrium configuration of neutral molecules at the intersection point. Available types of such transitions are schematically shown in Figure 1.4: the left part of the figure is related to the positive ion formation, and the right part deals with negative ions. As seen in the figure, the state of a formed ion is greatly dependent on the shift of equilibrium positions of atoms under ionization. If the shift is small or absent, the formed ions are stable (Figure 1.4(a)). If the shift is large, vibrationally excited ions are formed under ionization (Figure 1.4(b)), and the energy of the vibrational excitation may approach or even exceed the dissociation limit (Figure 1.4(c)). Therefore, the chemoionization may

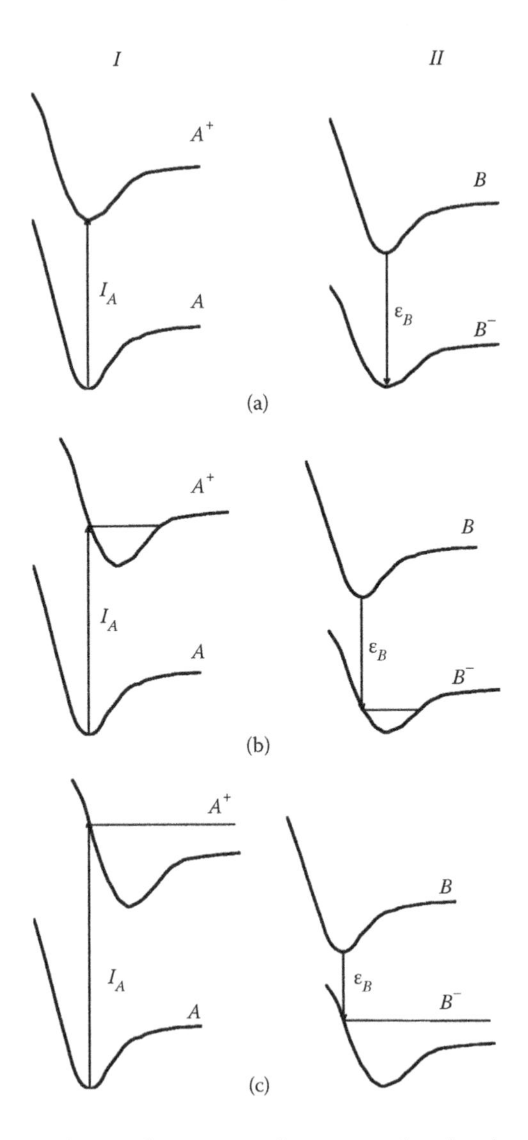

Figure 1.4 Scheme of vertical transitions from neutral molecule to ion.

be accompanied by the dissociation of either A or B, which should be taken into account when designing models for simulating the $[A^+B^-]$ adduct.

Second, the produced ions are subjected to a Coulomb interaction at the intersection point, the energy of which is $\approx I_A - \varepsilon_B$ by an order of magnitude.[57] As a result, the ions or products of their dissociation can form molecular compounds of the $M = A^{\delta+}B^{\delta-}$ and $M = (A_1 + A_2)^{\delta+}B^{\delta-}$ types, where A_1 and A_2 are dissociation products of, say, a donor in accordance with scheme Ic in Figure 1.4. The latter is an example of particular harpoon reactions observed within beams of neutral atoms and molecules.[55,57,58] Similar reactions with participation of the C_{60} fullerene will be discussed in Chapter 7.

1.3 Odd electrons approach as a basic concept of nanoscience

A peculiar aspect of nanomaterial science concerns a common essence joining topics that seem absolutely different at first glance. Thus, if radicals are well-accepted characteristics for molecular chemistry, dangling bonds and magnetic electrons are typical terms for surface science and magnetism. However, all the features are of the same origin and are connected with *odd electrons* of atoms that compose either molecules or surfaces and magnetic solids. The term stands for the difference between the number of atom valence electrons and the number of chemical bonds formed by the atom.[62]

The very fact of the odd electron availability is absolutely necessary for an atomic system to be peculiar, but it is not enough. The peculiarity implementation is directly dependent on the electrons' behavior, mainly on their coupling. Thus, in the case of carboneous substances, such as ethylene or benzene and to some degree other aromatic molecules, odd electrons are covalently bonded, while silicoethylene as well as siliceous aromatic molecules do not exist at all. The bare carbon surfaces are nonconductive and nonmagnetic, while silicon surfaces are conductive and magnetic. Differences in magnetic properties of bulk traditional magnetic solids and their surfaces or nanosize clusters, as well as different magnetic behaviors of solids composed of molecular nanocomplexes of the same structure but differing by transition metal atoms (say, Ni and Co), etc., follow from the difference in coupling the available odd electrons.

Historically, theoretical approaches to the phenomena have been developed in different ways. Thus, the bonding quantum theory forms the grounds for molecular chemistry. At the same time, a widely accepted defect-state approach to surfaces on the basis of bulk solid-state physics is mainly used in surface science. Particular theoretical approaches concentrated on the exchange and superexchange interaction of electrons are characteristic for the solid-state magnetism consideration. However, if the same origin of events is taken into account, one can suggest a unified theoretical or computational approach to all the phenomena, making possible their consideration on the same conceptual basis as well as on the same computational footing.[62]

References

1. Hoffman, R. 1971. Interaction of orbitals through space and through bonds. *Acc. Chem. Res.* 4:1–9.
2. Salem, L., and Rowland, C. 1972. The electronic properties of diradicals. *Ang. Chem. Intern. Ed.* 11:92–111.
3. Noodleman, L. 1981. Valence bond description of antiferromagnetic coupling in transition metal dimers. *J. Chem. Phys.* 74:5737–42.
4. Illas, F., Moreira, I. de P. R., de Graaf, C., and Barone, V. 2000. Magnetic coupling in biradicals, binuclear complexes and wide-gap insulators: A survey of *ab initio* wave function and density functional theory approaches. *Theor. Chem. Acc.* 104:265–72.
5. Garat, P. J. 1986. *Aromaticity*. New York: Wiley.
6. Minkin, V. I., Glukhovtsev, M. N., and Simkin, B. Y. 1994. *Aromaticity and antiaromaticity. Electronic and structural aspects.* New York: John Wiley & Sons.
7. Schleyer, P. v. R., and Jiao, H. 1996. What is aromaticity? *Pure Appl. Chem.* 68:209–18.
8. Schleyer, P. v. R., ed. 2001. Aromaticity. *Chem. Rev.* 101(5).
9. Schleyer, P. v. R., ed. 2005. Delocalization—Pi and sigma. *Chem. Rev.* 105(10).
10. Hückel, E. 1931. Quantum-theoretical contributions to the benzene problem. I. The electron configuration of benzene and related compounds. *Z. Physik* 70:204 and 72:310–5.
11. Kroto, H. W., Heath, J. R., O'Brien, S. C., Curl R. F., and Smalley, R. E. 1985. C_{60}: Buckminsterfullerene. *Nature* 318:162–63.
12. Bühl, M., and Hirsh, A. 2001. Spherical aromaticity of fullerenes. *Chem. Rev.* 101:1153–83.
13. Elser, V., and Haddon, R. C. 1987. Icosahedral C_{60}: An aromatic molecule with a vanishingly small ring current magnetic susceptibility. *Nature* 325:792–94.
14. Taylor, R., and Walton, D. R. M. 1987. The chemistry of fullerenes. *Nature* 325:685–92.
15. Hirsh, A. 1993. The chemistry of the fullerenes: An overview. *Ang. Chem. Intern. Ed.* 32:1138–41.
16. Hirsh, A. 1994. *The chemistry of fullerenes.* Stuttgart: Thieme.
17. Hirsh, A. 1999. Principles of fullerene reactivity. *Top. Curr. Chem.* 199:1–65.
18. Haddon, R. C., Schneemeyer, L. F., Waszczak, J. V., Glarum, S. H., Tycko, R., Dabbagh, G., Kortan, A. R., Muller, A. J., Mujsce, A. M., Rosseinsky, M. J., Zahurak, S. M., Makhija, A. V., Thiel, F. A., Raghavachari, K., Cockayne, E., and Elser, V. 1991. Experimental and theoretical determination of the magnetic susceptibility of C_{60} and C_{70}. *Nature* 350:46–48.
19. Haddon, R. C. 1993. Chemistry of the fullerenes: The manifestation of strain in a class of continuous aromatic molecules. *Science* 261:1545–50.
20. Fowler, P. 1991. Aromaticity revisited. *Nature* 350:20–21.
21. Krygowski, T. M., and Ciesielski, A. 1995. Local aromatic character of C_{60} and C_{70} and their derivatives. *J. Chem. Inf. Comp. Sci.* 35:1001–3.
22. Haddon, R. C. 1995. Magnetism of the carbon allotropes. *Nature* 378:249–54.
23. Steiner, E., Fowler, P. W., and Jenneskens, L. W. 1972. Counter-rotating ring currents in coronene and corannulene. *Ang. Chem. Intern. Ed.* 11:92–111.
24. Wudl, F. 1992. The chemical properties of buckminsterfullerene (C_{60}) and the birth and infancy of fulleroids. *Acc. Chem. Res.* 25:157–61.

25. Cioslowski, J. 1993. Heats of formation of higher fullerenes from *ab initio* Hartree-Fock and correlation energy functional calculations. *Chem. Phys. Lett.* 216:389–93.
26. Liu, S., Lu, Y.-J., Kappes, M. M., and Ibers, J. A. 1991. The structure of the C_{60} molecule: X-ray crystal structure determination of a twin at 110 K. *Science* 234:408–10.
27. Hedberg, K., Hedberg, L., Bethune, D. S., Brown, C. A., Dorn, H. C., Johnson, R. D., and De Vries, M. 1991. Bond lengths in free molecules of buckminsterfullerene, C_{60}, from gas-phase electron diffraction. *Science* 234:410–12.
28. Krygowski, T. M., and Cyrański, M. K. 2001. Structural aspects of aromaticity. *Chem. Rev.* 101:1385–420.
29. Cyrański, M. K., Krygowski, T. M., Katritzky, A. R., and Schleyer, P. V. R. 2002. To what extent can aromaticity be defined uniquely? *J. Org. Chem.* 67:1333–38.
30. de Proft, F., and Geerlings, P. 2001. Conceptual and computational DFT in the study of aromaticity. *Chem. Rev.* 101:1451–64.
31. Chen, Z., and King, R. B. 2005. Spherical aromaticity: Recent work on fullerenes, polyhedral boranes, and related structures. *Chem. Rev.* 105:3613–42.
32. Yoshida, M., and Aihara, J.-I. 1999. Validity of the weighted HOMO-LUMO energy separation as an index of kinetic stability for fullerenes with up to 120 carbon atoms. *Phys. Chem. Chem. Phys.* 1:227–30.
33. Poater, J., Duran, M., and Sola, M. 2004. Analysis of electronic delocalization in buckminsterfullerene (C_{60}). *Int. J. Quant. Chem.* 98:361–66.
34. Noodleman, L., and Davidson, E. R. 1986. Ligand spin polarization and antiferromagnetic coupling in transition metal dimers. *Chem. Phys.* 109:131–43.
35. Benard, M. 1979. A study of Hartree-Fock instabilities in $Cr_2(O_2CH)_4$ and $Mo_2(O_2CH)_4$. *J. Chem. Phys.* 71:2546–56.
36. Ellis, D. E., and Freeman, A. J. 1968. Model calculations for the study of direct and superexchange interactions. *J. Appl. Phys.* 39:424–25.
37. Hay, P. J., Thibeault, J. C., and Hoffman, R. 1975. Orbital interactions in metal dimer complexes. *J. Am. Chem. Soc.* 97:4884–99.
38. Davidson, E. R., and Clark, A. E. 2007. Analysis of wave functions for open-shell molecules. *Phys. Chem. Chem. Phys.* 9:1881–94.
39. Sadoc, I., Boer, R., and de Graaf, C. 2006. Role of charge transfer configurations in $LaMnO_3$, $CaMnO_3$, and $CaFeO_3$. *J. Chem. Phys.* 126:134709.
40. Slater, J. C. 1930. Cohesion in monovalent metals. *Phys. Rev.* 35:509–29.
41. Löwdin, P.-O. 1960. Expansion theorems for the total wave function and extended Hartree-Fock schemes. *Rev. Mod. Phys.* 32:328–34.
42. Sheka, E. F. 2003. Violation of covalent bonding in fullerenes. In *Lecture notes in computer science, computational science—ICCS2003*, ed. P. M. A. Sloot, D. Abramson, A. V. Bogdanov, J. Dongarra, A. Y. Zomaya, and Y. E. Gorbachev, 386–98. Part II. Berlin: Springer.
43. Sheka, E. F. 2004. Odd electrons and covalent bonding in fullerenes. *Int. J. Quant. Chem.* 100:375–87.
44. Sheka, E. F. 2004. Fullerenes as polyradicals. *Centr. Eur. J. Phys.* 2:160–82.
45. Sheka, E. F., and Zayets, V. A. 2005. The radical nature of fullerene and its chemical activity. *Russ. J. Phys. Chem.* 79:2009–14.
46. Sheka, E. F. 2006. 'Chemical' portrait of fullerenes. *J. Struct. Chem.* 47:593–99.
47. Sheka, E. F. 2007. Chemical susceptibility of fullerenes in view of Hartree-Fock approach. *Int. J. Quant. Chem.* 107:2803–16.

48. Sheka, E. F., and Chernozatonskii, L. A. 2007. Bond length effect on odd electrons behavior in single-walled carbon nanotubes. *J. Phys. Chem.* C 111:10771–80.
49. Sheka, E. F., and Chernozatonskii, L. A. 2010. Broken symmetry approach and chemical susceptibility of carbon nanotubes. *Int. J. Quant. Chem.* 110: 1466–80.
50. Sheka, E. F., and Chernozatonskii, L. A. 2010. Broken spin symmetry approach to chemical reactivity and magnetism of graphenium species. *J. Exp. Theor. Phys.* 110:121–32.
51. Mulliken, R. S. 1950. Structures of complexes formed by halogen molecules with aromatic and with oxygenated solvents. *J. Am. Chem. Soc.* 72:600–8.
52. Mulliken, R. S., and Person, W. B. 1969. *Molecular complexes*. New York: J. Wiley & Sons.
53. Forster, R. 1969. *Organic charge-transfer complexes*. New York: Academic Press.
54. Kaplan, I. G. 1986. *Theory of intermolecular interaction*. Amsterdam: Elsevier.
55. Evans, M. G., and Polanyi, J. 1939. Notes on the luminescence of sodium vapour in highly dilute flames. *Trans. Faraday Soc.* 35:178–85.
56. Yarkony, D. R. 1996. Current issues in nonadiabatic chemistry. *J. Phys. Chem.* 100:18612–28.
57. Herschbach, D. R. 1966. Reactive scattering in molecular beams. In *Advances in chemical physics*, ed. J. Ross, 319–93. New York: Wiley & Sons.
58. Leonas, V. B., and Kalinin, A. P. 1977. Ionization study at slow collisions of atomic particles. *Soviet Physics Uspekhi* 121:561–92.
59. Sheka, E. F. 2004. Intermolecular interaction in C_{60}-based donor acceptor complexes. *Int. J. Quant. Chem.* 100:388–406.
60. Morokuma, K. 1971. Molecular orbital studies of hydrogen bonds. III. C=O····H—O hydrogen bond in $H_2CO\cdots H_2O$ and $H_2CO\cdots 2H_2O$. *J. Chem. Phys.* 55:1236.
61. Kitaura, K., and Morokuma, K. 1976. A new energy decomposition scheme for molecular interactions within the Hartree-Fock approximation. *Int. J. Quant. Chem.* 10:325–40.
62. Sheka, E. F. 2007. A new aspect in computational nanomaterial science: Odd electrons in molecular chemistry, surface science, and solid state magnetism. *Mater. Sci. Forum* 555:19–27.

chapter two

Grounds of computational science of fullerenes

2.1 Unrestricted broken symmetry approach: Basic relations

The unrestricted broken symmetry (UBS) approach formulated by Noodleman[1] is focused on the determination of the energies of pure spin states on the basis of either unrestricted Hartree-Fock (UHF) or unrestricted density functional theory (UDFT) calculations. This was an obvious response to the requirements for magnetic properties of molecular metal dimers to be described. The energy of pure spin states with different total spin S for systems with an even number of magnetic electrons is subordinated to a simple relation

$$E^{PS}(S) = E^{PS}(0) - S(S+1)J \tag{2.1}$$

where $E^{PS}(0)$ presents the energy of the pure-spin singlet state, while J is the Heisenberg exchange integral, or *magnetic coupling constant*. A possibility to obtain the constant, which is the main characteristic of the magnetic behavior of the molecular compounds, was the main goal of the approach. Within the framework of the UBS approach, J can be determined as[1]

$$J = \frac{E_B(0) - E_B(S_{\max})}{S_{\max}^2} \tag{2.2}$$

Here $E_B(0)$ and $E_B(S_{\max})$ are the energy of singlet and $S_{\max}$ spin states obtained by either UHF or UDFT schemes using $|\psi_B\rangle$ (see Equation 1.1). Since any $S_{\max}$ spin state is presented by a single configuration (see discussion of configuration V in regard to Figure 1.1), the $|\psi_B\rangle$-based calculation always provides the pure spin state in this case. For n pairs of magnetic electrons $S_{\max} = n$. According to Noodleman,[1] the energy of a pure-spin singlet state is expressed as

$$E^{PS}(0) = E_B(0) + S_{\max} J \tag{2.3}$$

while the energies of higher spin states, following Equations 2.1 and 2.3, are determined as

$$E^{PS}(S) = E_B(0) - \left[S(S+1) - S_{\max} \right] J \tag{2.4}$$

Here

$$J = J_F + J_{AF} \tag{2.5}$$

and

$$J_F = \frac{1}{n^2} \sum_{i,j} \left(a_i b_j \left| \frac{1}{r_{12}} \right| a_i b_j \right) \tag{2.6}$$

$$J_{AF} = \frac{1}{n^2} \sum_{i=1}^{n} S_{\bar{a}_i \bar{b}_i} \left[\left(a_i \left| h \right| b_i \right) + \left(a_i b_i \left| \frac{1}{r_{12}} \right| a_i a_i \right) \right] \tag{2.7}$$

J_F and J_{AF} are ferromagnetic (positive) and antiferromagnetic (negative) coupling constants, $S_{\bar{a}_i \bar{b}_i}$ is the overlap of the nonorthogonal magnetic orbitals $\bar{a}_i = a_i + c_i b_i$ and $\bar{b}_i = b_i + c_i a_i$ (see Equation 1.1), h is one-electron Hamiltonian, and r_{12} determines the distance between electrons of atoms A and B in the dimer.[1] The sign of magnetic coupling J indicates which kind of electronic interaction, exchange or Coulombic, dominates.

The approach formalism does not put any restriction on the number of magnetic electrons and atoms. However, in the case of molecular complexes with transition metal atoms, modern semiempirical unrestricted HF cannot be used as the basis for the UBS approach due to parametrization problems with d electrons of the metals.[2] *Ab initio* unrestricted HF calculations are very time-consuming. That is why the majority of the available UBS calculations on the species have been performed by using one of the UDFT computational schemes, thus revealing advantages and disadvantages of the technique. The schemes differ in a concrete form of exchange-correlation potential that generally consists of a few members, among which the main contributors form a series:

$$E^{XC} = C_1 E_X^{HF} + C_2 E_X^{Slater} + C_3 \Delta E_X^{Becke88} + C_4 E_C^{VWN} + C_5 \Delta E_C^{LYP} \tag{2.8}$$

Here the first and second terms in the right-hand part indicate Hartree-Fock and Slater exchange functionals, respectively. The third and fourth terms represent Becke's exchange corrections involving the density gradient and Vosko, Wilk, and Nusair correlation functional, and the last term is the correlation correction of Lee, Yang, and Parr (LYP), which includes the density gradient. C_i ($i = 1, \ldots, 5$) are the mixing coefficients.

The functional complexity both facilitates and complicates UDFT (as well as RDFT) application. Thus, since the obtained results are obviously method dependent, changing mixing coefficient C_i can provide any beforehand decided result.[3,4] That is why the predominant majority of computational results *postfactum* describe the available experimental data. Due to this, the predictive ability of the technique is rather weak and is always based on some reference empirical data that assist in adapting the functional in use to the considered case. In contrast to UDFT, UHF solutions are strictly standard and retain a transparent single-determinant description, the restricted version of which led to the foundation of modern chemistry language. There was something behind when Noodleman called his approach "Valence bond description," presenting it in terms of a UHF solution, albeit mentioning its possible implementation in terms of UDFT. He has been working in the field of magnetic behavior of biomolecular complexes with transition metals and applying UBS DFT for many years.[5]

A higher transparency of the UHF solution can be seen through its better chance to "translate" the correlation incompleteness of the solution in convenient terms, so that it might become practically useful if properly designed. This concerns another aspect of the spin contamination of the broken symmetry solutions that can be coherently included in the UBS approach. The wave function $|\psi_B\rangle$ of both UHF and UDFT approaches corresponds to the state with the definite value of the spin projection, S_z, but does not, in general, correspond to the state with the definite value of the total spin, S, so that

$$\hat{S}_z|\psi_B\rangle = S_z|\psi_B\rangle \tag{2.9}$$

and

$$\hat{S}^2|\psi_B\rangle \neq S(S+1)|\psi_B\rangle \tag{2.10}$$

As a result, both UBS solutions are spin contaminated if only all electrons have not one direction, so that $S = S_{max}$. The contamination

$$C = \langle \hat{S}^2 \rangle - S(S+1) \tag{2.11}$$

is often substantial.

The inequalities in Equation 2.10 are tightly connected with the Löwdin symmetry dilemma[6] that is expressed as asymmetric electron densities of

UHF solution and asymmetric Local Spin Density Approximation (LSDA) Hamiltonian of UDFT with different exchange-correlation potentials for spin-up and spin-down orbitals. This feature exhibits the tendency of spin-up and spin-down electrons to occupy different portions of space. The asymmetry produces the appearance of an extra density function first described by Takatsuka, Fueno, and Yamaguchi more than thirty years ago[7] and named as the distribution of odd electrons:

$$D(r|r') = 2\rho(r|r') - \int \rho(r|r'')\rho(r''|r')dr'' \tag{2.12}$$

The function $D(R|r')$ trace,

$$N_D = trD(r|r') \tag{2.13}$$

was interpreted as the total number of such electrons.[8] The authors suggested the function $D(R|r')$ trace N_D to manifest the radical character of the species under investigation. Twenty-two years later Staroverov and Davidson changed the term to distribution of *effectively unpaired electrons*,[9,10] emphasizing a measure of the radical character that is determined by the N_D electrons taken out of the covalent bonding. Even in the paper of Takatsuka et al.[7] it was mentioned that the function $D(R|r')$ can be subjected to a population analysis within the framework of the Mulliken partitioning scheme. In the case of a single Slater determinant, Equation 2.13 takes the form[10]

$$N_D = trDS \tag{2.14}$$

where

$$DS = 2PS - (PS)^2 \tag{2.15}$$

Here D is the spin-density matrix $P = P^\alpha - P^\beta$, $P = P^\alpha + P^\beta$ is a standard density matrix in the atomic orbital basis, and S is the orbital overlap matrix (α and β mark different spin directions). The population of effectively unpaired electrons on atom A is obtained by partitioning the diagonal of the matrix DS as

$$D_A = \sum_{\mu \in A} (DS)_{\mu\mu} \tag{2.16}$$

so that

$$N_D = \sum_A D_A \tag{2.17}$$

Staroverov and Davidson[10] showed that atomic population D_A was close to the Mayer free valence index[11] F_A in the general case, while in the singlet state D_A and F_A are identical. Thus, a plot of D_A over atoms gives a visual picture of the actual radical electrons' distribution,[12] which in turn exhibits atoms with enhanced chemical reactivity.

The effectively unpaired electron population is definitely connected with the spin contamination of the UBS solution state, which in the case of the UBS HF scheme results in a straight relation between N_D and square spin, $\langle \hat{S}^2 \rangle$:[10]

$$N_D = 2\left(\left\langle \hat{S}^2 \right\rangle - \frac{(N^\alpha - N^\beta)^2}{4} \right) \tag{2.18}$$

where

$$\left\langle \hat{S}^2 \right\rangle = \left(\frac{(N^\alpha - N^\beta)^2}{4} \right) + \frac{N^\alpha + N^\beta}{2} - \sum_i^{N^\alpha} \sum_j^{N^\beta} \left| \left\langle \phi_i | \phi_j \right\rangle \right|^2 \tag{2.19}$$

Here ϕ_i and ϕ_j are atomic orbitals, and N^α and N^β are the numbers of electrons with spin α and β, respectively.

If UBS HF computations are realized via neglect-of-diatomic-differential-overlap (*NDDO*) approximation (the basis for AM1/PM3 semiempirical techniques),[13] a zero overlap of orbitals in Equation 2.15 leads to $S = I$, where I is the identity matrix. The spin-density matrix D (taking into account that $P^{2\alpha} = P^\alpha$ and $P^{2\beta} = P^{\beta}$[14]) assumes the form

$$D = (P^\alpha - P^\beta)^2 \tag{2.20}$$

The elements of the density matrices $P_{ij}^{\alpha(\beta)}$ can be written in terms of eigenvectors of the UHF solution C_{ik}:

$$P_{ij}^{\alpha(\beta)} = \sum_k^{N^{\alpha(\beta)}} C_{ik}^{\alpha(\beta)} * C_{jk}^{\alpha(\beta)} \tag{2.21}$$

The expression for $\langle \hat{S}^2 \rangle$ has the form[15]

$$\left\langle \hat{S}^2 \right\rangle = \left(\frac{(N^\alpha - N^\beta)^2}{4} \right) + \frac{N^\alpha + N^\beta}{2} - \sum_{i,j=1}^{NORBS} P_{ij}^\alpha P_{ij}^\beta \tag{2.22}$$

This explicit expression is the consequence of the Ψ-based character of the UBS HF. Since the corresponding coordinate wave functions are subordinated to the definite permutation symmetry, each value of spin S corresponds to a definite expectation value of energy.[16] In contrast, the electron density, ρ, is invariant to the permutation symmetry. The latter causes a serious spin multiplicity problem for the UBS DFT.[16,17] Additionally, the spin density, $D(R|r')$, of the UBS DFT depends on spin-dependent exchange and correlation functionals only and cannot be expressed analytically.[16] Since the exchange-correlation composition deviates from one method to the other, the spin density is not fixed and deviates alongside with the composition.

Within the framework of the *NDDO* approach, the total N_D and atomic N_{DA} populations of effectively unpaired electrons take the form[18]

$$N_D = \sum_A N_{DA} = \sum_{i,j=1}^{NORBS} D_{ij} \tag{2.23}$$

and

$$N_{DA} = \sum_{i \in A} \sum_{B=1}^{NAT} \sum_{j \in B} D_{ij} \tag{2.24}$$

Here D_{ij} present matrix elements of the spin-density matrix D.

First applied to fullerenes,[18–20] N_{DA} in the form of Equation 2.24 has actually disclosed the different chemical activities of atoms just visualizing a "chemical portrait" of the molecule. It was natural to rename N_{DA} as *atomic chemical susceptibility* (ACS). Similarly, N_D was renamed as *molecular chemical susceptibility* (MCS). Rigorously computed ACS (N_{DA}) is an obvious quantifier that highlights targets to be the most favorable for addition reactions of any type at each stage of the reaction, thus forming grounds for *computational chemical synthesis*. A high potentiality of the approach will be exemplified later by fluorination (Chapter 4), hydrogenation (Chapter 5), and cyanation and amination (Chapters 6 and 7) of fullerene C_{60}.

Opposite to UBS HF, UBS DFT does not suggest enough reliable expressions for either N_D or N_{DA}. The only known detailed discussion of the problem comparing UBS HF and UBS DFT results in a complete active space self-consistent field (CASSCF), and multireference configuration interaction (MRCI) ones concern the description of the diradical character of the Cope rearrangement transition state.[9] When UBS DFT calculations gave $N_D = 0$, CASSCF, MRCI, and UBS HF calculations gave 1.05, 1.55, and 1.45 e, respectively. Therefore, the experimentally recognized radical character of the transition state was well supported by the latter three techniques, with quite a small deviation in numerical quantities, while UBS DFT results just rejected the radical character of the state. Serious UBS DFT problems are known as well in relevance to $\langle \hat{S}^2 \rangle$ calculations.[21,22] These obvious shortcomings of the UDFT approach might be a reason why UBS DFT calculations of this kind are rather scarce. A few recent results concerning UBS DFT calculations applied to carbon nanotubes and graphene will be discussed in Chapters 11 and 12.

As seen above, currently the UBS HF approach seems to be the only alternative that may provide a quite accurate and reliable description of both the energy of the ground state and particular features of the odd electrons' behavior in fullerene molecules and other sp^2 nanocarbons on the same theoretical platform. An additional support in favor of the approach is its easy and fully transparent communication with widely used atom-matched "informative devices,"[23] which are related to such quantities as charge, spin density, strength of bonding (pairing) with other atoms, free valence, etc. The latter constitute the language of modern molecular science and are quite important in establishing both correlations between different objects and different responses of the considered object to various external actions. Obviously, the devices should have a clear physical meaning and should be properly calculated. Looking at a number of quantities or terms in use from this viewpoint, one must accept that those have been introduced into molecular science in terms that originated from the fundamental single-determinant HF approach.

2.2 *UBS approach realization in semiempirical calculation*

The pure-spin-state energy-oriented Noodleman's consideration[1] supplemented by the effectively unpaired electron population analysis[7–10,18] forms the ground for the modern UBS approach. Its efficacy has been repeatedly proven (see Han and Noodleman[5] and Davidson and Clark[12] and references therein). However, the approach application was mainly restricted to diradicals and molecular complex dimers containing ions of transition metals. Either *ab initio* UBS HF or UBS DFT calculations dominated in the studies. Looking at fullerenes (as well as carbon nanotubes and graphene) as desirable candidates to be analyzed within the approach, it becomes clear that the bulk of computations needed in these cases cannot be performed by using *ab initio* UBS HF, for which the molecules are too complex; neither can UBS DFT be used due to reasons explained in the previous section. At the same time, being composed of carbon atoms, fullerenes are excellent candidates for high-level semiempirical calculations, which are provided with perfect parameterization of computational versions.[24]

The computations, results of which are presented in the current book, have been performed by using the Austin model 1 (AM1) semiempirical method,[25] implemented in CLUSTER-Z1 codes.[26] Based on the *NDDO* approximation,[13] the codes are very efficient, accurate, and suitably enlarged by including calculation of square spin, $\langle \hat{S}^2 \rangle$ (Equation 2.19), as well as effectively unpaired electrons' population analysis in terms of N_D (Equation 2.23) and N_{DA} (Equation 2.24).[18] A comprehensive description of the codes is given elsewhere,[2] alongside their parallel versions. We shall use UBS HF inscription throughout the book, emphasizing that the UBS approach is applied on the basis of unrestricted HF calculations.

2.3 UBS HF approach testing

Analysis of the N_D values' behavior along the potential energy curve of diatomic molecules gives an excellent possibility to check the correctness of the UBS HF approach to the description of effectively unpaired electrons. Following the first calculations of this kind performed by Staroverov and Davidson,[10] the calculations included in the book were carried out for hydrogen, nitrogen, and oxygen molecules.[18,20,27] The obtained data for the ground states of the molecules (singlet H_2 and N_2, and triplet O_2) are plotted in Figure 2.1. As seen in the figure, a characteristic *S*-like character of the $N_D(R)$ dependence is common for all molecules. Each *S*-curve contains three regions: (I) $R \leq R_{cov}$, (II) $R_{cov} \leq R \leq R_{rad}$, and (III) $R \geq R_{rad}$ (the corresponding R_{cov} and R_{rad} values for the studied molecules are given in Table 2.1). R_{cov} marks the extreme distance that corresponds to the completion of the covalent bonding of the molecule electrons according to the multiplicity of its ground state and exceeding which indicates the onset of the molecule's radicalization. R_{rad} matches a start of a complete homolytic

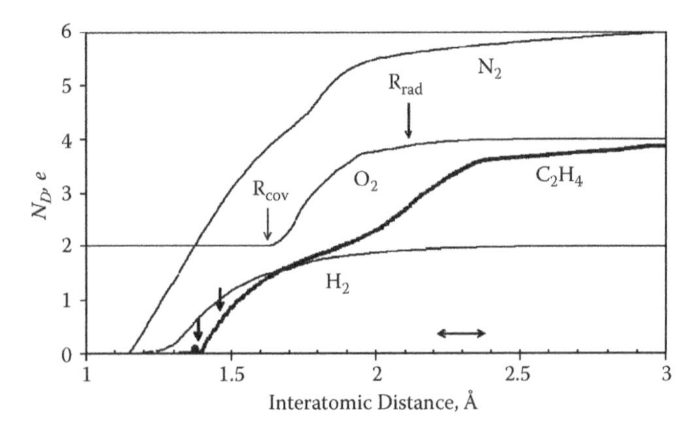

Figure 2.1 Molecular chemical susceptibility N_D of diatomic molecules and ethylene vs. the distance between nuclei.

Table 2.1 Critical distances R_{cov} and R_{rad} for simple molecules, *nm*

Molecules	R_{cov}, UBS HF[27]	R_{cov}, UBS DFT (B3-LYP)[28]	R_{rad}[27]
H_2	0.115	0.140	0.20
N_2	0.120	0.133	0.25
O_2	0.124	—	0.17
C_2H_4	0.1395	0.1565	0.30

bond cleavage followed by the formation of two free radicals (see the corresponding arrows indicating the quantities for the O_2 molecule in Figure 2.1). The intermediate region II with a continuously growing N_D value demonstrates convincingly the buildup of the molecule's radicalization caused by electron extraction from the covalent bonding as the corresponding interatomic bond is gradually stretched. It should be noted that region II only slightly depends on the chemical nature of the species and is rather narrow, occupying the interval of ~1.5Å in width.

The analysis of the $N_D(R)$ curves for hydrogen and nitrogen molecules shows that the obtained results fully satisfy predictions followed from the chemical intuition. The ground state of both molecules is singlet, and both curves were calculated for this state. As could be expected, there are no effectively unpaired electrons in region I that correspond to the covalent bonding of all molecule electrons. Consequently, unrestricted and restricted HF solutions coincide and correspond to pure spin states. The numbers of effectively unpaired electrons in region III are two and six for hydrogen and nitrogen, respectively. As seen in Figure 2.1, the $N_D(R)$ curves do approach the asymptotes corresponding to two and six electrons in this region. As for oxygen, the analysis of the results obtained in regions I and III shows a similar behavior, but for a triplet state. The N_D asymptote should approach 4 in this case, which we actually see in the figure.

As for benzenoid-based carboneous species (fullerenes, graphite, graphene, and carbon nanotubes), the characteristic two-atomic unit is connected with a particular C_2 pair, each atom of which has three neighbors, covalently bonded with the central atom by three σ bonds. We are interested in the fortune of the fourth odd electron, depending on the C-C distance. The analysis of the ethylene molecule dissociation evidently suits the case. The evolution of the total number of unpaired electrons along the potential curve for the molecule is shown in Figure 2.1 by a thickened curve. As seen in the figure, the $N_D(R)$ curve for the molecule is similar to S-curves of the diatomic molecules discussed above. Region I corresponds to covalently bound electrons with $N_D = 0$ until R reaches R_{cov} = 1.395 Å. On the other hand, when R reaches R_{rad} = 3.0 Å, N_D is approaching

4, showing a cleavage of the C-C bond and a total radicalization of the CH_2-CH_2 pair. The interval between 1.395 and 3.0 Å corresponds to the intermediate case when the odd electrons change their behavior from being totally bonded to totally free.

Supposing that the behavior of stretched ethylene simulates the variation of C-C bonds in benzenoid species correctly enough, one can conclude that the odd electron pairing in the species depends on the corresponding interatomic C-C distance in the molecules. This leads to a new numerical factor of their electronic structure and highlights three regions limited by R_{cov} and R_{rad} values that determine extremes (the highest in the first case and the lowest in the second) at which either the covalent bonding or the free radical formation is completed. The intermediate range between R_{rad} and R_{cov} may be attributed to species with an intermediate radical character. Therefore, the greater the number of long C-C bonds in the molecule, the greater the total expected number of effectively unpaired electrons. This point can be easily checked for planar $\{4n + 2\}$ aromatic molecules. As seen in Table 2.2, the molecules have practically the same set of C-C bonds but differ in the related bond number, particularly that of long bonds. The N_D values presented in the table show that larger numbers of long bonds in the molecules are accompanied by greater N_D values. This finding made it possible to suggest that the presence of long bonds in fullerenes, as well as extending the C-C bond lengths over R_{cov} in carbon nanotubes and graphene, causes a partial unpairing of odd electrons in the species, thus making them partially radicalized.

A strong support in favor of the UBS HF data obtained was found in the recent many-body configurational interaction calculations of polyacenes.[28] Applying *ab initio* density matrix renormalization group (DMRG) algorithms, the authors highlighted the radical character of the acenes

Table 2.2 Effectively unpaired electrons in polyacenes

	C-C bond length, Å[27]				N_D	
Molecules	Number of bonds				UBS HF [27]	DMRG [28]
Benzene	1.395				0	—
	6					
Naththalene	1.385	1.411	1.420	1.430	1.48	1.95
	4	2	4	1		
Anthracene	1.387	1.410	1.421	1.435	3.00	3.00
	4	6	4	2		
Tetracene	1.388	1.410	1.421	1.436	4.32	4.00
	4	8	6	3		
Pentacene	1.388	1.411	1.420	1.436	5.54	5.20
	4	10	8	4		

that is caused by the appearance of effectively unpaired electrons and that starts in naphthalene and strengthens when the acene size increases in full agreement with the UBS HF data for lower acenes, discussed above. In contrast, the UBS DFT approach rejects the radicalization in this case until the acene becomes quite long.[29] The DMRG approach also permits the determination of the total number of effectively unpaired electrons, N_D. When utilizing the algorithm for the quantity determination presented by Equation 2.23 in Section 2.2, the authors obtained N_D values that coincide with the relevant data obtained in the framework of the UBS HF approach based on the same algorithm (see Table 2.2). The observed fitting of DMRG and UBS HF is undoubtedly a strong support in favor of the ability of the latter to highlight the physical reality of a system of weakly interacting electrons.

Two arrows attached to the curve corresponding to ethylene in Figure 2.1 mark the bond length interval characteristic for fullerene C_{60} as well as for carbon nanotubes and graphene. C-C bonds of higher fullerenes also fill this interval. In all the cases the interval is located in the intermediate region above R_{cov}. A black dot at ~1.4 Å corresponds to the benzene molecule. The molecule takes the site at the boundary between aromatic and gradually radicalized species. Its N_D is practically zero, so that the molecule is assuming its right place in the aromatic world. It is worthwhile to mention in this connection that CASSCF/6-31G* calculations, which are usually expected to be closer to the exact solution than UBS HF, gave $N_{DA} = 0.15\ e$ on each carbon atom of the molecule, which results in $N_D = 0.9\ e$.[9] This finding calls into question whether benzene belongs to the aromatic series due to its evident partial radical character, followed by enhanced chemical reactivity. The conclusion fully contradicts the empirical reality. Therefore, *a priori* judgment in favor of any computational scheme should not be taken for granted and needs a thorough consideration in view of fitting (and predicting) reasonable data over a set of different physicochemical quantities.

A double-headed arrow in the region around 2.3 Å corresponds to an averaged interval of Si-Si valence bonds. The interval is located in the region where the radicalization is either very considerable or fully completed. This means that the formation of π electrons followed by the sp^2 configuration of odd electrons in siliceous species is highly unfavorable, so that the species are to be highly radicalized. We shall come back to the discussion of siliceous sp^2 nanomaterials in Chapter 3 (Section 3.4) and Chapter 14.

As follows from UBS HF calculations, $R_{cov} = 1.395$ Å is an extreme above which electron unpairing related to the benzenoid C-C bond may occur. Bond lengths of fullerenes and carbon nanotubes exceed the distance by ~0.1 Å. The value seems to be rather small, which makes the accuracy of the R_{cov} determination crucial and may question the use of the computational technique. As for UBS HF, the value is big enough to be reliably distinguished. A comparative study of the homolytic cleavage of the C-C bond in ethylene by different UBS DFT techniques has shown[30] that critical distances R_{cov} form a series

$$R_{cov}\text{: UBS HF} < \text{B3-LYP} < \text{B-P} < \text{S-VWN} \tag{2.25}$$

where B3-LYP, B-P, and S-VWN correspond to different DFT functionals used in the UBS calculations with a split valence plus polarization basis set. The difference between UBS HF and the nearest UBS DFT values exceeds 0.1 Å, so that no electron unpairing should be expected for fullerenes according to these UBS DFT data. However, the situation is fully similar to that discussed at the end of Section 2.1 in regards to N_D data for the diradical character of the Cope rearrangement transition state.[9] As in the current case, the B3-LYP UBS DFT calculations gave $N_D = 0$, while CASSCF, MRCI, and UBS HF calculations provided a value between 1.55 and 1.05*e* that suits experimental findings. These two cases convincingly show overpressing configurational interaction in the applied UBS DFT techniques.

2.4 UBS HF approach and fullerene nanoscience

The UBS HF building of fullerene nanoscience is seen as the following construction. Partially radicalized fullerene molecules, having extreme donor and acceptor characteristics, form the foundation of the building. Effectively unpaired electrons provide enhanced chemical reactivity of the molecules and lay the foundation of their enhanced magnetic behavior under particular conditions. Chemical reactivity stimulates and explains the high tendency of the molecules to addition reactions and, in connection with high donor–acceptor ability, promotes a very large class of such reactions, covering not only large fields of synthetic chemistry, but also bio- and medicinal chemistry. The reaction products are produced under the leadership of the numerical indicator N_{DA}. Thus, nanochemistry and nanomedicine form two walls of the building. Two other walls are connected with particular reactions related to self-aggregation of the molecules. The reactions result in the formation of either chemically bound oligomers at the R^{+-} minimum, the formation of which is subordinated to the N_{DA} indications, or weakly bound clusters at the R^{00} minimum due to the two-well structure of the ground-state potential energy term. The former composites are characterized by the dependence of odd electrons pairing on the size of the composites, which leads to the ability of their magnetization when achieving a certain size. The latter composites, being charge transfer complexes, are able to cause photostimulated spin-flip and to enhance the electromagnetic field, leading to the transformation $^3O_2 \rightarrow {}^1O_2$, on one hand, and to a number of enhanced spectral and nonlinear effects, on the other. These appearances of the fullerene aggregates exhibited as nanomagnetism and nanophotonics of fullerene form the remaining two walls. Closing the fullerene space, the roof of the building makes it possible to coherently put the building into a much more extended space filled by other representatives of a large class of benzenoid nanocarbons, such as carbon nanotubes, graphite, graphene, etc. The designed fullerene building is inhabited by a particular family. This book acquaints readers with the family members and motivates people to construct their own buildings with different families.

References

1. Noodleman, L. 1981. Valence bond description of antiferromagnetic coupling in transition metal dimers. *J. Chem. Phys.* 74:5737–42.
2. Berzigiyarov, P. K., Zayets, V. A., Ginzburg, I. Ya., Razumov, V. F., and Sheka, E. F. 2002. NANOPACK: Parallel codes for semiempirical quantum chemical calculations of large systems in the *sp*- and *spd*-basis. *Int. J. Quant. Chem.* 88:449–62.
3. Illas, F., Moreira, I. de P. R., de Graaf, C., and Barone, V. 2000. Magnetic coupling in biradicals, binuclear complexes and wide-gap insolators; a survey of *ab initio* function and density functional theory approaches. *Theor. Chem. Acc.* 104:265–72.
4. Adamo, C., Barone, V., Bencini, A., Broer, R., Filatov, M., Harrison, N. M., Illas, F., Malrieu, J. P., and Moreira, I. de P. R. 2006. Comment on "About the calculation of exchange coupling constants using density-functional theory: The role of the self-interaction error" [*J. Chem. Phys.* 123, 164110 (2005)]. *J. Chem. Phys.* 124:107101.
5. Han, W.-C., and Noodleman, L. 2008. Structural model studies for the high-valent intermediate Q of methane monooxygenase from broken-symmetry density functional calculations. *Inorg. Chim. Acta* 361:973–86.
6. Löwdin, P. O. 1969. Quantum theory of many-particle systems. *Adv. Chem. Phys.* 14:283–90.
7. Takatsuka, K., Fueno, T., and Yamaguchi, K. 1978. Distribution of odd electrons in ground-state molecules. *Theor. Chim. Acta* 48:175–83.
8. Takatsuka, K., and Fueno, T. 1978. The spin-optimized SCF general spin orbitals. II. The 2 ^2S and 2 ^2P states of the lithium atom. *J. Chem. Phys.* 69:661–69.
9. Staroverov, V. N., and Davidson, E. R. 2000. Diradical character of the Cope rearrangement transition state. *J. Am. Chem. Soc.* 122:186–87.
10. Staroverov, V. N., and Davidson, E. R. 2000. Distribution of effectively unpaired electrons. *Chem. Phys. Lett.* 330:161–68.
11. Mayer, I. 1986. On bond orders and valences in the *ab initio* quantum chemical theory. *Int. J. Quant. Chem.* 29:73–84.
12. Davidson, E. R., and Clark, A. E. 2007. Analysis of wave functions for open-shell molecules. *J. Chem. Phys.* 9:1881–94.

13. Dewar, M. J. S., and Thiel, W. 1977. Ground states of molecules. 38. The MNDO method. Approximations and parameters. *J. Am. Chem. Soc.* 99:4899–907.
14. Zaets, V. A. 2004. Private communication.
15. Zhogolev, D. A., and Volkov, V. B. 1976. *Methods, algorithms and programs for quantum-chemical calculations of molecules* [in Russian]. Kiev: Naukova Dumka.
16. Kaplan, I. 2007. Problems in DFT with the total spin and degenerate states. *Int. J. Quant. Chem.* 107:2595–603.
17. Davidson, E. 1998. How robust is present-day DFT? *Int. J. Quant. Chem.* 69:214–45.
18. Sheka, E. F., and Zayets, V. A. 2005. The radical nature of fullerene and its chemical activity. *Russ. J. Phys. Chem.* 79:2009–14.
19. Sheka, E. F. 2006. Chemical portrait of fullerenes. *J. Struct. Chem.* 47:593–99.
20. Sheka, E. F. 2007. Chemical susceptibility of fullerenes in view of Hartree-Fock approach. *Int. J. Quant. Chem.* 107:2803–16.
21. Wang, J., Becke, A. D., and Smith, V. H., Jr. 1995. Evaluation of $\langle \hat{S}^2 \rangle$ in restricted, unrestricted Hartree-Fock, and density functional based theory. *J. Chem. Phys.* 102:3477–80.
22. Cohen, A. J., Tozer, D. J., and Handy, N. C. 2007. Evaluation of $\langle \hat{S}^2 \rangle$ in density functional theory. *J. Chem. Phys.* 126:214104.
23. Szabo, A., and Ostlund, N. S. 1989. *Modern quantum chemistry*, rev. sections 3.8.5–3.8.7. 1st ed. New York: McGraw-Hill.
24. Chen, Zh., and Thiel, W. 2003. Performance of semiempirical methods in fullerene chemistry: Relative energies and nucleous-independent chemical shift. *Chem. Phys. Lett.* 367:15–25.
25. Dewar, M. J. S., Zoebisch, E. G., Healey, E. F., and Stewart, J. J. P. 1985. Development and use of quantum mechanical molecular models. 76. AM1: A new general purpose quantum mechanical molecular model. *J. Am. Chem. Soc.* 107:3902–9.
26. Zayets, V. A. 1990. *CLUSTER-Z1: Quantum-chemical software for calculations in the s,p-basis.* Kiev: Institute of Surface Chemistry, National Academy of Science of Ukraine.
27. Sheka, E. F., and Chernozatonskii, L. A. 2007. Bond length effect on odd electrons behavior in single-walled carbon nanotubes. *J. Phys. Chem. C* 111:10771–80.
28. Hachmann, J., Dorando, J. J., Aviles, M., and Chan, G. K. 2007. The radical character of the acenes: A density matrix renormalization group study. *J. Chem. Phys.* 127:134309.
29. Poater, J., Bofill, J. M., Alemany, P., and Sola, M. 2005. Local aromaticity of the lowest-lying singlet states of [*n*]acenes (*n*) 6–9. *J. Phys. Chem. A* 109:10629–32.
30. Bauernschmitt, R., and Ahlrichs, R. 1996. Stability analysis for solutions of the closed shell Kohn–Sham equation. *J. Chem. Phys.* 104:9047.

chapter three

Fullerene C_{60} in view of the unrestricted broken symmetry Hartree–Fock approach

3.1 Introduction

Fullerene C_{60} is related to those rare molecules whose structure had been predicted before the molecule was discovered experimentally. The first suggestion was made by Jones in 1966, who assumed that intercalation of pentagon defects into a plain graphite layer (graphene, according to today's nomination) consisting of perfect hexagons would result in the transformation of this layer into a closed hollow shell that presents a giant carbon molecule.[1] In 1970 Osawa published a short communication[2] showing a possibility of an existing C_{60} molecule in the form of truncated icosahedron, consisting of sixty atoms. The following year Yoshida and Osawa discussed possible aromatic properties of the molecule.[3] Two years later Bochvar and Galpern[4] as well as Stankevich and coworkers[5] performed calculations of the molecule electronic structure. This calculation, repeated a few years later by Davidson,[6] showed that the closed-hollow-cell C_{60} molecule is characterized by a large distance between the energies of the highest occupied (HOMO) and lowest unoccupied (LUMO) molecular orbitals, which points to a chemical stability.

The first discovery of the molecule is usually connected with Kroto, Curl, and Smalley (KCS),[7] although a year earlier its presence among carbon clusters was heralded.[8] Not knowing the above-cited results, KCS suggested a truncated icosahedron shape of the molecule, consisting of thirty-two faces, sixty apices (carbon atoms), and ninety ribs. Supposing a full equivalence of atom positions, the symmetry of the polyhedron as a geometrical replica for equilibrium positions of the molecule atoms was attributed to the point group symmetry I_h. The beauty of the molecule is so impressive that it is worthwhile to remember words said by Sir Harrold Kroto in his Nobel lecture on December 7, 1996, p. 75:[9] "The molecule's most delightful property lies in the inherent charisma which arises from its elegantly simple and highly symmetric structure that is quite unlike any other. It is this charisma that has stimulated delight and fascination for chemistry in young and old alike." These words encouraged Katz to write a beautiful scientific fiction: "*Fullerenes, Carbon Nanotubes and Nanoclusters. Pedigree of Shapes and Concepts.*"[10]

Fullerene C_{60} is the main object of discussion presented in the current book. However, the concepts laying the foundation of the consideration address the benzenoid structure and odd electron nature of the molecule, as well as its high D–A characteristics, but not the number of atoms. That is why the discussions and, particularly, conclusions can obviously be expanded over the whole fullerene family. And only a predominant majority of the performed investigations, both computational and empirical, distinguish the molecule from other members of the family. In what follows, addressing fullerene C_{60}, or simply C_{60}, means consideration of concrete results related to the molecule, while addressing fullerene without specification points to features common to the whole family.

Two sides of the UBS HF approach, covering the ability to determine spin-pure electronic states and odd electron behavior, offer a large set of characteristics that describe the ground and high spin states of the molecules, among which we select the following:

Structure and symmetry
Energy of pure spin states, $E^{PS}(0)$
Total spin as $\langle \hat{S}^2 \rangle$
Total number of effectively unpaired electrons N_D; molecular chemical susceptibility (MCS)
N_D distribution over the molecule's atoms (N_{DA}); atomic chemical susceptibility (ACS)
Ionization potential (IP)
Electron affinity (EA)

3.2 Structure and symmetry

Symmetry is understood here to be the maximal point group compatible with the molecular graph. The I_h symmetry suggestion has laid the foundation for a doubtless view on the molecule shape symmetry that has been, until now, supported by the absolute majority of the fullerene community. The number of contradictions to this paradigm when interpreting some of the molecule properties are usually considered in terms of "a slight lowering of the molecule symmetry from the ideal one under particular conditions." In the current chapter we advocate a new consideration of the molecule property, showing that there are some reasons underlying these contradictions, so that the symmetry problem of fullerene C_{60} is not so simple.

Table 3.1 Experimental data on C-C bond lengths of the C_{60} molecule, Å

Experiment[a]	Single C-C Bonds (p)	Double C-C Bonds (h)	Reference
EGD (I_h)	1.455 (6)	1.398 (10)	11
ND (powder) (I_h)	1.452 (66)	1.391 (63)	12
XRD (I_h)	1.452 (66)	1.391 (63)	13
^{13}C NMR (I_h)	1.450 (15)	1.400 (15)	14
QCh (UBS HF AM1) (C_i)	1.464 (13)	1.391 (32)	15
QCh (RHF AM1) (I_h)	1.463 (3)	1.385 (0.4)	15

[a] Uncertainties of physical experiments and dispersions of the C-C bond distributions provided by UBS HF AM1 calculations are given in parentheses.

3.2.1 C_{60} shape symmetry: Structural experiments

As accepted, the most reliable atomically mapped description of the molecule structure follows from the electron gas diffraction (EGD).[11] Actually, the diffraction pattern is well fitted when suggesting that the molecule shape is a truncated icosahedron of the point group symmetry I_h, which is formed by two types of C-C bonds, ones of a substantial double bond character and h = 1.398 (10) Å in length, and ones that are p = 1.455 (6) Å long and have a prevalent single bond character. C-C bonds separating two hexagons are double bonds (h), while the pentagon C-C bonds (p) are single. Under the same suggestion, a good fitting to experimental data related to neutron diffraction (ND) from C_{60} powder[12] and x-ray data (XRD) for the C_{60} crystal[13] was obtained. The available data concerning C-C bond length are collected in Table 3.1. Analyzing the data, one can conclude that EGD provides the most accurate bond length determination, while its accuracy is only 10^{-2} Å.[11,16] At the same time, ND and XRD methods do not allow determining the value better than 10^{-1} Å.

3.2.2 C_{60} shape symmetry: Quantum chemical calculations

The modern Born-Oppenheimer approach to quantum chemistry can provide accuracy in bond length determination not worse than of 10^{-5} Å. That is why quantum chemical calculations can be regarded as a highly accurate structure determination technique under conditions when an extended set of calculated results fits experimental findings of the object under consideration. Starting from Chang et al.[17] and Weaver,[18] the molecule has been repeatedly and thoroughly studied computationally (see, to name a few, Lee et al.,[19,22] Nagase,[20] and Slanina and Lee[21] and references therein). In some sense, C_{60} turned out to be a proving ground for testing different computational techniques, from the simplest to the most sophisticated. However, all calculations were based on the aromaticity concept that revealed itself in a closed-shell or restricted approximation. All calculations have shown I_h symmetry of the truncated icosahedron structure to correspond to the lowest potential energy minimum in the singlet state. However, as was discussed in Section 2.3 and shown in Figure 2.1, lengths of the molecule-long C-C bonds exceed the region that provides a complete covalent bonding of odd electrons, so that a remarkable difference, ΔE^{RU} (see Equation 1.1), in the energies of restricted and unrestricted solutions should be expected, highlighting the nonstability of the restricted solutions and positioning a stable pure spin and physically real ground state of the molecule at lower energy. This is so, and Table 3.2 presents data covering both restricted Hartree-Fock (RHF) and unrestricted broken symmetry Hartree-Fock (UBS HF) calculations.[15] To show that the same happens for other fullerenic structures, data for C_{70} and Si_{60} molecules are included in the table.

As seen from the table, spin-contaminated energies of UBS HF solutions and pure spin energies of all three molecules are much lower than those of the restricted solutions. The ΔE^{RU} values constitute 1.53, 3.85, and 22.9% the RHF heats of formations for C_{60}, C_{70}, and Si_{60}, respectively. The energy lowering not only exhibits the instability of the restricted solution, but can be followed by the symmetry lowering as well. A tight connection between the energy lowering and the molecule symmetry has been thoroughly discussed (see, for example, Touless[25] and Benard[26]). Since it is commonly believed that the energy lowering is generally accompanied with a descent of symmetry (which adds the geometry and/or symmetry instability to the spin one) that is why lowering I_h symmetry of the RHF solution of C_{60} to C_i symmetry of the UBS HF one is not unexpected.

Since the real state of the molecule is spin pure and since its energy is lower than that of the spin-contaminated UBS HF one, the real symmetry of the molecule cannot be higher than the symmetry of the UBS HF state. That is why the UBS HF symmetry of the species is suggested for pure spin states in Table 3.2. A final decision on the real symmetry of the molecule might be put on the shoulders of experiments. However, this way is rather complex and ambiguous. The situation is not particularly typical for fullerene species and is well known for other cases, such as a discussion about D_{6h} or D_{3h} symmetry of the benzene molecule.

In the case of C_{60}, the statement that the point group symmetry of the molecule is C_i seems to drastically contradict common opinion. Actually, a beautiful geometrical truncated-icosahedron shape does not allow one even to think about a molecule symmetry lower than that for I_h. And this intuitive expectation is proved by direct structural experiments and quantum chemical calculations. But any experimental proof has never been an absolute one, providing a rigid *yes/no* answer concerning physical object symmetry due to unavoidable uncertainties inherited in experimental

Table 3.2 Characteristics of singlet ground state of fullerenes[a]

	Solution/data	C_{60}	C_{70}	Si_{60}
RHF	Heat of formation,[b] kcal/mol	970.180	1061.136	1295.967
	Symmetry	I_h	D_{5h}	C_i
	Number of effectively unpaired electrons, N_D, *e*	0		0
	Total spin, $\langle \hat{S}^2 \rangle$	0	0	0
	Ionization potential,[c] IP, eV	9.64	9.14	8.00
	Electron affinity,[c] EA, eV	2.95	3.27	3.38
UBSHF	Heat of formation,[b] kcal/mol	955.362	1020.226	1011.722
	Symmetry	C_i	D_{5h}	C_1
	Number of effectively unpaired electrons, N_D, *e*	9.84	14.40	63.52
	Total spin, $\langle \hat{S}^2 \rangle$	4.92	7.20	31.76
	ΔE^{RU}, kcal/mol	14.818	40.910	284.245
	Ionization potential,[c] IP, eV	9.87 (8.74[1])	9.87	8.98
	Electron affinity,[c] EA, eV	2.66 (2.69[2])	2.73	2.70
Pure spin state	Heat of formation, kcal/mol	899.580	963.176	994.597
	Symmetry	C_i	D_{5h}	C_1
	Number of effectively unpaired electrons, N_D, *e*	0	0	0
	Total spin, $\langle \hat{S}^2 \rangle$	0	0	0
	$\Delta E^{RPS\,d}$, kcal/mol	70.600	97.960	314.903

[a] Data were obtained by using the HF AM1 semiempirical version.

[b] Molecular energies are presented throughout the book by heats of formation, ΔH, determined as

$$\Delta H = E_{tot} - \sum_A \left(E^A_{elec} + EHEAT^A \right).$$

Here $E_{tot} = E_{elec} + E_{nuc}$, while E_{elec} and E_{nuc} are the electron and core energies. E^A_{elec} and $EHEAT^A$ are the electron energy and heat of formation of an isolated atom, respectively.

[c] Here ionization potential (IP) and electron affinity (EA) correspond to energies of HOMO and LUMO, respectively, just inverted by sign. Experimental data for relevant orbitals of C_{60} are taken from Weaver et al.[23] (1) and Wang et al.[24] (2).

[d] ΔE^{RPS} presents the difference between the heats of formation of RHF and pure spin states.

data. The latter is related even to such sophisticated techniques as quantum chemical calculations, where, as seen from Table 3.2, application of the restricted approach results in lifting the molecule symmetry.

As for structural computational experiments, let us consider what it means to lower the C_{60} symmetry from I_h (RHF) to C_i (UBS HF). As it turns out, the UBS HF solution supports the molecule truncated icosahedron shape formed by two groups of C-C bonds. Moreover, similar to the RHF solution, the UBS HF one supports the difference in the character of the bonds. Thus, the short bonds have close to the double bond character, which might be characterized by the Wiberg bond index.[27] In both cases the average Wiberg index for the bonds is 1.494, similar to that in benzene. As for long bonds, the average Wiberg index of 1.10 clearly evidences the single bond character. Therefore, the symmetry lowering is not connected with changing either the molecule shape or C-C bond character and is concerned with some delicate quantitative characteristics. Such is the case, indeed, and the feature concerns the scattering of C-C bond lengths. Figure 3.1(a) presents the dispersion of C-C bond distribution for two molecule structures that correspond to RHF and UBS HF solutions, while Figure 3.1(b) reveals the bond transformation. As seen in the figure, if the

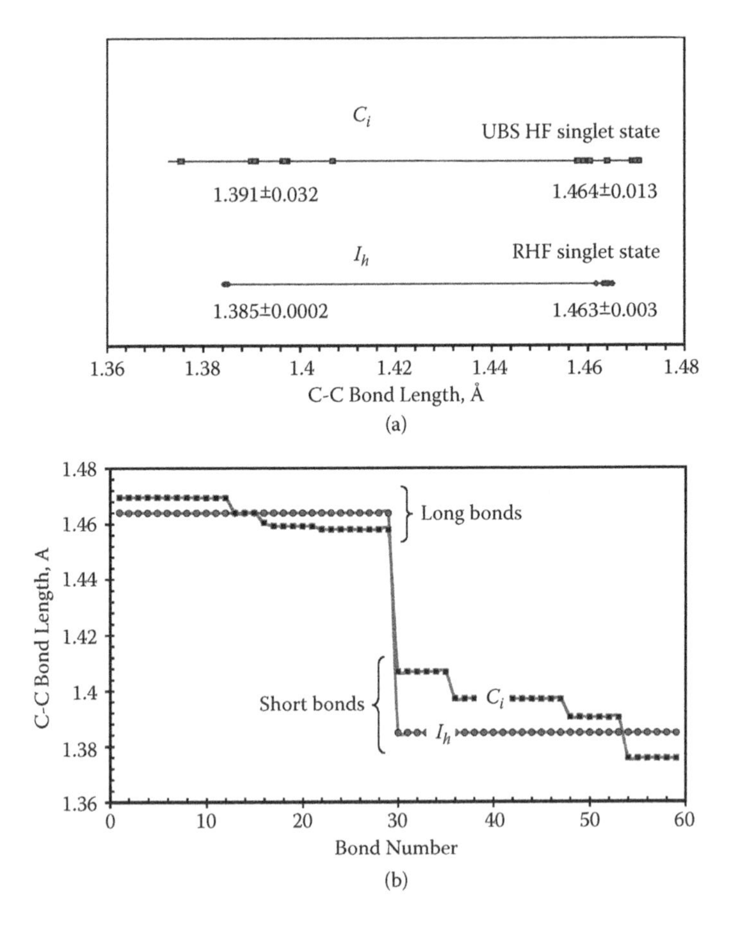

Figure 3.1 Correlation between C-C bonds of C_{60} fullerene related to restricted and unrestricted HF calculations. (a) Bond dispersion. (b) Bond transformation. (a. From Sheka, E. F., *Int. J. Quant. Chem.*, 107, 2803–16, 2007.)

average values of the bond lengths practically coincide for long bonds and slightly differ for short ones, the corresponding dispersions caused by splitting bonds of both kinds into four groups differ many times (four times and sixteen times for long and short bonds, respectively). This large difference in the bond length dispersion strongly actualizes the question concerning uncertainty of structural experiments performed, since the two symmetry-different structures can be distinguished only if the accuracy of the bond length determination is better than 10^{-2} Å. The requirement is just on the limit of the modern experiment accuracy, so that it is really not definitely discriminative for the case.

3.2.3 C_{60} *shape symmetry: Optical spectra*

Electronic optical spectra are another source of information related to the point group symmetry. An exhausted and detailed analysis of highly structure-sensitive low-energy $S_0 \leftrightarrow S_1$ electronic optical spectra of C_{60} forced Orlandi and Negri[28] to conclude that in spite of a high-symmetry pattern of the spectra as a whole, some peculiarities, such as a weak pure electronic 0_0^0 transition in both absorption and fluorescence spectra, the vibronic series in the latter spectrum based on *g* symmetry vibrations, absolutely allowed a pattern of the phosphorescence spectrum, the "silent" modes in Raman and IR spectra and others are inconsistent with the high symmetry of the molecule and cast some suspicion on the point.

The conclusions made on the basis of spectral studies as well as discriminative inability of structural experiments could seemingly be ponderable arguments in favor of the point group symmetry C_i, although a high-symmetry pattern of the molecule spectra requires an additional explanation. This fact, as well as other physical reality, shows that the symmetry problem of fullerene C_{60} cannot be solved in terms of exact symmetry.

Further evidence of the problem actuality was obtained when studying optical spectra of fullerene C_{60} monoderivatives. Figure 3.2 presents three pairs of specular absorption/luminescence spectra belonging to molecules C_{60} and two of its monoderivatives dissolved in crystalline toluene. The spectra manifest the Shpol'skii effect, which concerns the fine structure of vibronic spectra of solute molecules in crystalline matrices under particular conditions.[29] The monoderivatives differ by added atomic groups, shown in Figure 3.3. A high-symmetry pattern of the pristine fullerene spectra in Figure 3.2(c) is obvious: a very low intensity of the pure electronic 0_0^0 bands in both spectra and the dominance of vibronic series based on nontotally *u* symmetric vibrations that is just typical for a Herzberg-Teller (HT) pattern. This pattern is characteristic for high-symmetry molecules with forbidden or slightly allowed electron transitions, such as spectra of benzene and naphthalene.[33] It is usually accepted as convincing experimental evidence of high symmetry of the molecular

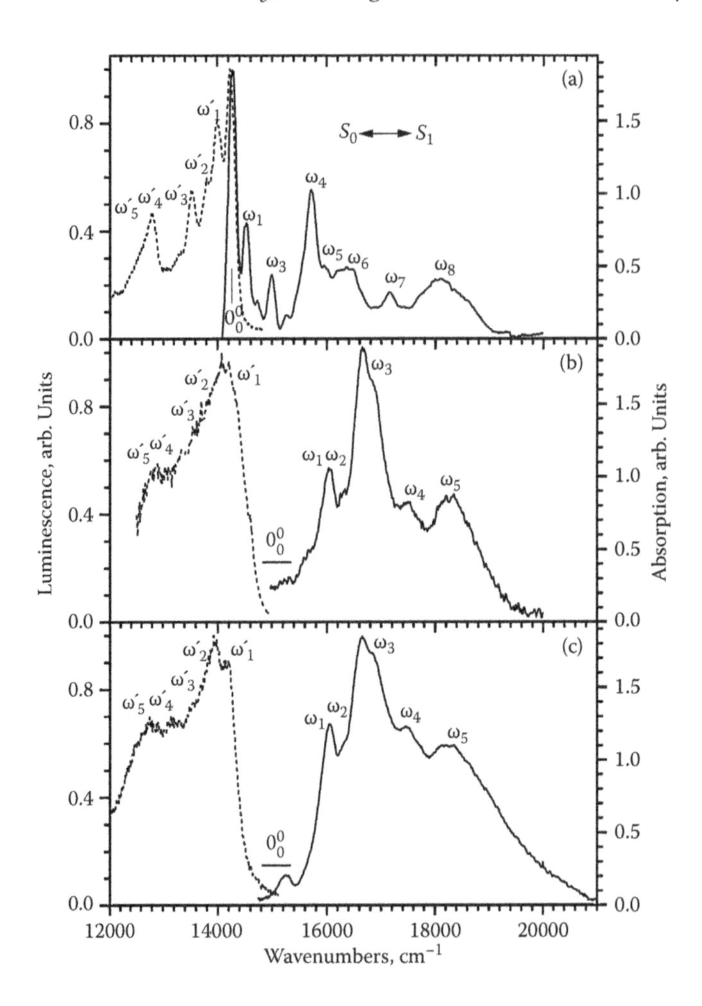

Figure 3.2 Specular background-free absorption (solid lines) and luminescence (dotted lines) spectra (see details of the spectra plotting in Section 10.2) of I (a), II (b), and C_{60} (c) in crystalline toluene (T = 80 K). For details of vibrational analysis, see Razbirin et al.[30] Symbol 0_0^0 marks the position of a pure electronic transition that is quite certain for I and fills some regions for II and C_{60} due to the multiplet structure. Structures of I and II are presented in Figure 3.3.

object under study. If one expects the pattern for C_{60} based on its I_h symmetry, UBS HF calculations show that both derivatives are fully nonsymmetric (C_1), which does not imply any forbidden electronic transitions. Consequently, their spectra should drastically differ from that of pristine fullerene. As it turns out, the reality does not follow the expectation completely. Actually, as seen in the figure, the spectra of species I in Figure 3.2(a) are absolutely different from those of pristine fullerene in Figure 3.2(c), while the spectra of species II in Figure 3.2(b) are quite similar to the latter. The difference between spectra of C_{60} and I concerns not only the detailed vibrational structure of the spectra, which is expected, but the shape of the total spectra. Thus, a high-symmetry HT pattern of the C_{60} is substituted by the Franck-Condon (FC) pattern that is typical for resolved electronic transitions and is characterized by intense pure electronic 0^0_0 bands and vibronic series based on g symmetric vibrations only.[34] In contrast to the case, spectra of species II preserve the HT pattern and a close similarity to the spectra of C_{60}. Therefore, in spite of a low symmetry, molecule II is characterized by a high-symmetry pattern of its electronic spectra.

The finding clearly shows that the problem concerning the symmetry of the C_{60} skeleton is not simple. The notations C_i and C_1 themselves do not mean much, since the matter is evidently not about exact yes/no symmetries but about how much they both differ from, say, the I_h one. The situation is not unique in the molecular world, but the exclusive position of the fullerene draws particular attention to the problem. A concept based on the continuous symmetry approach seems to be the most suitable tool for the problem elucidation.

3.2.4 *Continuous symmetry concept*

A highly heuristic view that symmetry can be and, in many instances, should be treated as a continuous gray property, and not as a black or white one that exists or does not exist, was introduced by Zabrodski, Peleg, and Avnir (ZPA) in 1992.[35] Speaking about continuous symmetry, the ZPA approach does not restrict itself by the object shape, but concerns mass or electronic distributions as well. This means that *continuous symmetry* should be determined not only for the object shape, but also for other massifs of data that provide the description of molecular properties on a matrix-element level.[36,37] However, usually the shape symmetry is considered as a basic one that strictly governs all other object properties.

As thoroughly analyzed in Estrada and Avnir,[38] the strict codex of perfect symmetry is particularly devastating in the case of small deviations from symmetry that force a jump in the symmetry description, the magnitude of which is totally out of proportion from that deviation. At the same time, the continuous symmetry concept makes it possible to replace a conventional question. "What is the object symmetry under such or other conditions?" by another one, "How much does the symmetry of the object deviate from an ideal or reference one under such or other conditions?" If the answer to the latter can be quantified, one can expect to be able to substitute usually met statements like: "The NH group perturbs the electron system of the C_{60} molecule to a small extent" (with respect to the intensity of the lowest electronic transition in Figure 3.2(b)) by some quantitative structure-property correlations.

The continuous symmetry approach can satisfy these expectations to the most extent.

Mathematically, the approach is based on the distance function formulation that makes it possible to determine a set of useful quantitative characteristics. The first is the *continuous symmetry measure* (CSM) that is a distance function commonly employed in symmetry analysis. If M is a structure composed of n vertices (atoms) in an original configuration, Q_i, and G is any symmetry point group, the amount of G symmetric shape in M is defined as[39]

$$Sy(M,G)=\frac{1}{nD^2}\sum_{i=1}^{n}\left\|Q_i-P_i\right\|^2 \tag{3.1}$$

Here P_i are the searched corresponding points of the nearest G perfectly symmetric configuration, and D is a size normalization factor (the rms of all distances from the center of mass to the vertices), making Sy size invariant. The distance $Q_i - P_i$ is squared to avoid sign limitation, a common practice in distance function formulations. The bounds of Sy are 0 (Q_i coincides with P_i; M is perfectly symmetric) and 1 (the nearest symmetric structure coalesces onto a center point, the distance to which is 1). To make the values more sensitive to the geometry deviation, the latter are usually scaled by 100.

The *continuous symmetry number* (CSN) is the next distance function, which measures the deviation from rotational symmetry. Originally, they were introduced as

$$\sigma=\sum_{k}r_k \tag{3.2}$$

to correct evaluation of rotational entropy.[38] Here k numbers proper C_n rotations and r_k is determined as

$$r_k=1-Sy_k \tag{3.3}$$

where Sy_k evaluates the degree of the deviation a specific rotation operation k from perfection and is expressed as

$$Sy_k(M,k)=\frac{1}{nD^2}\sum_{i=1}^{n}\left\|Q_i-P_{ik}\right\|^2 \tag{3.4}$$

The distance function Sy_k should be determined for each of k symmetry elements of the point group G within the interval from 0 to 1. Obviously, the last three equations can be expanded over improper rotations S_n, inversion S_2, mirror symmetry C_s, and chirality C_h, thus allowing determination of the *continuous symmetry level* (CSL), σ_{cont}.[40] When the set of symmetry elements of the point group G is formed from the above elements, CSL σ_{cont} describes an absolute contribution of the G symmetry in the studied structure, while $\eta_{cont} = \sigma_{cont}/\sigma_{classic}$, where $\sigma_{classic}$ equal to k is a perfect measure of the G symmetry, presents a relative contribution.

A complex of programs created at the Hebrew University of Jerusalem[41] offers a great possibility of determining a consistence of shapes of two structures by the SHAPE program, finding a selected point group symmetry contribution in a studied structure by the SYMMETRY program, and evaluating the chirality of the structure by the CHIRALITY program. Let us look at pristine fullerene C_{60} and its derivatives from the viewpoint of continuous symmetry.

3.2.5 *Continuous symmetry of fullerene C_{60} and its monoderivatives*

The first application of the continuous symmetry concept to fullerene C_{60} concerned the symmetry of its anions from C_{60}^{-1} to C_{60}^{-6}.[38] Classically, the structural change induced in C_{60} upon charging to C_{60}^{-1} is associated with an abrupt drop in symmetry from I_h to D_{3h}, which leads to a grossly overestimated rotational entropy change and to CSN σ_{cont} according to Equation 3.2, limited by proper rotations, equal to either 60 or 6, respectively. Similar abrupt changes without any reasonable order are also obtained for the rest anions toward C_{60}^{-6}. At the same time, changes in CSNs, calculated for rotations involved in the rotational subgroup of the point group I_h, for all anions (σ_{cont} lies in the interval from 58.73 to 58.72) constitute only 2.2% instead of the 90% expected classically. This means that the anion C_{60} skeleton preserves the I_h-ness to a great extent.

Let us address the continuous symmetry measure to answer two questions: (1) How much C_i structure of C_{60} fullerene that follows from the UBS HF solution deviates from the I_h structure of the RHF solution, and (2) how much I_h symmetry contributes to the structure of various C_{60} monoderivatives as a whole and to their C_{60} skeletons in particular? Programs SHAPE and SYMMETRY give quite an exhaustive answer to the first question. The shape symmetry analysis of C_{60}_RHF and C_{60}_UBS HF structures shows that when the former structure is compared with itself, $Sy(M,G)$ is equal to 0. When it is compared with the second one, $Sy(M,G)$ = 0.011951 (0 to 100 scale). $Sy(M,G) = 0$ trivially points to the structure's identity, while nonzero $Sy(M,G) = 0.011951$ evidences a deviation of the C_i structure from the I_h one. However, the deviation is small enough to be accepted as practically negligible.

Eight symmetry elements of the I_h point group are given at the top of Table 3.3. Below the corresponding Sy_k values determined according

Table 3.3 Continuous symmetry measures Sy_k related to the symmetry elements of I_h point group[a]

Molecule	E	C5	C3	C2	i	S10	S6	σ (Ch)
C60_RHF	0	0	0	0	0	0	0	0
C60_UBS HF	0	**0.000093**	**0.000085**	**0.000054**	**0.000056**	**0.00011**	**0.0001**	**0.000056**
I	0	**0.097742**	**0.058939**	**0.374478**	**0.945437**	**0.47872**	**0.42051**	**0.033549**
I_sk	0	0.000231	0.000186	0.000128	0.000132	0.00024	0.00023	0.000132
II	0	**0.040622**	**0.034247**	**0.0007**	**0.064343**	**0.14461**	**0.06438**	**0.025276**
II_sk	0	0.000094	0.000086	0.000057	0.00058	0.0001	0.00011	0.000056
III	0	**0.041919**	**0.050337**	**0.036587**	**0.363893**	**0.39697**	**0.39599**	**0.0193**
III_sk	0	0.000102	0.000092	0.000067	0.00006	0.00011	0.00011	0.00006
IV	0	**0.056269**	**0.032465**	**0.032465**	**0.987355**	**0.35785**	**0.35683**	**0**
IV_sk	0	0.000215	0.000159	0.000121	0.000121	0.00023	0.00022	0.000121
V	0	**0.101112**	**0.068744**	**0.042874**	**0.877885**	**0.63536**	**0.5452**	**0.159748**
V_sk	0	0.000225	0.000165	0.000126	0.000127	0.00024	0.00023	0.000127

[a] Equilibrium structures of species molecules are given in Figure 3.3. The addition of sk marks a sixty-atom fullerene skeleton of the corresponding derivatives.

Table 3.4 Classical and continuous symmetry levels of fullerene C_{60} and its monoderivatives[a]

Molecule	Symmetry	σ Classic	σ Continuous		
			$E + C_n$	$i + S_n + \sigma$	Total
C60_RHF	I_1	120	**60**	**60**	**120**
C60_UBS HF	C_i	2	**59.995**	**59.994**	**119.99**
I	C_i	1	50.858	38.652	89.51
I_sk	C_i	1	**59.989**	**59.987**	**119.976**
II	C_i	1	58.33	54.798	113.1279
II_sk	C_i	1	**59.995**	**59.994**	**119.989**
III	C_i	1	57.438	41.899	99.338
III_sk	C_i	1	**59.995**	**59.994**	**119.989**
IV	C_i	1	57.513	43.288	100.801
IV_sk	C_i	1	**59.99**	**59.988**	**119.978**
V	C_i	1	55.555	30.573	86.129
V_sk	C_i	1	**59.989**	**59.988**	**119.977**

[a] See footnote to Table 3.3.

to Equation 3.4 within the SYMMETRY program are listed. The values related to C_{60}_RHF (I_h) equal zero, as should be expected. Sy_k values for C_{60}_UBS HF (C_i) differ from zero starting from the fourth or fifth digit after the point. However, the deviation is actually small, so that the calculated CSL σ_{cont} of the structure (see Table 3.4) constitutes 119.99 instead of $\sigma_{classic}$ = 120 and η_{cont} = 99.99%. Therefore, the symmetry analysis shows that the C_i structure is practically of I_h symmetry. The high continuous symmetry of the molecule provides high-symmetry patterns of all the symmetry-sensitive experimental recordings. At the same time, its deviation from the exact I_h symmetry may be used to explain the inconsistencies of experimental recordings from the exact high-symmetry ones.

As for fullerene derivatives, the continuous symmetry measure analysis makes it possible to trace changes in both the derivative as a whole and its C_{60} skeleton when an atomic group is added to the pristine fullerene. Let us look at the relevant data for the structures shown in Figure 3.3, which are presented in Tables 3.3 and 3.4. The data related to the molecules themselves are in bold. As seen from the table, all additions considerably disturb the molecule structures, so that their deviations from the I_h one are quite large. Actually, σ_{cont} in Table 3.4 changes from 113.128 to 86.129, resulting in η_{cont} lying in the interval from 94.3 to 71.8%. In its turn, the expansion of the changes over symmetry elements in terms of Sy_k in Table 3.3 forms the ground for a new *symmetry language* for the change description, as well as for its comparative analysis with respect to the varied structure and composition of the added groups. This makes it possible to substitute a

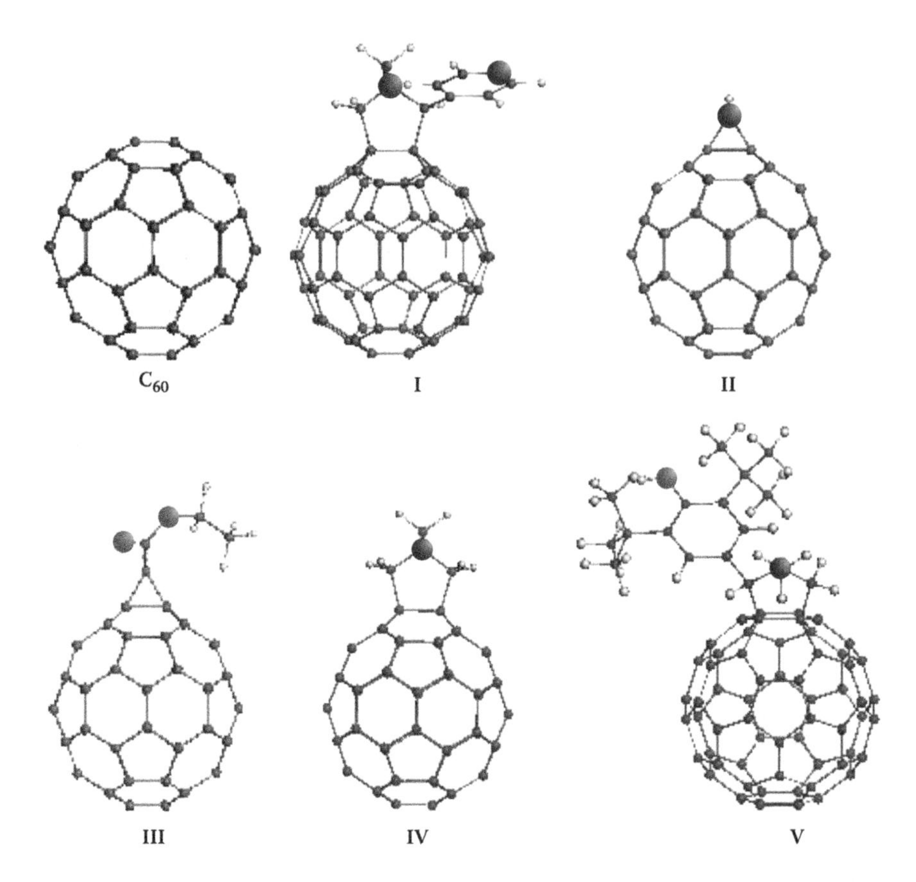

Figure 3.3 Equilibrium structures of fullerene C_{60}, N-methyl-2(4-pyridine)-3,4-[C_{60}]fulleropyrrolidine (I),[30] fullerene azyridine [C_{60}] (II), ethyl ester of [C_{60}]fullerene acetic acid (III),[31] N-methyl-3,4-[C_{60}]fulleropyrrolidine (IV),[30] and N-methyl-2-(3,5-di-*tert*-butyl-4-hydroxyphenyl)-[C_{60}] fulleropyrrolidine (V).[32] Carbon atoms are not shown. Big blue and red balls mark nitrogen and oxygen atoms. Carbon and hydrogen atoms are shown by small dark and white balls, respectively. Big dark gray and light gray balls mark nitrogen and oxygen atoms. UBS HF singlet state.

usually met typical sentence such as "CH_3 unit causes a rather weak effect on the benzene ring structure" with another one: "C_s symmetry of toluene molecule preserves 77% symmetry with respect to C_6 operation, 80% of C_3, 98% of C_2, 92.3% of inversion, 100% of C_s, thus conserving ~92% of D_{6h} symmetry in total." Similarly, the data presented in Table 3.3 may be used to describe C_1 symmetry of the discussed derivatives. It is clear from the table that concrete content of this description is different for the molecules characterized by different addends, in spite of the fact that the latter are added to the same atoms of the pristine fullerene cage.

In contrast to the whole molecules, their C_{60} skeletons having C_1 exact symmetry in all cases preserve (99.99 ÷ 99.98)% of I_h symmetry, as follows from Tables 3.3 and 3.4. Therefore, the same molecule may produce both low-symmetry and high-symmetry experimental patterns, depending on which atoms are involved in the empirical response. If those are skeleton atoms only, one obtains a high-symmetry patterned response. If addend atoms are mostly involved, a low-symmetry patterned response will be obtained. Let us consider this effect exemplified by optical electronic spectra of the studied derivatives.

3.2.6 *Continuous symmetry view on optical electronic spectra of fullerene C_{60} and its derivatives*

We shall consider electronic spectra corresponding to optical transitions between the ground and excited electronic states produced within the first HOMO-LUMO gap.[28] In the framework of single-determinant HF approximation, the atomic function composition of HOMO and LUMO will govern the atomic-sensitive characteristic of the wave functions of excited states. Table 3.5 presents a percentage contribution of addend atoms in HOMOs and LUMOs of the studied derivatives. As seen from the table, the addend atoms do not contribute to LUMOs in all cases. As for HOMO contributions, the latter is zero for molecule II, constitutes 0.9 ÷ 1.4% for molecules III, IV, and I, and absolutely dominates at the level of 97.1% for species V. The obtained data allow for making certain conclusions concerning optical spectra of the molecules:[40]

1. Spectra of molecule II are governed by C_{60} skeleton atoms only, so that the spectra pattern should be similar to that of the pristine fullerene.
2. Spectra of molecules I, III, and IV are initiated by both C_{60} skeleton atoms and addends, so that the whole molecule is responsible for the spectra intensity. Suggesting that the intensity would correlate with

Table 3.5 Percentage contribution of the addend atoms into HOMOs and LUMOs (UBS HF singlet state)

Molecule	HOMO	LUMO
I	1.4	0
II	0	0
III	0.9	0
IV	1.0	0
V	97.1	0

the deviation of the molecule structure from I_h symmetry (taken as a high-symmetry (HS) measure of the C_1-ness in terms of I_h symmetry), the intensity should increase when going from molecule IV (16%) to III (17%), and then to I (25%). At the same time, the intensity of the spectra, which correspond to the excitation of the lowest excited state, should be compared with that of pristine fullerene due to the low contribution of the addend atoms to the relevant HOMOs.
3. Spectra of molecule V should have a spectral-allowed pattern and be of the highest intensity due to both a considerable HS measure of the C_1-ness (19.9%) and a high addend atom contribution to the HOMO.

These predictions have been supported by direct calculations of the oscillator strength related to optical dipole transitions to the first low-lying excited states. The data obtained in the framework of the ZINDO/S method implemented in the GAUSSIAN package for molecular structures (Figure 3.3) are given in Table 3.6. The table lists five lowest excited states that contribute to low-frequency optical absorption spectra of the species.[28] As seen from the table, both I_h and C_i structures of C_{60} are characterized by zero oscillator strength of electronic transitions in this region. This might be interpreted that in spite of a formal allowance of the transitions for the C_i molecule, its wave functions are in fact highly symmetrical enough to provide zero dipole matrix elements related to the transitions. This appears to be consistent with both the shape and symmetry analysis discussed earlier.

The remaining five molecules are characterized by allowed transitions in the region, of which the summary intensity forms a series V > I ≥ IV > III > II. It should be noted that when summary intensities of the transitions of molecules V, I, and IV are comparable, those of species III and II are two to three times less. The series is fully synchronous with the one presenting the deviation of the molecule symmetry from the I_h one, as follows from Table 3.4. Actually, following the disclosed tendency V > I ≥ IV > III, one can see a synchronous directed changing in the spectra pattern of molecules V, I, IV, and III, presented in Figure 3.4. The spectra of the first three molecules have a clearly exhibited FC vibronic pattern, and their intensity somewhat decreases downward, accompanied by the simultaneous extension of the spectra vibronic structure of the luminescence spectra. Thus, in the case of molecule V, the zero-vibration (pure electronic) 0_0^0 band dominates in the spectrum, while its vibrational repetitions are rather scarce. The same happens in the case of molecules I and IV, with the domination of the 0_0^0 band slightly decreased. In contrast to these cases, spectra of molecule III, whose intensity is twice that of those of the above-mentioned molecules, are of a different shape; the 0_0^0 band in both absorption and luminescence spectra constitutes an ordinary part of the extended vibronic series, which is a mixture of the FC and HT series in

Table 3.6 Calculated energies of the lowest excited states and oscillator strengths of the lowest electronic transitions (ZINDO/S)

Molecule	Energy, eV	Oscillator strength
C_{60}-RHF	2.2394	0
	2.2414	0
	2.2429	0
	2.3156	0
	2.3163	0
	Σ	**0**
C_{60}-UBS HF	2.0785	0
	2.1456	0
	2.1458	0
	2.2217	0
	2.2856	0
	Σ	**0**
I	1.9712	0.0098
	1.9898	0.0004
	2.1897	0.0001
	2.1897	0.0001
	2.2506	0.0003
	Σ	**0.0107**
II	2.0120	0
	2.0490	0.0043
	2.1440	0
	2.225	0.0003
	2.3	0.0001
	Σ	**0.0047**
III	2.0548	0.0002
	2.1073	0
	2.1417	0.0061
	2.2226	0
	2.3314	0.0001
	Σ	**0.0064**

(continued on next page)

Table 3.6 (continued) Calculated energies of the lowest excited states and oscillator strengths of the lowest electronic transitions (ZINDO/S)

Molecule	Energy, eV	Oscillator strength
IV	2.0487	0.0004
	2.0782	0.0001
	2.0840	0.0096
	2.1926	0
	2.3559	0.0002
	Σ	**0.0103**
V	2.0445	0.0007
	2.0636	0.0022
	2.0778	0.0090
	2.1839	0.0003
	2.3538	0.0003
	Σ	**0.0125**

favor of the former, so far starting with a rather weak but still clearly vivid 0_0^0 band.[31] This dependence of spectra vibronic structure on the intensity of electronic transition is a typical feature in molecular spectroscopy of large molecules, where weak electronic transitions are presented by extended series of vibronic bands, opposite to strong allowed transitions. This is connected with the fact that in the former case there is a comparable chance for a number of different vibrations to be exhibited, including both *g* and *u* symmetric vibrations. In the latter case, the vibrational repetitions are provided by a limited number of *g* symmetric vibrations only, the series extension of which is determined by the shift of the equilibrium positions of atoms under the electronic excitation. For large molecules, the latter is usually small, which results in a short vibronic series. Spectra of molecule V in Figure 3.4(a) clearly demonstrate the latter tendency, particularly for the luminescence partner. Lowering the intensity for molecules I and IV produces a vivid extension of the vibronic structure of their spectra (Figure 3.4(b) and (c)), while further lowering the intensity of the electronic transitions for molecule III causes a substantial enrichment of the vibronic structure, transforming it into a mixture of FC and HT series, so that it becomes quite similar to a rich vibronic HT structure of electronic transitions of the pristine fullerene C_{60} spectrum.

Figure 3.5 presents spectra of molecules III and C_{60} in more detail based on fine-structured Shpol'skii's spectra at 2K. Unfortunately, due to a very strong tendency to clusterization (see, for example, Sheka et al.[41]),

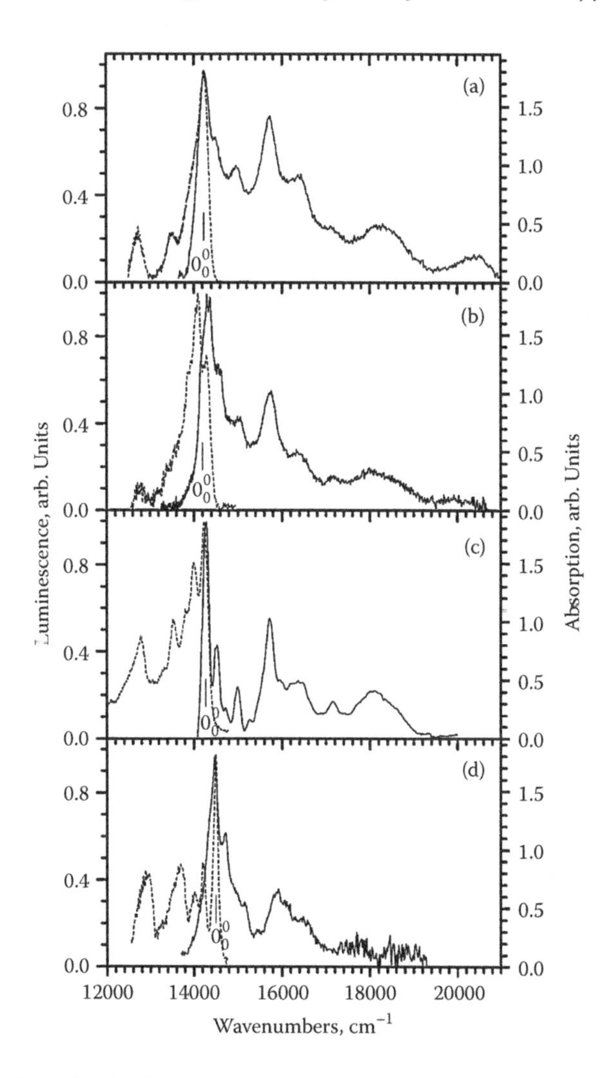

Figure 3.4 Specular background-free absorption (solid lines) and luminescence (dotted lines) spectra of (a) V,[32] (b) I,[30] (c) IV,[30] and (d) III[31] in crystalline toluene (T = 80 K). Symbols 0_0^0 mark positions of pure electronic transitions.

molecules II cannot be distributed in the toluene matrix as individual entities, which prevents them from obtaining their fine-structured vibronic spectra similar to those of molecules III and C_{60}, so that even at low temperature its spectra consist of rather broad bands (Figure 3.5(c)). As clearly seen in Figure 3.5(a) and (b), spectra of both molecules present a mixture of FC and HT series, with the difference that when the FC series is the most intense in the former case, the HT series dominates in the latter. Lowering the absolute intensity when passing from molecule III to C_{60} inhibits the

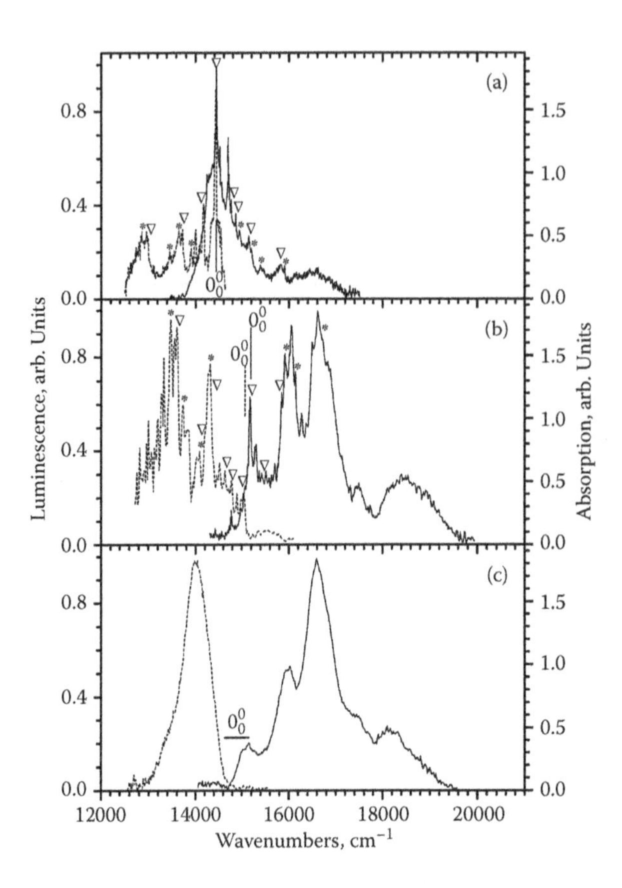

Figure 3.5 Specular background-free Shpolskii's absorption (solid lines) and luminescence (dotted lines) spectra of (a) III,[31] (b) C_{60},[30] and (c) II in crystalline toluene (T = 2 K). 0_0^0 marks positions of pure electronic transitions. Stars and triangles mark vibronic bands related to the HT and FC series, respectively. The series separation is based on the vibrational analysis of the spectra given in Razbirin et al. [30,31]

intensity of the FC series. The same happens when going from molecule III to II since the spectra of the latter are just overlapping those of C_{60}.

Based on this finding, it might be said that the summary strength of the electronic transitions at the level between 0.005 and 0.006, which is characteristic for II and III (Table 3.6), can be considered a limit for a transition from a predominantly forbidden HT spectra pattern to a predominantly allowed FC one. In the studied case, this corresponds to 10 to 15% deviation from the highest I_h symmetry. Therefore, a (0.006 – 0.005)/(10–15)% relationship seems to be a bordering condition when a change in the structure symmetry of a fullerene molecule will cause a qualitative reconstruction of its optical spectra that allows establishing a lifting or lowering of the molecule symmetry. Similar quantitative relationships can be established for other structure-sensitive experimental techniques, such as nuclear magnetic resonance (NMR), infrared (IR), and Raman spectroscopy. Evidently, quantitative expressions for bordering cases might be different for different techniques.

3.3 *Total spin* $\langle \hat{S}^2 \rangle$

Within the framework of the UBS HF approach, the total spin, presented by its squared value, $\langle \hat{S}^2 \rangle$, is described by Equation 2.22 and differs markedly from the exact value $S(S+1)$ when the spin contamination of the UBS HF solution is significant. Therefore, the ratio

$$\varsigma(\%) = \frac{\langle \hat{S} \rangle^2 - S(S-1)}{S(S-1)} \tag{3.5}$$

may serve as a measure of the spin contamination of the corresponding spin states. Figure 3.6 presents all three quantities, namely, $\langle \hat{S}^2 \rangle$, $S(S+1)$, and relative spin deviation, ς, calculated for both C_{60} and Si_{60} fullerenes in accordance with Equations 2.22 and 3.5.[42] As seen in the figure, UBS HF solutions for C_{60} and Si_{60} are spin contaminated in different ways: if the spin contamination in C_{60} vanishes at total spin $S = 12$, then for Si_{60} the same is observed at $S > 20$, so that a majority of high spin states of the molecule are spin contaminated within the framework of the UBS HF

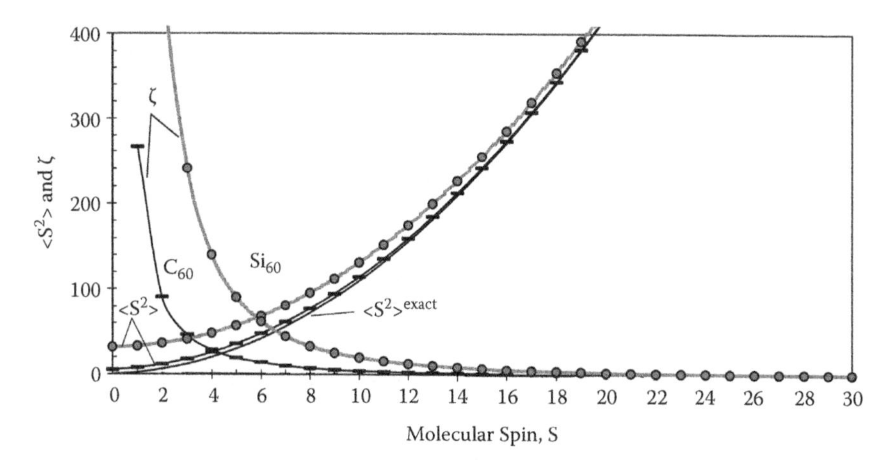

Figure 3.6 Total spin, $\langle \hat{S}^2 \rangle$, and its relative deviation, ζ, for C_{60} and Si_{60} in different spin states. Curves with filled dots are related to Si_{60}, while those with dashes are attributed to C_{60}. The solid curve depicts $\langle \hat{S}^2 \rangle = S(S+1)$. UBS HF singlet state. (From Sheka, E. F. *Int. Journ. Quant. Chem.* 100, 375–87, 2004.)

approach. Another difference concerns the singlet state where the contamination is the largest. For C_{60} the ζ value reaches 266.85, while for Si_{60} it is ~6 times larger and constitutes 1,529.55. The data reflect a drastic difference in the odd electrons' behavior of these two species, which first was mentioned when discussing Figure 2.1. This lays the foundation for different chemical and physical properties of the species, and we shall come back to a comparative analysis of the species not once throughout the book. As for C_{70}, its behavior is similar to that of C_{60}, both qualitatively and quantitatively.

3.4 Chemical reactivity of fullerenes C_{60} and C_{70}

A large part of C-C distances in both fullerene C_{60} and C_{70} exceeds the critical value R_{cov} = 1.395Å, so that a significant number of effectively unpaired electrons N_D looks quite natural.[43] Listed in Table 3.2, that constitutes 9.84 *e* and 14.40 *e* for C_{60} and C_{70} molecules, respectively, or about 15 to 20% of the total number of odd electrons. The fractions show a measure of a partial radicalization of both molecules, while N_D describes their molecular chemical susceptibility (MCS). In contrast, N_D = 63.52 for Si_{60}, which means that all sixty odd electrons are unpaired, causing a complete radicalization of the molecule. The latter should not seem to be strange in view of Si-Si distances filling the interval of 2.25 to 2.35 Å (see Figure 2.1). On the other hand, the radicalization explains the failure in producing Si_{60} fullerene in practice.

The distribution of atomic chemical susceptibility (ACS), quantified by N_{DA} in accordance with Equation 2.24, over atoms of Si_{60} and C_{60} fullerenes is shown in Figure 3.7. In complete agreement with the above reasoning, the N_{DA} values evidence the availability of one free electron on each of the Si_{60} atoms, whereas these values for C_{60} vary from 0.029 to 0.271 *e*. Similarly, for C_{70}, N_{DA} fills the interval 0.029 to 0.323 *e*. It is natural to compare ACS values with those of atomic free valence, V_A^{free}. Within the framework of UBS HF approximation, the quantity is defined as

$$V_A^{free} = N_{val}^A - \sum_{B \neq A} K_{AB} \tag{3.6}$$

Here N_{val}^A is the number of valence electrons of atom *A*, and

$$\sum_{B \neq A} K_{AB}$$

presents a sum over the generalized bond index, $K_{AB} = |P_{AB}|^2 + |D_{AB}|^2$. Here, the first term is the Wiberg bond index and the second term is

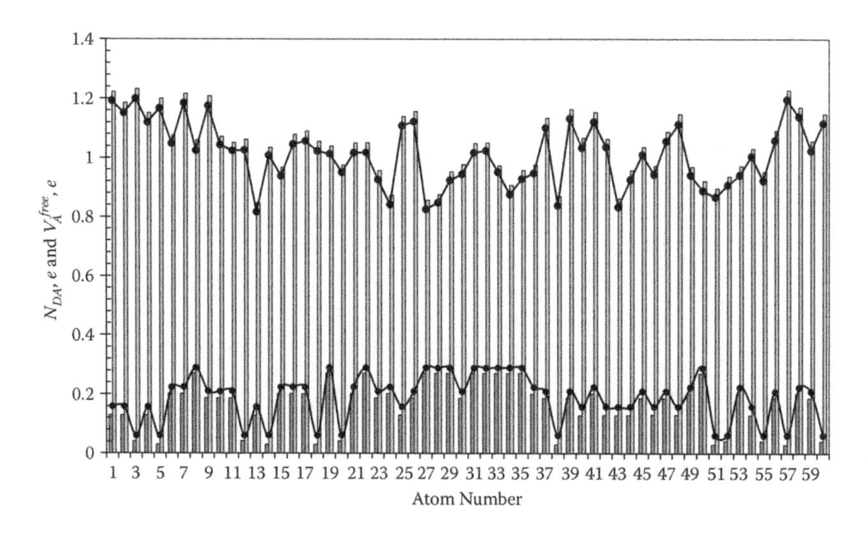

Figure 3.7 Atomic chemical susceptibility, N_{DA} (histograms), and free valence, V_A^{free} (curves with dots), over atoms of C_{60} and Si_{60} fullerenes. (From Sheka, E. F., *Int. J. Quant. Chem.*, 107, 2803–16, 2007.)

determined by the spin density matrix. The V_A^{free} distributions over the atoms of the molecules are shown by solid lines with dots in Figure 3.7. Excellent agreement of N_{DA} and V_A^{free} values show that the former is actually a quantitative measure of the atom chemical reactivity and can serve as a quantitative pointer of the target atoms on the fullerene cage, to which atom-atom contacts are the most active in addition reactions. Thus, the values distribution over molecule atoms forms a unique ACS molecular map. This opens a transparent methodology of a successive algorithmic *computational synthesis* of any fullerene-cage-based derivatives, just selecting the cage target atoms by the largest N_{DA} value. A large feasibility of this methodology will be shown later, in Chapters 4 to 7. Let us first look at the ACS maps of the C_{60} and C_{70} molecules.

3.4.1 *Chemical portrait of fullerene C_{60}*

Effectively unpaired electrons of 9.84 by total number are distributed over the molecule atoms, as shown in Figure 3.8. The ACS map in Figure 3.8(a) corresponds to an arbitrary numbering of atoms from the input/output files (like that in Figure 3.7). However, if N_{DA} are aligned by the value in a Z → A manner, the distribution takes a more ordered view, as seen in Figure 3.8(b). The map is divided into five groups with twelve atoms in each, which are characterized by the same N_{DA} value. Atoms of each group form six identical pairs consisting of two carbon atoms coupled via

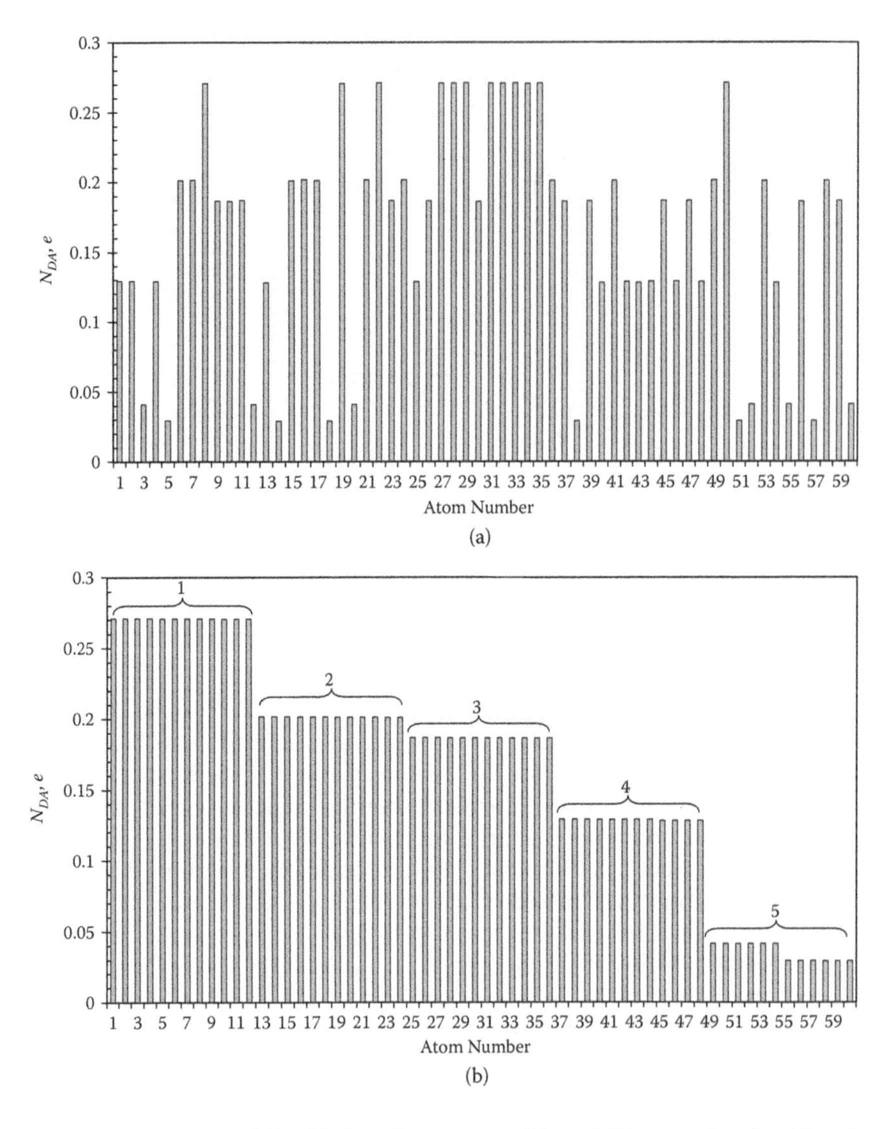

Figure 3.8 N_{DA} map of C_{60}: (a) data from output file and (b) same data but Z → A aligned by value.

a short C-C bond. Spin densities (the matter is about extra spin density caused by the spin contamination of the UBS HF solution; see details in Chapter 2) on the two atoms in any pair are equal by value and opposite by sign. Therefore, the total spin density over the molecule is zero, which is consistent with zero total spin of the singlet state. According to the ACS map, the C_{60} molecule consists of six identical C_{10} compositions formed by five pairs in accordance with the total number of the groups. Distributing

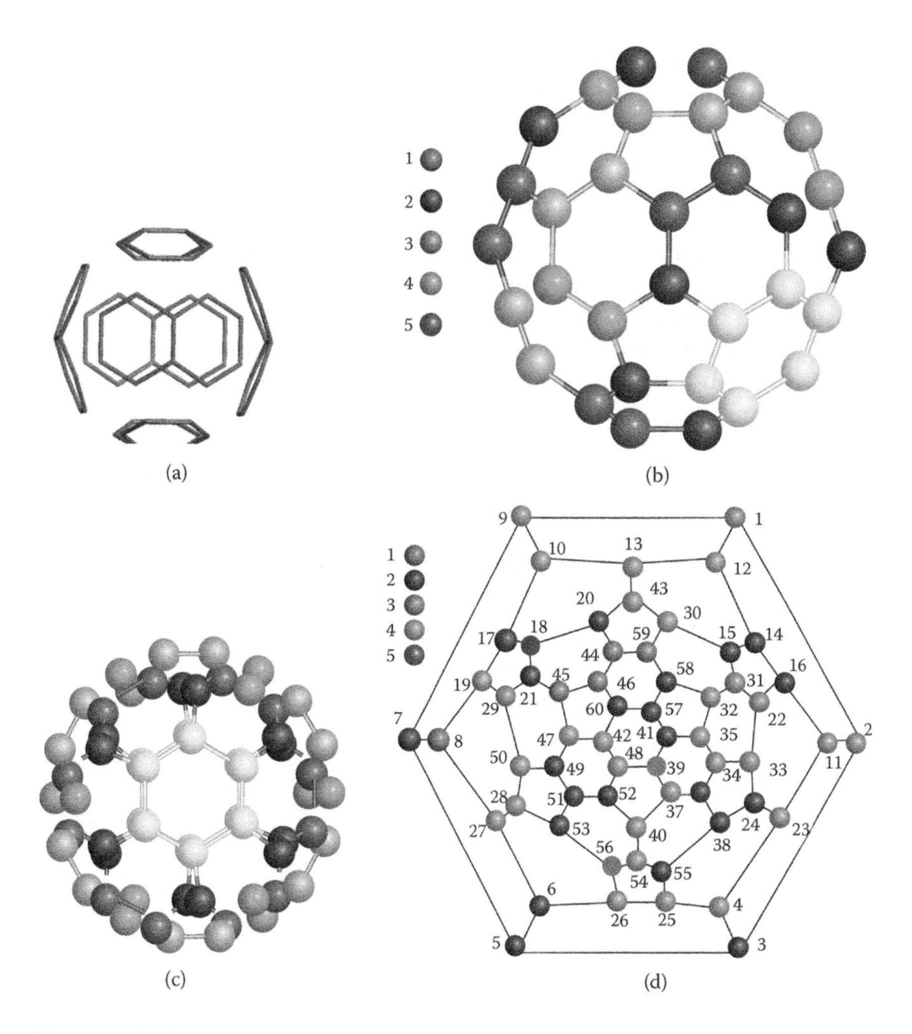

Figure 3.9 Chemical portrait of C_{60}. (a) 6^*C_{10} composition. (b and c) Two projections of colored C_{60} with different colors marking atoms with different ACS. Figures point to different atom groups in Figure 3.8. (d) C_{60} Schlegel diagram corresponding to colors and atom numbering in (b) and (c).

atoms over six fragments following the map shown in Figure 3.8(a), one gets a 6^*C_{10} configuration consisting of six identical naphthalene cores, shown in Figure 3.9(a). Applying different coloring to atoms of different N_{DA} values, one obtains a *chemical portrait* of the C_{60} molecule[43,44] in the singlet UBS HF state (Figure 3.9(b) and (c)). The picture unexpectedly favors one of the hypotheses of C_{60} molecule formation from mutually bonded C_5 carbene chains.[43] Additionally, the Schlegel diagram is presented in

Figure 3.9(d), of which the colored spots correspond to those on three-dimensional views of the molecule, while the atom numbering is identical to that in Figure 3.8(a).

3.4.2 Chemical portrait of fullerene C_{70}

The ACS map of the molecule[43,44] shown in Figure 3.10 presents the distribution of 14.39 unpaired electrons and shows much less contrast in comparison with that of C_{60}. Nevertheless, as previously, the N_{DA} distribution points to a well-defined grouping. Contrary to the previous case, one cannot distinguish a unique basic structural element, besides a benzenoid cycle C_6, multiplying which one might compose the molecule structure. The D_{5h} symmetry of the molecule in the UBS HF singlet state (see Figure 3.11(a)) governs the molecule structure decomposition into three five-benzenoid fragments (Figure 3.11(b) and (c)). Two twenty-atom fragments I in Figure 3.11(b) are formed by conjugated benzenoids and look like a five-lobe flower each. Five benzenoids mutually coupled via a single C-C bond of fragment II in Figure 3.11(c) form a thirty-atom closed rarefied chain bracelet. The highest N_{DA} values are concentrated just in this area. It is clearly seen in Figure 3.11(d) where a chemical portrait of C_{70} is presented, but in terms of spin density that fully correlates with N_{DA} values.

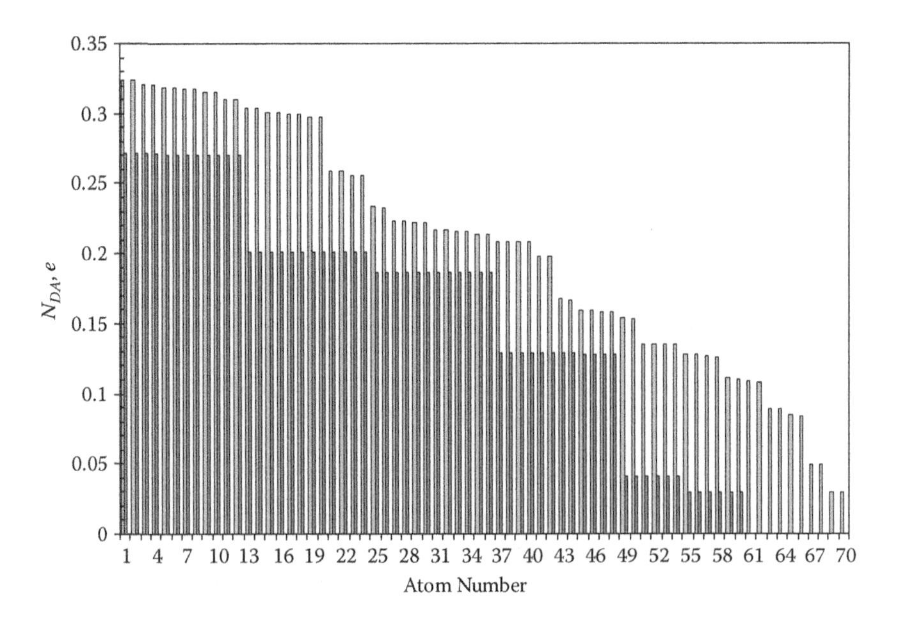

Figure 3.10 A comparative view of the N_{DA} maps of C_{60} (dark gray) and C_{70} (light gray); the data are aligned in a Z → A manner.

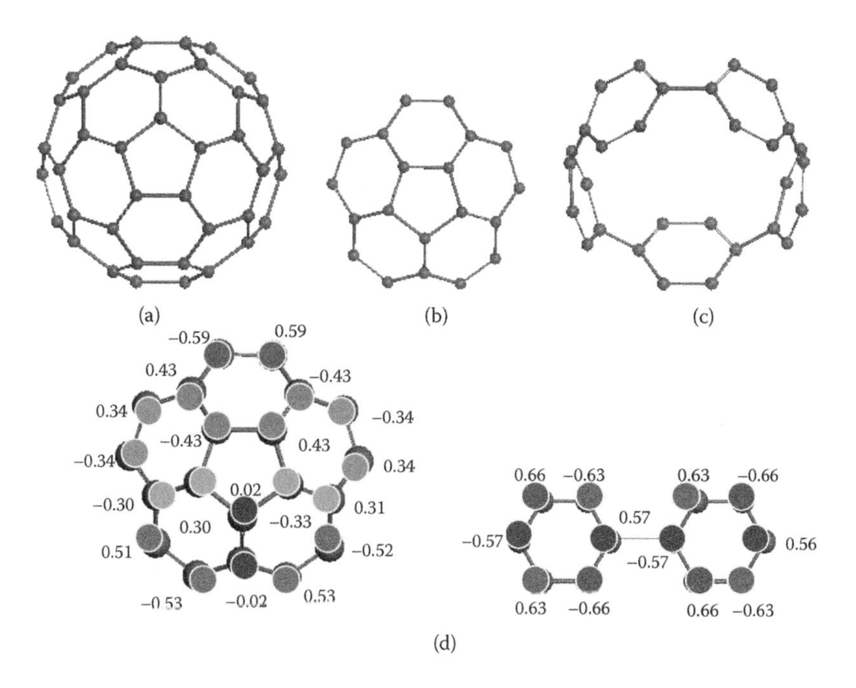

Figure 3.11 (a) C_{70} fullerene. Five-membered fragments I (b) and II (c). (d) Chemical portrait of fragments I and II. Digits point to spin density. UBS HF singlet state.

3.5 C_{60} isomers

Until now we have considered monoisomeric molecules C_{60} and C_{70} possessing only the structures presented in Figures 3.9 and 3.11. However, actually a great number of high-symmetry fullerenes can be assembled with high probability from nucleus clusters of a "good" symmetry.[46] Nevertheless, a monoisomeric approach dominates in the modern science of fullerenes, which implies that only one I_h isomer of C_{60} and a D_{5h} one of C_{70} is obtained in practice in macroscopic quantity. There is nothing to tell how important this monoisomeric approach is for computational simulations.

The approach might have been fully accepted if there were no particular inconsistencies of empirical results with I_h symmetry of C_{60}. Actually, the reality is full of inconsistencies, and some of them are discussed in Sections 3.2.4 and 3.2.5 related to optical spectra of C_{60}. Besides this, there are others facts concerning NMR spectra of the fullerene derivatives, ambiguity of structural data related to polyderivatives, differences in structure of homologous C_{60}-based derivatives, such as, say, $C_{60}F_6$ and $C_{60}Cl_6$, and many others. We shall face some of these in Chapters 4 and 6. Since the computational science of fullerene presented in the current book

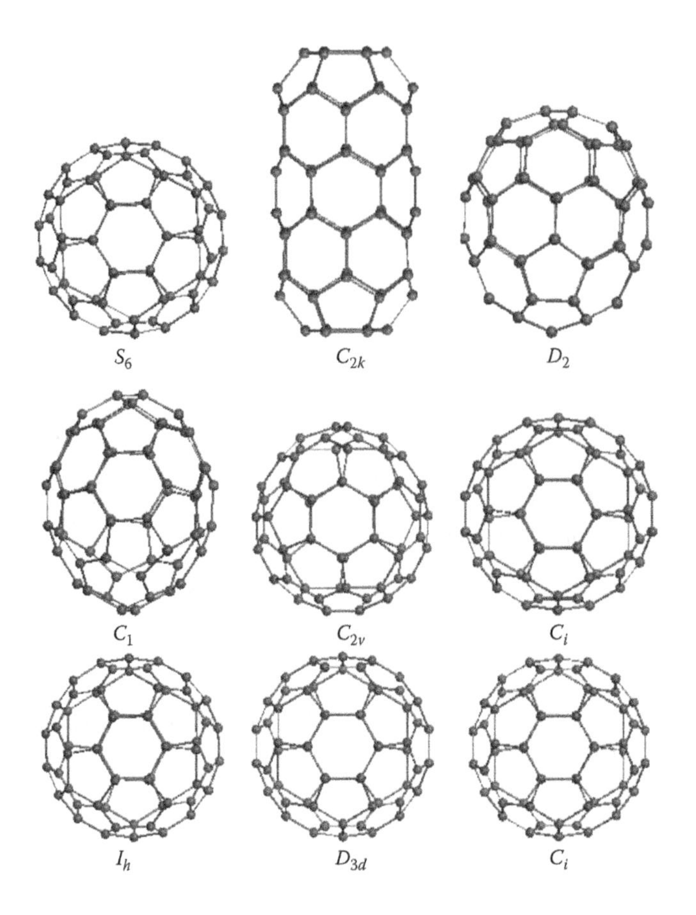

Figure 3.12 Equilibrium structures of C_{60} isomers. UBS HF singlet state.

is based on a monoisomeric approach, we have to evaluate at least how big the inaccuracy connected with this restriction might be.

Let us look at results of one of a few detailed investigations of the fullerene C_{60} isomerism presented in Tomlin et al.[45] The authors considered a few C_{60} isomers, among which there were molecules of I_h, D_{2d}, C_2, C_{2v}, and D_5 symmetry, and showed that some of them could be produced in due course of fullerene synthesis. We will follow their example and will consider a similar set. Figure 3.12 presents equilibrium structures of the relevant molecules resulting from UBS HF calculations, while Table 3.7 lists total energies and numbers of effectively unpaired electrons. Among nine structures in Figure 3.12, only a peculiar C_{2h} one presents a small carbon nanotube, while the others are truncated icosahedrons. As follows from the table, all isomers are partially radicalized—the more so the higher the total energy. Besides high energy, big N_D values are inhibiting factors toward the isomer formation since C_{60} clusters are too reactive, so that the matter

Table 3.7 Energies (E) and numbers (N_D) of effectively unpaired electrons of C_{60} isomers[a]

Isomer	E, kcal/mol	N_D, e
S_6	2017.93	36.22
C_{2h}	1194.29	23.58
D_2	1041.19	18.22
C_1	1038.00	19.85
C_{2v}	994.32	12.17
C_i	961.80	9.58
I_h	955.36	9.87
D_{3d}	955.37	9.88
C_i	955.36	9.87

[a] Equilibrium structures in UBS HF AM1 calculations. The isomer notations follow those in Figure 3.12.

can be only about the remaining four species, three of which, with the lowest energy, are isoenergetic and N_D isonumbered. Isomer C_i, included in the group, we described in the previous sections of this chapter in detail. Two others correspond to different initial structures.

To characterize these isomers and exhibit their difference, Figure 3.13 presents the distribution of a fixed set of C-C bonds. As clearly seen, the set of each isomer consists of eight groups of bonds that are identical by

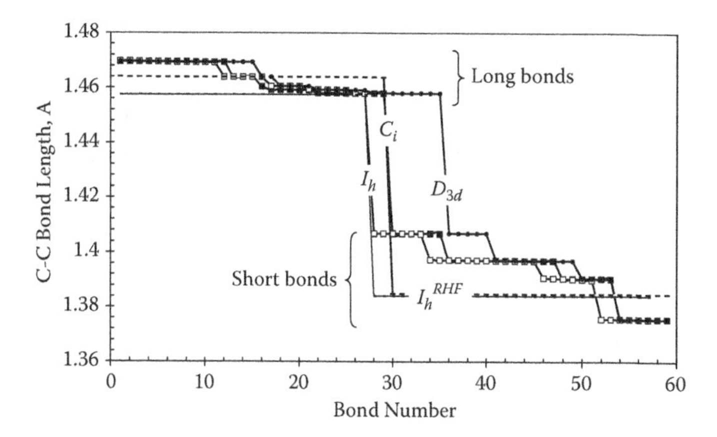

Figure 3.13 Curves I_h, C_i, and D_{3d} present the distributions of C-C bond lengths of the relevant C_{60} isomers. UBS HF singlet state. Two I_h^{RHF} curves correspond to results of RHF calculations of I_h (solid curve with dashes) and C_i and D_{3d} (dotted curve with dashes) isomers. Both UBS HF and RHF curves for I_h isomer are shifted to the left to avoid a complete overlapping data plotting for I_h and C_i isomers.

Table 3.8 Symmetry analysis of C_{60} isomers

C_{60} isomer	Shape $Sy(M,G)$	Symmetry	
		σ_{cont}	I_h percentage[a]
I_h	0	120	100
D_{3d}	0.000387	118.359	98.63
C_i	0. 011951	119.990	99.99

[a] I_h isomer presents the reference structure for comparison.

length for all isomers. The number of bonds within each group is the same for I_h and C_i isomers while different for the D_{3d} one. The bond grouping in comparison with the only short and long bonds forming two sets of the RHF solution is typical not only for C_i isomer (see Figure 3.1), but for I_h and D_{3d} isomers as well.

The difference in the number of bonds within the groups may be attributed to the difference in the bond assembly in isomer structures. At the same time, the picture presented in the figure forces one to think about the close resemblance between the three structures, as well as the closeness of their actual symmetries. Table 3.8 lists results of the symmetry analysis of the three structures in terms of the SHAPE and SYMMETRY programs, as described in Section 3.2.4. As seen from the table, all three isomers are practically I_h symmetric. There is, nevertheless, a very important difference in their structures. The latter concerns the pentagon unit framing: if in the case of the I_h and C_i isomers only long single bonds form the frame, one short double bond is involved in the framing in the D_{3d} isomer. As we shall see in Section 4.4.5, the feature might explain some peculiarities connected with the formation of polyderivative C_{60} fluorides.

Besides the discussed three isomers, the next by energy C_i and C_{2v} isomers shown in Figure 3.12 should be taken into account. Because there is not a big difference in the total energy, they might play an important role under particular conditions. Sure, a small set of C_{60} isomers considered does not cover the problem exhaustively. However, a definite glance at the problem as a whole we have nevertheless obtained.

3.6 *Concluding remarks*

In view of these isomer circumstances, we have to make a choice of a standard reference C_{60} structure between the three isoenergetic isomers to continue in a monoisomeric way of the result presentation later in the book. Long-bond framing of pentagons that have been exhibited experimentally for any commercial C_{60} substance might serve as the leading argument that allows rejecting the D_{3d} isomer, giving us a choice between the I_h and C_i ones. It should be emphasized that the matter is about the I_h isomer that is characterized by the dispersed structure in accordance with the bond distribution shown in Figure 3.13, but not about the I_h^{RHF} structure that is obtained in all restricted calculations. Isomers I_h and C_i are nondistinguishable by structure testing and possess practically identical chemical portraits. Nevertheless, a preference can be given to the C_i one due to the following factors: (1) the C_i molecule is evidently in better relations with extended spectroscopy of fullerene itself and its derivatives, and (2) the RHF modification of its structure in Figure 3.13 is fully consistent with a predominant majority of the available calculations, mainly restricted ones, concerning C_{60}. We understand that this choice may be considered arbitrary, but are pretty sure that the main results and conclusions presented in the following chapters do not change when the C_i isomer is substituted with I_h or even a D_{3d} one. The best way to support this suggestion is to consider the computational synthesis of C_{60}-based derivatives that offer a lot of points of independent comparison with experiments. Let us start this way with addition reactions concerning the fluorination of the pristine fullerene that might lead to manifestation of the D_{3d} isomer of the pristine molecule when forming high C_{60} fluorides.

References

1. Jones, D. E. H. 1966. Ariadne. *New Scientist* 32:245.
2. Osawa, E. 1970. Superaromaticity [in Japanese]. *Kagaku* (Kyoto) 25:854–63.
3. Yoshida, Z., and Osawa, E. 1971. *Aromaticity* [in Japanese], 174–78. Kyoto: Kagaku Dojin.
4. Bochvar, D. A., and Galpern, E. G. 1973. Hypothetical systems: Carbododecahedron, s-icosahedron, and carbo-s-icosahedron [in Russian]. *Dokl. Akad. Nauk SSSR* 209:610–12.
5. Stankevich, I. V., Nikerov, M. V., and Bochvar, D. A. 1984. The structural chemistry of crystalline carbon: Geometry, stability, and electronic spectrum. *Russ. Chem. Rev.* 53:640–55.
6. Davidson, R. A. 1981. Spectral analysis of graphs by cyclic automorphism subgroups. *Theor. Chim. Acta* 58:193–231.
7. Kroto, H. W., Heath, J. R., O'Brien, S. C., Curl R. F., and Smalley, R. E. 1985. C_{60}: Buckminsterfullerene. *Nature* 318:162–63.
8. Rohlfing, E. A., Cox, D. M., and Kaldor, A. 1984. Production and characterization of supersonic carbon cluster beams. *J. Chem. Phys.* 81:3322–30.
9. Kroto, H. 2003. Symmetry, space, stars, and C_{60}. Nobel lecture on December 7, 1996. In *Nobel lectures, chemistry 1996–2000*, ed. I. Grenthe. Singapore: World Scientific Publishing Co. 44–79.
10. Katz, E. A. 2008. *Fullerenes, carbon nanotubes and nanoclusters. Pedigree of shapes and concepts* [in Russian]. Moscow: IKK Publishing Co.
11. Hedberg, K., Hedberg, L., Bethune, D. S., Brown, C. A., Dorn, H. C., Johnson, R. D., and De Vries, M. 1991. Bond lengths in free molecules of buckminsterfullerene, C_{60}, from gas-phase electron diffraction. *Science* 234:410–12.

12. Leclercq, F., Damay, P., Foukani, M., Chieux, P., Bellisent-Funnel, M. C., Rassat, A., and Fabre, C. 1993. Precise determination of the molecular geometry in fullerene C_{60} powder: A study of the structure factor by neutron scattering in a large momentum-transfer range. *Phys. Rev. B.* 48:2748–58.
13. Slovokhotov, Yu. L., Moskaleva, I. V., Shil'nikov, V. I., Valeev, E. F., Novikov, Yu. N., Yanovski, A. I., and Struchkov, Yu. T. 1996. Molecular and crystal structure of C_{60} derivatives: CDS statistics and theoretical modeling. *Mol. Mater.* 8:117–124.
14. Yanonni, C. S., Bernier, P. P., Bethune, D. S., Meijer, G., and Salem, J. K. 1991. NMR determination of the bond lengths in C_{60}. *J. Am. Chem. Soc.* 113:3190–92.
15. Sheka, E. F. 2007. Chemical susceptibility of fullerenes in view of Hartree-Fock approach. *Int. J. Quant. Chem.* 107:2803–16.
16. Hedberg, L., Hedberg, K., Boltalina, O. V., Galeva, N. A., Zapolskii, A. S., and Bagryantsev, V. F. 2004. Electron-diffraction investigation of the fluorofullerene $C_{60}F_{48}$. *J Phys. Chem. A* 108:4731–36.
17. Chang, A. H. H., Ermler, W. C., and Pitzer, R. M. 1991. Carbon molecule (C_{60}) and its ions: Electronic structure, ionization potentials, and excitation energies. *J. Phys. Chem.* 95:9288–91.
18. Weaver, J. H. 1992. Fullerenes and fullerides: Photoemission and scanning tunneling microscopy studies. *Acc. Chem. Res.* 25:143–49.
19. Lee, B. X., Cao, P. L., and Que, D. L. 2000. Distorted icosahedral cage structure of Si_{60} clusters. *Phys. Rev. B* 61:1685–87.
20. Nagase, S. 1993. Theoretical study of heteroatom-containing compounds. From aromatic and polycyclic molecules to hollow cage clusters. *Pure Appl. Chem.* 65:675–82.
21. Slanina, Z., and Lee, S. L. 1994. A comparative study of C_{60}, Si_{60}, and Ge_{60}. *Fullerene Sci. Technol.* 2:459–69.
22. Lee, B. X., Jiang, M., and Cao, P. L. 1999. A full-potential linear-muffin-tin-orbital molecular-dynamics study on the distorted cage structures of Si_{60} and Ge_{60} clusters. *J. Phys. Condens. Matter* 11:8517–21.
23. Weaver, J. H., Martins, J. L., Komeda, T., Chen, Y., Ohno, T. R., Kroll, G. H., Troullier, N., Haufler, R., and Smalley, R. E. 1991. Electronic structure of solid C_{60}: Experiment and theory. *Phys. Rev. Lett.* 66:1741–44.
24. Wang, X.-B., Ding, C.-F., and Wang, L.-S. 1999. High resolution photoelectron spectroscopy of C60–. *J. Chem. Phys.* 110:8217–20.
25. Touless, D. J. 1961. *The quantum mechanics of many-body systems.* New York: AP.
26. Benard, M. 1979. A study of Hartree–Fock instabilities in $Cr_2(O_2CH)_4$ and $Mo_2(O_2CH)_4$. *J. Chem. Phys.* 71:2546–56.
27. Wiberg, K. B. 1968. Application of the Pople-Santry-Segal CNDO method to the cyclopropylcarbinyl and cyclobutyl cation and to bicyclobutane. *Tetrahedron* 24:1083–96.
28. Orlandi, G., and Negri, F. 2002. Electronic states and transitions in C_{60} and C_{70} fullerenes. *Photochem. Photobiol. Sci.* 1:289–308.
29. Shpol'skii, E. V. 1962. Problems of the origin and structure of quasi-line spectra of organic compounds at low temperatures [in Russian]. *Soviet Physics-Uspekhi* 77:321.
30. Razbirin, B. S., Starukhin, A. N., Nelson, D. K., Sheka, E. F., and Prato, M. 2007. Optical spectra and covalent chemistry of fulleropyrrolidines. *Int. J. Quant. Chem.* 107:2787–802.

31. Razbirin, B. S., Sheka, E. F., Starukhin, A. N., Nelson, D. K., Troshin, P. A., and Lyubovskaya, R. N. 2009. Shpolski effect in optical spectra of frozen solutions. *Phys. Sol. State* 51:1315–19.
32. Sheka, E. F., Razbirin, B. S., Starukhin, A. N., Nelson, D. K., Degunov, M. Yu., Fazleeva, G. M., Gubskaya, V. P., and Nuretdinov, I. A. 2009. Influence of the structure of fullerene molecules on their clusterization in the crystalline matrix. *Phys. Sol. State* 51:2193–98.
33. Broude, V. L., Rashba, E. I., and Sheka, E. F. 1985. *Spectroscopy of molecular excitons*. Berlin: Springer.
34. Osadko, I. S. 1979. Study of electron-vibrational interaction by structural optical spectra of impurity centers. *Physics-Uspekhi*. 22:311–48.
35. Zabrodski, H., Peleg, S., and Avnir, D. 1992. Continuous symmetry measures. *J. Am. Chem. Soc.* 114:7843–51.
36. Grimm, S. 1998. Continuous symmetry measures for electronic wavefunctions. *Chem. Phys. Lett.* 297:15–22.
37. Avnir, D., and Dryzhun, Ch. 2009. Generalization of the continuous symmetry measure: The symmetry of vectors, matrices, operators and functions. *Phys. Chem. Chem. Phys.* 11:9653–66.
38. Estrada, E., and Avnir, D. 2003. Continuous symmetry numbers and entropy. *J. Am. Chem. Soc.* 125:4368–75.
39. Sheka, E. F., Razbirin, B. S., Nikitina, E. A., and Nelson, D. K. 2010. Continuous symmetry of C_{60} fullerene and its derivatives. http://arXiv.org/abs/1005.1829.vl. [cond-mat.mes-hall].
40. SHAPE, SYMMETRY, and CHIRALITY programs. 2010. http://www.csm.huji.ac.il/new/.
41. Sheka, E. F., Razbirin, B. S., Starukhin, A. N., Nelson, D. K., Degunov, M. Yu., Troshin, P. A., and Lyubovskaya, R. N. 2009. Fullerene-cluster amplifiers and nanophotonics of fullerene solutions. *J. Nanophot. SPIE* 3:033501.
42. Sheka, E. F. 2004. Odd electrons and covalent bonding in fullerenes. *Int. J. Quant. Chem.* 100:375–87.
43. Sheka, E. F. and Zayets, V. A. 2005. The radical nature of fullerene and its chemical activity. *Russ. J. Phys. Chem.* 79:2009–14.
44. Sheka, E. F. 2006. 'Chemical' portrait of fullerenes. *J. Struct. Chem.* 47:593–99.
45. Tomilin, F. N., Avramov, P. V., Varganov, S. A., Kuzubov, A. A., and Ovchinnikov, S. G. 2001. Possible scheme of synthesis-assembling of fullerenes. *Phys. Sol. State* 43:973–81.
46. Fowler, P. V., and Manolopoulos, D. E. 1995. *Atlas of fullerenes*. Oxford: Clarendon Press.

chapter four

Nanochemistry of fullerene C_{60}

Stepwise computational synthesis of fluorinated fullerenes $C_{60}F_{2k}$

4.1 Introduction

In view of general regularities in intermolecular interactions (IMIs) of fullerene with other compounds, described in Section 1.2, which are complicated by the D–A interaction, starting the computational chemistry of fullerenes, we have to clearly define:

1. Which product of the fullerene (A) + reactant (B) dyad, either $[A^+B^-]$ adduct or $[A^0B^0]$ complex, is under consideration?
2. Which type of the IMI potential is characteristic for the dyad?
3. How big is the shift of equilibrium positions of atoms in both reactant ions under ionization, and should we expect a dissociation of either of the components?
4. Which algorithm should be used when looking for atomic contacts between reactants?

In this chapter we are interested in the interaction between fullerene C_{60} and fluorine gas consisting of F_2 molecules.[1] Let us start from the C_{60} + F_2 dyad and answer the above questions.

1. At the level of UBS HF AM1 calculations, both types of final products with two modifications of the $[A^+B^-]$ fluorides are formed (see Figure 4.1).
2. Coupling energies E_{cpl}^{+-} of the two $[A^+B^-]$ fluorides are –89.29 kcal/mol for $[C_{60}^{+0.2}F_2^{-0.2}]$ and –70.61 kcal/mol for $[C_{60}^{+0.1}F_1^{-0.1}]$. Energy E_{cpl}^{00} for $[C_{60}^0F_2^0]$ constitutes –0.008 kcal/mol. With these energetic parameters, the IMI potential of the dyad is obviously of type 1, presented in Figure 1.2(b). Complex $[C_{60}^0F_2^0]$ possesses all the characteristics of a charge transfer complex, where C_{60} donates the electron, while the F_2 molecule accepts it, in the energy region $E_{gap} = I_A - \varepsilon_B = 9.64$ eV, which is 2.43 eV bigger than the relevant E_{gap} for C_{60} itself due to a

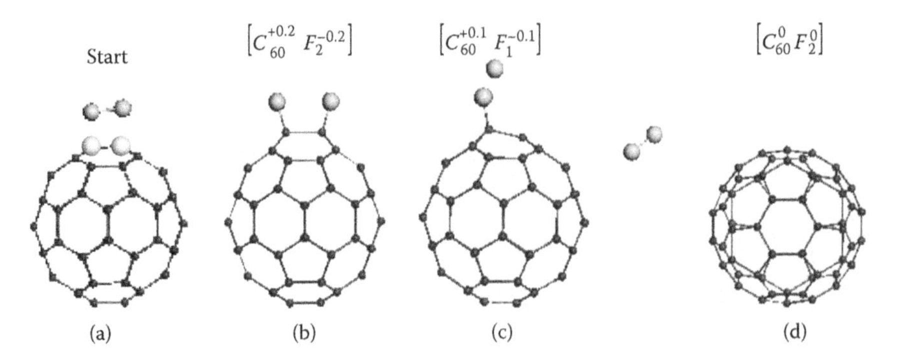

Figure 4.1 Attachment of one fluorine molecule to the C_{60} cage. (a) Starting geometry. (b) Adduct $[C_{60}^{+0.2}F_2^{-0.2}]$. (c) Adduct $[C_{60}^{+0.1}F_1^{-0.1}]$. (d) Complex $[C_{60}^{0}F_2^{0}]$. Target atoms of the C_{60} cage are shown by white. Fluorine atoms are presented by light gray balls. UBS HF AM1 singlet state. (From Sheka, E. F., arXiv:0904.4893 (cond-mat.mes-hall) (online May 1, 2009).)

much smaller electron affinity (EA) (0.23 eV) of the F_2 molecule. The corresponding B_2 band should lie far in the UV region.

3. The shift of the equilibrium position of atoms can be evaluated via changing bond lengths between atoms under ionization. In the case of cation C_{60}^{+}, the shift does not exceed 0.03 Å, while for anion F_2^{-} it is quite large and constitutes +0.208 Å. This means that the molecule can be dissociated in an ionized state. A large interatomic distance in the F_2 molecule (1.427 Å) is the reason for the two-mode behavior of the $[C_{60}^{+}F_2^{-}]$ fluoride. The adduct modification depends on the orientation of the F_2 molecule axis with respect to C-C bond that joins carbon contact atoms. When the axis is parallel to the bond, both fluorine atoms are attached to the fullerene cage (Figure 4.1(b)). If the axis is inclined to the bond, only one fluorine atom attaches to the cage, while the other one goes away (Figure 4.1(c)). The latter is possible only if the F_2 molecule is dissociated when forming fullerene fluoride. The distance between fluorine atoms exceeds both C-F and C-C standard bond lengths (1.378 and 1.391 Å, respectively) and puts the contact carbon atoms in nonequivalent positions in regards to the two fluorine atoms when one of them is farther from the cage than the other. The nonequivalence grows when the F_2 molecule is inclined to the C-C bond.
4. According to the concept of odd electron–enhanced reactivity of fullerene atoms discussed in Section 3.4, the algorithm of the contact carbon atoms' selection is in following the highest values of the atomic chemical susceptibility (ACS) of fullerene atoms, which is considered a quantitative indicator of the atom's chemical reactivity. Two light gray atoms on the C_{60} cage in Figure 4.1(a) belong to those

of six equivalent pairs, joined by a short C-C bond and belonging to group 1. These atoms are white in color in Figure 3.9(b) and, according to the N_{DA} map, possess the highest reactivity.

The above exhibits a general manner of tackling the computational chemistry of fullerene. Actually, in any particular case, the main attention can be shifted toward either $[A^+B^-]$ adducts or $[A^0B^0]$ complexes, depending on the fullerene partners and requirements dictated by practice. Thus, in the case of fluorine, hydrogen, cyan, chlorine, and other small molecules, the former products are of current interest in practice. On the other hand, complex amine and carboxylic entities and other partners with D–A characteristics similar to those of fullerene make both products actual. The two cases cover the main features of fullerene chemistry, which is extremely large. In the current book, the first community will be presented by the computational chemistry of fluorinated (current chapter), hydrogenated (Chapter 5), cyano, NH, and chlorine derivatives of C_{60} (Chapter 6), while complicated amines of different structure will be presented as the second community (Chapter 7). Particular attention to $[A^0B^0]$ complexes will be given in Chapters 9 and 10 in regards to the clustering of C_{60} and its derivatives in molecular solutions.

4.2 Background and the problem formulation

Fluorinated fullerenes, $C_{60}F_{2k}$, are in a particular place in the nanochemistry of fullerenes due to the extremely high concentration of both experimental and computational efforts aimed at synthesis, characterization, and simulation of the species. Undoubtedly, successes achieved in the field make up the best platform for exhibition of the main peculiarities of the computational nanochemistry of fullerenes due to the (1) simplicity of the addition reaction, (2) large size of a set of possible polyderivatives, (3) clear setup of the isomerism problem, (4) large quantum chemical computations' practice that provides a number of computational techniques to be compared, and (5) extended explored experimental field. These factors, taken together, provide a unique situation for a reliable computational approach to the fullerene nanochemistry to be originated and proved. Given below is such an approach that turns out to be a ground of the computational synthesis of fullerene derivatives of the $[A^+B^-]$ type in general. To make the exposition transparent, the presentation is done in a manner of successive consideration of stepwise reactions, each of which is shown in detail.

4.2.1 A historical background

A fluorinated fullerene C_{60} decade started with the first synthesis in 1991 (see reviews[2–7]) and ended by a generalizing theoretical approach to the fluorinated fullerenes characterization in 2003,[6] and without any doubt is one of the best examples of the power and ability of modern chemistry to produce, characterize, and describe at the microscopic level a new family of fascinating chemicals. It is especially impressive since the number of species, hidden under the general formula $C_{60}F_{2k}$ (with $k = 1, \ldots, 30$), is definitely countless if all possible isomers at each particular k are taken into account. The first breaking through the problem was made by experimentalists who showed that (1) species with an even number of fluorine atoms could be observed only, (2) not all [$k = 1, \ldots, 30$]-fluorinated products but only a restricted k-set of them could be produced and identified in practice, and (3) a very limited number of isomers, from one to three, at these particular k were revealed. Thus, species with chemical formula $C_{60}F_{18}$, $C_{60}F_{36}$, or $C_{60}F_{48}$ dominate in the production list, while products from $C_{60}F_2$ to $C_{60}F_{20}$ have been identified as minor additions only.[6] Mass spectrometry, infrared (IR), and ^{19}F and ^{3}He nuclear magnetic resonance (NMR) spectroscopy manifested themselves as reliable tools for the species identification.

At the same time, quantum chemical simulations have faced the many-fold isomerism problem in full measure. Thus, to suggest a convincing atomic structure of the products produced experimentally, one has to choose from one to three isomers[7] among, say, $\sim 6 \cdot 10^{14}$ structural isomers of C_{60} F_{36} and $\sim 2 \cdot 10^{10}$ ones of C_{60} F_{48}, and so on.[8] To make computations feasible, one has to restrict the isomer number to units. Obviously, it might be possible if some global regularities, which govern the fluorination process, can be exhibited. To start going along the way, it was necessary to answer the following questions:

1. To what kind of chemical reactions does the fullerene fluorination belong?
2. What is the pathway for fluorination of fullerene C_{60} and does fluorination of $C_{60}F_2$ to, say, $C_{60}F_{48}$ follow a single pathway and occur in regular steps?
3. How are target (contact) carbon atoms of the fullerene cage chosen *a priori* to any successive step of fluorination?
4. How can these measures restrict the number of possible isomers?

Answering the first question, the computing community has considered fullerene fluorination a radical reaction[9] of the one-by-one addition of fluorine atoms, where the first addition drastically violates a double C-C bond of the fullerene cage, while the next addition completes the transformation of the bond into a single one. This statement is consistent with the conclusion made above, that F_2 molecule dissociation precedes the fluorine atom attachment to the fullerene cage.

Experimental data, particularly the exhausted study of the electronic structure and chemical bonding of fluorinated fullerenes by X-ray Photoelectron Spectroscopy (XPS), Near Edge X-ray Absorption Fine Structure (NEXAFS), Ultraviolet Photoelectron Spectroscopy (UPS), and vacuum-UV absorption,[10] give a positive answer to the second question, so that fluorination of $C_{60}F_2$ to $C_{60}F_{48}$ follows a single pathway and occurs in regular steps.

As for the choice of target atoms for the subsequent steps, a computing scientist had to solve the dilemma of either to accept a full equality of the fullerene C_{60} atoms with respect to chemical reactivity or to look for their regioselectivity. Obviously, the first suggestion, which has been accepted by the majority of the computational community until now, is absolutely impotent in the solution of the above isomerism problem. At the same time, the very fact of a scarce number of isomers obtained is strong evidence of the inherited selectivity of the cage atoms, so that to proceed with the problem solution, one has to accept the presence of some governing factors. The first suggestion concerned the selection between 6,6 double bonds joining two hexagons and 6,5 single bonds framing pentagons in favor of the former. The next suggestion deals with the separation between carbon sites of the subsequent addition of fluorine atoms either via bonds (1,2-addition, adjacent carbon sites) or via space (1,3- and 1,4-additions, widely spaced carbon sites). Matsuzawa et al.[11] were first to consider the point proving that 1,2-addition is more preferable for fluorine. In a decade, the conclusion was confirmed by Jaffe.[6] The next question concerns the succession of the 1,2-additions. Following the concept of a supposed increasing of aromaticity of the fullerene hexagons caused by fluorine addition, generalized by Taylor,[5,7] a contiguous F_2 addition was suggested where the preference is done to C-C double bonds adjacent to the prior addition sites.

The above issues are quite important for designing model structures, but they should be nevertheless qualified as general recommendations only. The main problem concerning the number of isomers attributed to each of the $C_{60}F_{2k}$ species still remains. The way out of the situation was suggested by Clare and Kepert,[12] who, tackling a similar problem with the $C_{60}H_{2k}$ family, introduced the symmetry factor into the problem and proposed to restrict the isomer number by those possessing a threefold symmetry axis. The suggestion was based on the experimental report on the C_{3v} symmetry of the crown-shaped $C_{60}H_{18}$ species.[13] That was spread over $C_{60}F_{18}$ as well,[12] and later on the high-symmetry controlling preference was expanded over $C_{60}F_{36}$ and $C_{60}F_{48}$ adducts too.[14–17] However, under these assumptions, the number of possible isomers is still quite large and constitutes, say, 2,695 for $C_{60}F_{36}$ species.[18] The next step toward decreasing the isomer number has implied concrete suggestions concerning the molecules point group symmetry. Thus, based on a crown-shaped structure of

$C_{60}F_{18}$ of the C_{3v} symmetry,[12] the consideration of T, C_3, S_6, and D_{3d} structures for $C_{60}F_{36}$ has allowed for decreasing the isomer number to sixty-three[15]; similarly, ninety-four isomers have been selected in the case of $C_{60}F_{48}$.[14]

The logic of this action is a clear example of the strict symmetry codex dictate based on the paradigm that the symmetric situation is easier to formulate theoretically, and that exact symmetry is superior and sets the rules (see a detailed discussion of the problem in Section 3.2). The justification of the high-symmetry assumption was addressed by the ^{19}F NMR spectra of $C_{60}F_{18}$,[19] $C_{60}F_{36}$,[20] and $C_{60}F_{48}$,[21,22], the relevant experimental recordings of which actually show a high-symmetry pattern. However, as was discussed in Section 3.2.3, this is necessary but not enough to distinguish an exact and continuous symmetry, since exact nonsymmetric but continuously symmetric compositions may provide a high-symmetry pattern of experimental data as well. Moreover, thoroughly analyzing both ^{19}F NMR and Two-Dimensional Correlation Spectroscopy (2D COSY) data, one has to conclude that in the case of $C_{60}F_{36}$ and $C_{60}F_{48}$ species, the high-symmetry pattern of the corresponding spectra is not absolutely perfect oppositely of $C_{60}F_{18}$. Therefore, the problem of the assignment of the most stable isomers still remains unsolved.

This deadlock situation in computational simulations has been a result of the restricted formalism forming the ground of computations performed.[13–18] The latter are unable to provide any inherent regioselectivity of pristine fullerene atoms, so that one has to look for an artificial one similar to that discussed above. But, as shown in the previous chapters, the restricted formalism is not suitable for fullerenes and should be substituted by an unrestricted one. Under these conditions, the situation acquires a spontaneous solution since the regioselectivity of pristine fullerene atoms is just a natural consequence of a partial radicalization of the molecule and is exactly quantified as ACS distribution over the atoms. Taking ACS (see Equation 2.24) as a quantitative pointer of the readiness of each atom to enter the chemical reaction, one is able to make a definite choice of targets at each stage of the reaction. Let us trace how it works in the case of a stepwise computational synthesis of fluorinated fullerenes of any composition considered in the framework of the UBS HF (AM1 version) approach.[1]

4.2.2 ACS algorithm of computational synthesis of fullerenes' derivatives

The initial ACS distribution over atoms of pristine C_{60} is shown in Figure 3.8. According to the figure, the first step of any addition chemical reaction involves atoms of the highest ACS, or high-rank N_{DA} atoms, of group 1. There are twelve identical atoms that form six short C-C bonds belonging to six identical naphthalene-core fragments (see white atoms in Figure 3.9). One may choose any of the pairs to start the reaction of attaching any addend to the fullerene cage. When the first adduct, $C_{60}R_1$, is formed, the cage and, consequently, the ACS distribution are reconfigured and the reaction proceeds around the cage atoms related to the $C_{60}R_1$ adduct, with the biggest N_{DA} values thus resulting in the formation of the second adduct, $C_{60}R_2$. A new ACS map of the adduct reveals the sites for the next addition step, and so on. The reaction prolongation is controlled by the molecular chemical susceptibility, N_D, and the reaction is terminated when N_D reaches zero. Following this methodology, a complete list of fluorinated species from $C_{60}X_2$ to $C_{60}X_{60}$ can be synthesized. Let us look at the process in detail, which makes it possible to vividly present the occurrence of addition reactions on the fullerene cage.

4.3 *Reactions of C_{60} fluorination*

4.3.1 *Start of C_{60} fluorination*

When starting fluorination of C_{60}, one fluorine molecule is placed in the vicinity of the selected atoms of group 1 (31 and 32 in this case) (Figure 4.1(a)), and a full optimization of the complex geometry in the singlet state is performed. As occurred, the fluorine molecule is willingly attached to the cage; however, two adducts, $[C_{60}^{+0.2}F_2^{-0.2}]$ (II, Figure 4.1(b)) and $[C_{60}^{+0.1}F_1^{-0.1}]$ (I, Figure 4.1(c)), are possible, depending on the fluorine molecule's orientation with respect to the chosen C-C bond, as was discussed in Section 4.1.

Figure 4.2 presents the ACS maps of the pristine C_{60} cage (light gray bars) and of the cage after formation of adducts I and II (black bars) following the atom numeration in the output file. Crosses mark initial target atoms 31 and 32. As seen in the figure, attaching either one or two fluorine atoms changes the initial map considerably, but differently in the two cases. When one atom is attached, the remaining target atom, 31, which is adjacent to the first target atom, 32, dominates on the adduct $C_{60}F_1$ ACS map (Figure 4.2(a)). The picture clearly evidences a readiness of the C_{60} cage to complete the reaction by adding a further fluorine atom to atom 31 via 1,2-addition. When two atoms are attached to the cage, the N_{DA} values become zero at target atoms 31 and 32, and star-marked atoms 18, 20, 38, and 55 become the most active (Figure 4.2(b)).

Following this indication, let us complete the adduct $[C_{60}^{+0.2}F_2^{-0.2}]$ formation by adding the second fluorine molecule to atom 31 of $[C_{60}^{+0.1}F_1^{-0.1}]$, as shown in Figure 4.3(a). In due course of the structure optimization, a wished adduct $[C_{60}^{+0.2}F_2^{-0.2}]$ ($C_{60}F_2$) and a fluorine molecule (Figure 4.3(b)) are formed. This way the formed adduct $C_{60}F_2$ is fully identical to that presented in Figure 4.1(b), which is confirmed by a complete identity of their N_{DA} maps. Therefore, independently of whether either one-stage

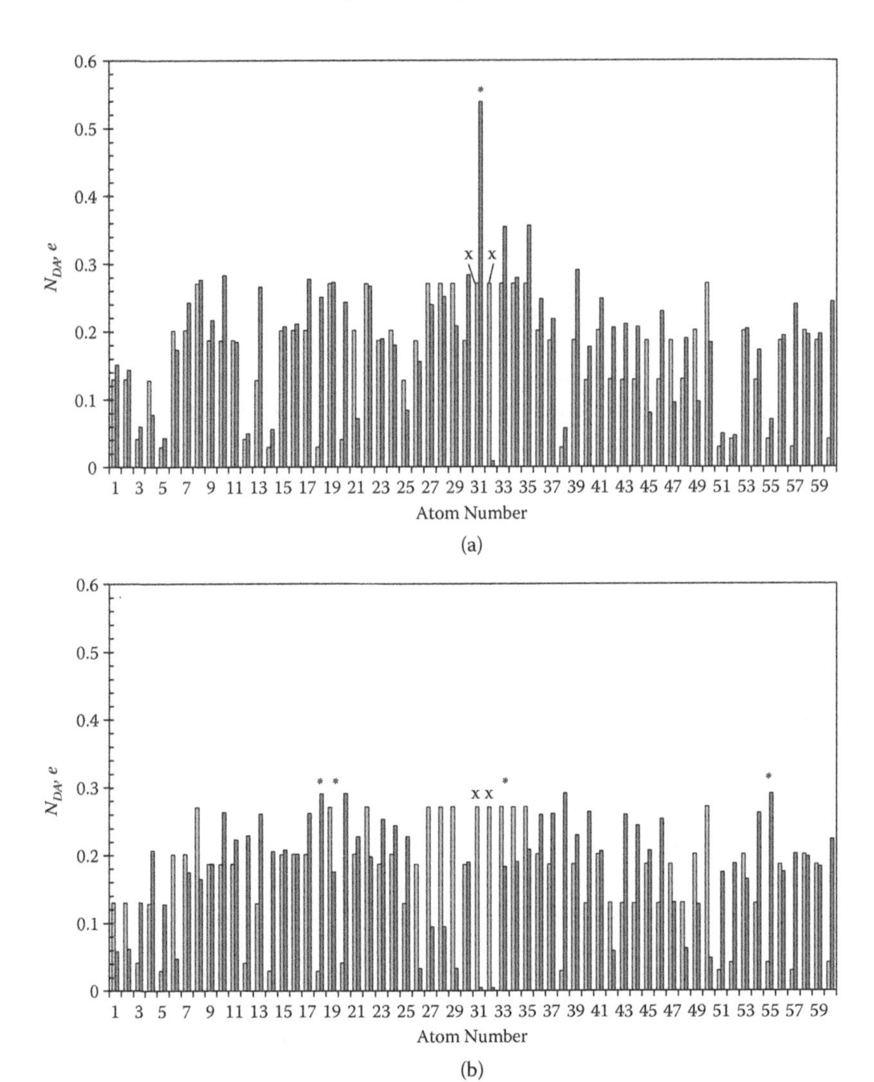

Figure 4.2 ACS map of the C_{60} cage of the adducts $[C_{60}^{+0.1}F_1^{-0.1}]$ (a) and $[C_{60}^{+0.2}F_2^{-0.2}]$ (b). Light-color bars represent the map of the pristine C_i isomer (see Section 3.5) of the C_{60} molecule. (From Sheka, E. F., arXiv:0904.4893 (cond-mat.mes-hall).

(Figure 4.1(c)) or two-stage (Figure 4.3(b)) processes of the F_2 attachment to the fullerene cage occur, the same final adduct $C_{60}F_2$ is formed. Obviously, a two-stage reaction should prevail in practice. It is radical by nature and follows a qualitative scheme suggested by Rogers and Fowler.[23]

The next step of the reaction is governed by the prevalence of atoms 18, 20, 38, and 55 on the ACS map of $C_{60}F_2$ (see Figure 4.2(b)). These atoms form

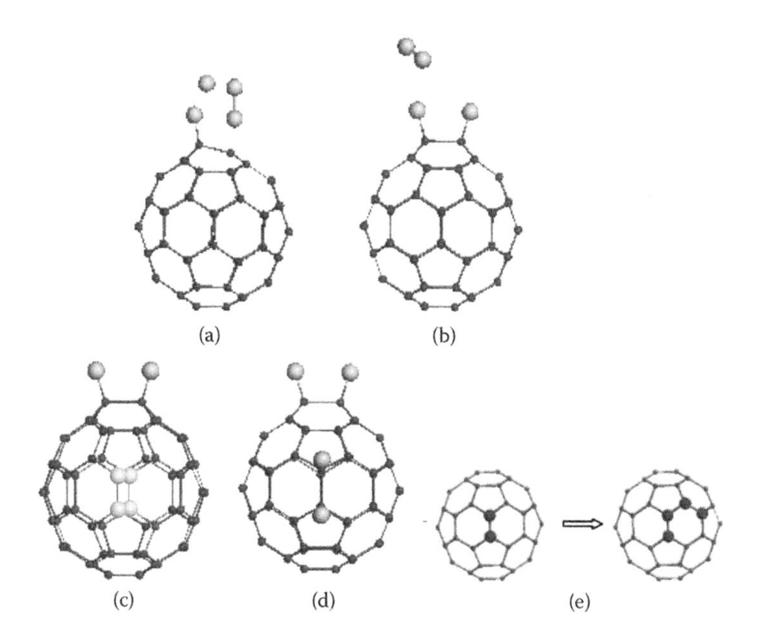

Figure 4.3 Attachment of the second fluorine molecule to the C_{60} cage. (a) Starting geometry. (b) The adduct $[C_{60}^{+0.2}F_2^{-0.2}]$ ($C_{60}F_2$) formation. (c) Atoms 18, 20, 38, and 55 of the C_{60} cage of $C_{60}F_2$ are shown by light coloring. (d) Adduct $C_{60}F_4$. (e) A scheme of contiguous F_2 addition following Taylor's suggestion. (From Taylor, R., *Chem. Eur. J.*, 7, 4074–84, 2001).

two identical pairs of short C-C bonds located in the equatorial plane with respect to first two target atoms (see lightly colored atoms in Figure 4.3(c)). One of these pairs is taken as a couple of targets, and the procedure of attaching F_2 to the pair atoms repeats the one described above. Consequently, a molecule $C_{60}F_4$ is formed (Figure 4.3(d)). The ACS map is calculated for the product to select target atoms for the next attaching. As seen from Figure 4.3, a subsequent F_2 addition is not contiguous (see Figure 4.3(e)), as was suggested by Taylor.[5]

Therefore, a successive one-by-one addition of fluorine atoms to the fullerene cage results in the formation of two adducts related to two reactions: $C_{60}F_{2K} + F = C_{60}F_{2K+1}$ and $C_{60}F_{2K+1} + F = C_{60}F_{2(K+1)}$ ($k = 1, 2, \ldots, 30$). Each step is controlled by the fullerene cage ACS map of the preceding adducts, namely, $C_{60}X_{2K}$ and $C_{60}F_{2K+1}$, respectively. Actually, when the difference between the high-rank N_{DA} values is not quite pronounced, every step should be additionally complicated by expanding calculations over a restricted set of isomers, which are pointed out by a set of high-rank values on the ACS map. A final choice of the most stable species is subordinated therewith to the preference of the structure with the least total energy.

4.3.2 $C_{60}F_2$–$C_{60}F_8$ adducts

Let us consider the synthesis of fluorinated adducts in the working regime exemplifying the procedure for $C_{60}F_2$-$C_{60}F_8$ adducts. Chart 4.1 presents high-rank N_{DA} values of the species from C_{60} to $C_{60}F_8$. The N_{DA} data are ordered in the Z → A manner, and only a small part of the N_{DA} list involving the high-rank data is shown. Fluorinated adducts are marked FN, where N numbers the attached fluorine atoms. The atom numbering used below is presented on the Schlegel diagram of the fullerene C_{60} cage in Figure 3.9(d).

Cage atom 32 was chosen as the first to be attacked by fluorine. As a result, atom 31 heads the list of the N_{DA} data of adduct F1, and obviously

Chart 4.1 F1 to F8 Fluorination

F0-C60		F1 (32)		F2 (31)		F3 (38)		F4 (55)	
Atom number	NDA	Atom number	NDA	Atom number	NDA	Atom number	NDA	Atom number	NDA
32	0.27072	31	0.53894	38	0.29094	55	0.51918	42	0.30143
31	0.27077	35	0.35590	20	0.29086	40	0.33620	22	0.30002
		33	0.35406	55	0.29049	4	0.31575	48	0.30000
		39	0.29065	18	0.29041	20	0.29535	2	0.29971
		30	0.28358	10	0.26369	18	0.29026	1	0.29812
		10	0.28304	40	0.26368	36	0.28537	35	0.29782

F5 (42)		F6 (48)		F7-1 (60)		F8-1 (59)	
Atom number	NDA	Atom number	NDA	Atom number	NDA	Atom number	NDA
48	0.52989	60	0.34210	59	0.51853	22	0.34686
39	0.35424	66	0.32535	57	0.37960	24	0.32989
22	0.34896	22	0.28518	39	0.37006	54	0.29221
36	0.34284	1	0.28104	46	0.36312	27	0.28239
35	0.34226	39	0.27708	60	0.34821	11	0.28026
49	0.34028	44	0.27568	22	0.34667	36	0.27552
		35	0.27411			35	0.27273
		59	0.26879			23	0.27116
		15	0.26840			28	0.27074
		2	0.26778			33	0.26836
		26	0.26445			25	0.26625
		12	0.26395			6	0.26346
		57	0.28360			44	0.26253
		5	0.25955			53	0.26195

Source: Sheka, E. F., arXiv:0904.4893 (cond-mat.mes-hall).

Note: In parentheses are given the numbers of the C_{60} cage atom to which the current fluorine atom is attached.

points to the place of the next attack. The head of the N_{DA} list of F2 involves two pairs of one-bond-connected atoms, 38 and 55, and 20 and 18 (the atoms' location is shown in Figure 4.3(c)). Shadowing is used to distinguish the pairs. Those are fully equivalent, and as shown by calculations, adducts series started from each of them are equivalent as well. Let us proceed with atom 38 to form F3, whose high-rank N_{DA} list immediately highlights the 1,2 pairing atom 55. The N_{DA} list of F4 is opened by one-bond-connected atoms 42 and 48 and includes atom 22, the pairing atom to which is located farther in the list. Continuing addition by attacking atom 42, one obtains F5, with atom 48 possessing the highest N_{DA}. The N_{DA} list of F6 is typical for the majority of succeeding addition events and is headed by two atoms with comparable N_{DA} values, while the corresponding 1,2 one-bond-connected atoms are shifted into the depth of the list. This is a typical case where a study of a few isomers is needed. Based on the energy analysis, a preference to isomer F8-1 occurred quite reliably.

Atomic structures of the F2K species for $k = 1, 2, 3$, and 4 are shown in Figure 4.4. As seen in the figure, neither F4 nor F6 and F8 follow the contiguous scheme of the addition suggested by Taylor.[5,7] It is not a hypothetical growing of the aromaticity of adjacent bonds, but the redistribution of the density of effectively unpaired electrons over the cage atoms in due course of fluorination that is the governing factor of the reaction pathway.

The total energy of adducts as well as their main geometric parameters are presented in Table 4.1. The latter concern C*-F and C*-C* bond lengths, where C* marks cage atoms bound to a fluorine one. Changing the total energy due to successive fluorination is shown in Figure 4.5, alongside the coupling energy, E_{cpl}, needed for the addition of every next pair of fluorine atoms on the pair number k. Suppose that the two-stage reaction occurs in the gaseous fluorine; then the coupling energy is determined as

$$E_{cpl} = \Delta H_{2k} - \Delta H_{2(k-1)} - \Delta H_{F_2} \tag{4.1}$$

Here ΔH_{2k} and $\Delta H_{2(k-1)}$ are the heats of formation (total energies) of F2K and F2(K – 1) products, while ΔH_{F_2} is the heat of formation of a fluorine molecule equal to –22.485 kcal/mol. The data fill the interval from –91 to –83 kcal/mol. A large value of the coupling energy tells suggests tight binding of the fluorine with the fullerene cage.

As seen in Figure 4.5(a), the total energy remarkably increases by absolute value, remaining negative when fluorination proceeds; this obviously favors polyaddition. That occurs experimentally, and $C_{60}F_{18}$, $C_{60}F_{36}$, and $C_{60}F_{48}$ are usually the main products obtained.[5] As shown by the Sussex[5] and Moscow[3] research groups, the fluorination yield greatly depends on

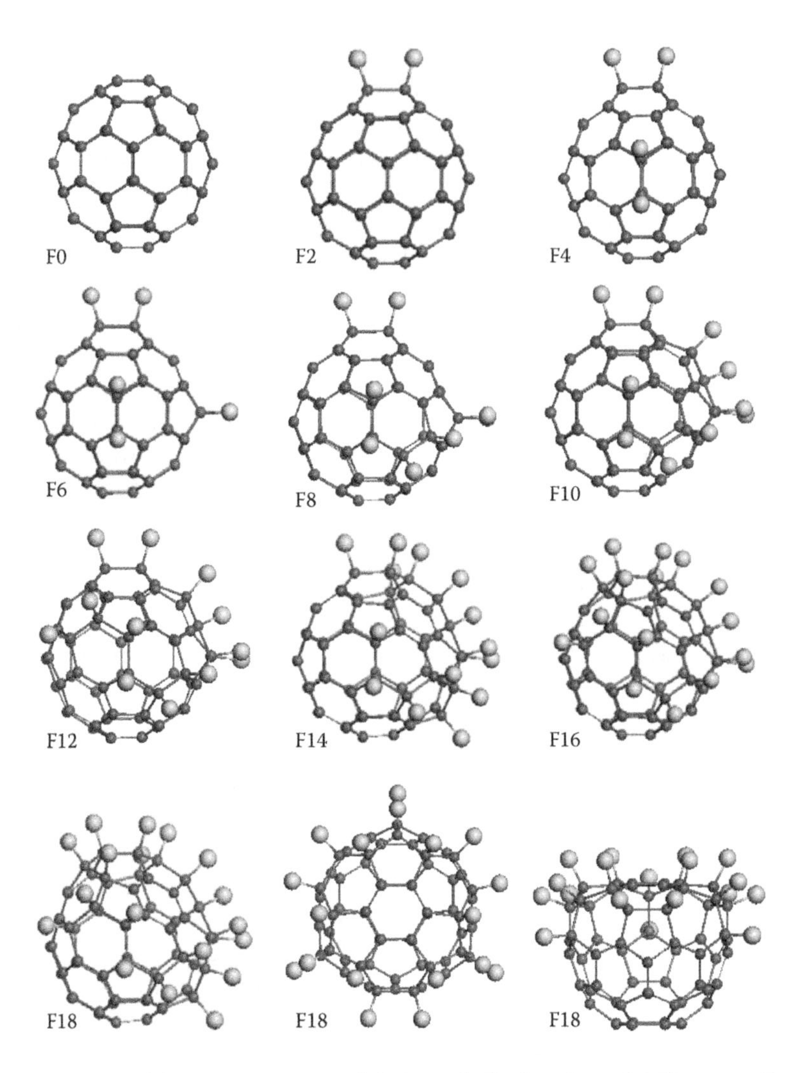

Figure 4.4 Equilibrium structures of $C_{60}F_0$ to $C_{60}F_{18}$ fluorinated fullerenes. (From Sheka, E. F., arXiv:0904.4893 (cond-mat.mes-hall).

the reaction conditions, and only after particular measures were taken did the authors succeed in separating lower fluorinated species, $C_{60}F_2$, $C_{60}F_4$, $C_{60}F_6$, and $C_{60}F_8$,[24,25] confirming these adducts' stability.

The top panel in Figure 4.5(b) presents the evolution of the molecular chemical susceptibility N_D under fluorination. As a total, the value gradually decreases when fluorination proceeds, thus demonstrating the N_D behavior as a "pool of chemical reactivity" that is being worked out in due course of the reaction proceeding. At the very beginning at low k, N_D slightly increases due to increasing the number of elongated C-C bonds

Table 4.1 Geometric parameters, symmetry, and total energy of fluorinated fullerenes $C_{60}F_{2k}$: UBS HF AM1 singlet state

	F1	F2	F3	F4	F5	F6	F7	F8	F10	F12	F14
R(C* – F),[a] Å	1,378	1,382	1,382, 1,378	1,382	1,382, 1,377	1,382	1,383–1,381	1,384–1,381	1,383–1,381	1,383–1,381	1,387–1,380
R(C* – C*),[b] Å	1,52[c]	1,61	1,52[c]	1,62, 1,61	1,52[c]	1,61	1,52[c]	1,61–1,60	1,61–1,60	1,61–1,58	1,60–1,58
ΔH,[c] kcal/mol	903,60	843,58	795,50	729,99	677,70	622,58	570,35	517,16	414,70	305,92	195,77
Symmetry		C_{2v}		C_s		C_1		C_1	C_1	C_1	C_1
	F16	**F18**	**F20**	**F22**	**F24**	**F26**	**F28**	**F30**	**F32**	**F34**	**F36**
R(C* – F),[a] Å	1,385–1,378	1,384–1,377	1,385–1,376	1,384–1,375	1,385–1,375	1,385–1,374	1,386–1,376	1,385–1,374	1,393–1,373	1,393–1,376	1,402–1,375
R(C* – C*),[b] Å	1,60–1,57	1,58–1,57	1,59–1,57	1,59–1,57	1,59–1,57	1,59–1,57	1,58–1,57	1,59–1,57	1,59–1,57	1,59–1,57	1,59–1,57
ΔH,[c] kcal/mol	86,80	–24,54	–132,87	–248,07	–355,61	–459,33	–552,60	–642,47	–738,80	–822,14	–914,95
Symmetry	C_s	C_{3v}	C_1	C_1	C_1	C_s	C_1	C_1	C_1	C_1	C_1
	F38	**F40**	**F42**	**F44**	**F46**	**F48**	**F50**	**F52**	**F54**	**F56**	**F60**
R(C* – F),[a] Å	1,402–1,375	1,402–1,376	1,402–1,381	1,402–1,381	1,402–1,383	1,408–1,387	1,408–1,387	1,410–1,387	1,412–1,387	1,412–1,385	1,412 (1,357)
R(C* – C*),[b] Å	1,59–1,57	1,59–1,57	1,59–1,57	1,59–1,57	1,59–1,57	1,59–1,57	1,59–1,57	1,59–1,57	1,59–1,57	1,59–1,57	1,585–1,565
ΔH,[c] kcal/mol	–995,77	–1069,62	–1120,36	–1184,68	–1237,16	–1293,30	–1338,08	–1356,10	–1371,49	–1387,42	–1410,63
Symmetry	C_1	C_1	C_1	C_1	C_1	C_1	C_1	C_s	C_s	C_s	I_h

Source: Sheka, E. F., arXiv:0904.4893 (cond-mat.mes-hall).

[a] C* marks the cage atom to which fluorine is added.
[b] C* – C* marks a pristine short bond of the cage to which a pair of fluorine atoms is added.
[c] See footnote b to Table 3.2.

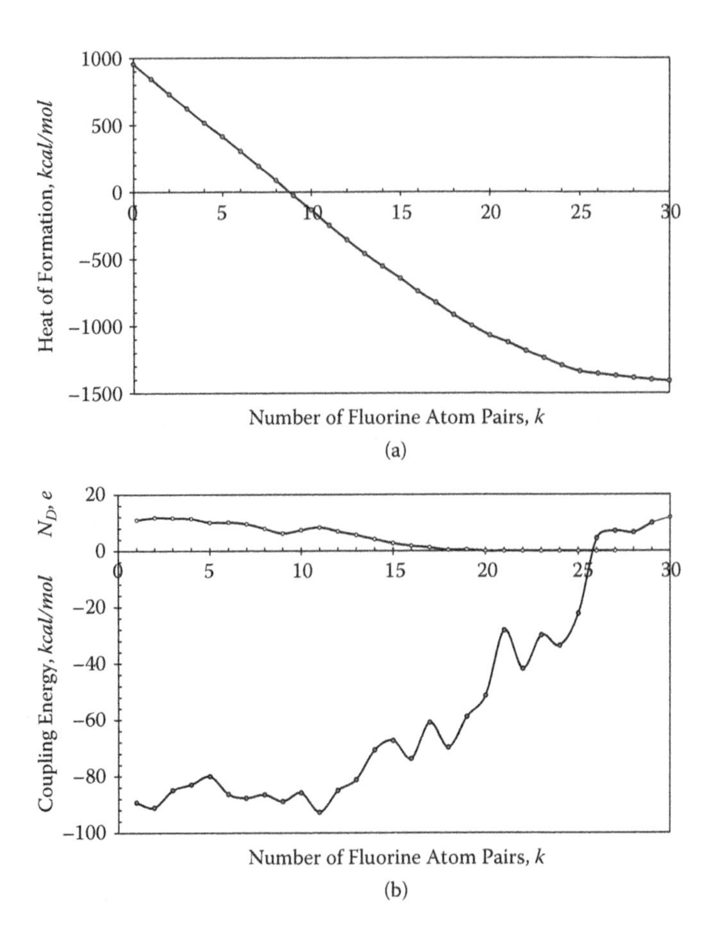

Figure 4.5 Evolution of the total energy (a), molecular chemical susceptibility, N_D, and coupling energy, E_{cpl} (b), at growing the number of fluorine atom pairs, k. (From Sheka, E. F., arXiv:0904.4893 (cond-mat.mes-hall).)

caused by the cage structure sp^2-sp^3 reconstruction under fluorination (see Section 4.5 for details).

4.3.3 $C_{60}F_{10}$–$C_{60}F_{18}$ adducts

The detailed procedure of the computational fluorination of the C_{60} cage via successive steps of 1,2-additions discussed above is retained throughout a full cycle of $C_{60}F_2$ to $C_{60}F_{60}$ fluorination.[1] A concise synopsis related to the formation of products from $C_{60}F_{10}$ to $C_{60}F_{18}$, which required a study of eleven isomers, is presented in Chart 4.2. The content of each cell involves the isomer name; the cage atom numbers; which were taken from the

Chart 4.2 Scheme of F8 to F18 fluorination

<table>
<tr><td></td><td></td><td></td><td></td><td>F8-1
60 and 57
517.162</td><td></td><td></td><td></td></tr>
<tr><td></td><td></td><td></td><td></td><td>F10-1
22 and 33
414.700</td><td></td><td></td><td></td></tr>
<tr><td>F12-1
54 and 40
310.952</td><td></td><td></td><td></td><td>F12-2
52 and 51
305.920</td><td></td><td></td><td>F12-3
53 and 56
311.329</td></tr>
<tr><td></td><td></td><td></td><td>F14-1
58 and 59
198.577</td><td></td><td></td><td>F14-2
53 and 56
203.669</td><td></td></tr>
<tr><td></td><td></td><td>F16-1
53 and 56
87.656</td><td></td><td></td><td>F16-2
40 and 54
86.799</td><td></td><td></td></tr>
<tr><td></td><td>F18-2
24 and 23
−18.287</td><td></td><td>F18-3
16 and 11
−8.705</td><td></td><td>F18-1
24 and 23
−24.538</td><td></td><td></td></tr>
</table>

Source: Sheka, E. F., arXiv:0904.4893 (cond-mat.mes-hall).

high-rank part of the N_{DA} list of the preceding isomer and addition, which provided the current isomer formation; and the isomer total energy. Isomers with the least total energy are marked in bold.

Equilibrium structures of adducts from the $C_{60}F_{10}$-$C_{60}F_{18}$ series that correspond to those of the least total energy are shown in Figure 4.4. Drawn in the same projection, the pictures allow vividly exhibiting a consequent deformation of the fullerene cage caused by fluorination until F18 is formed. The last row of panels in the figure presents the F18-1 isomer in different projections to highlight its crown-like structure of C_{3v} symmetry, which is well known experimentally.[19,26,27] Total energy and geometric parameters of adducts are given in Table 4.1. Data plotting in Figure 4.5(b) show that both N_D quantity and coupling energy do not differ much within the series and retain close to the biggest values, just disclosing high chemical reactivity of the individual adduct and a strong tendency to further fluorination.

Experimental evidence of the C_{3v} symmetry of $C_{60}F_{18}$ follows from the ^{19}F NMR spectrum[19] and is presented in Figure 4.6. The spectrum is a beautiful example of a perfect high-symmetry pattern when only four bands for eighteen fluorine atoms are located over practically a zero-point background. The relation between integral intensities of the bands (see in parentheses in the figure) points to three groups of atoms, each of which covers six identical atoms. Decisive structural information was obtained from a homonuclear ^{19}F-^{19}F shift correlation (COSY) experiment. The connectivity

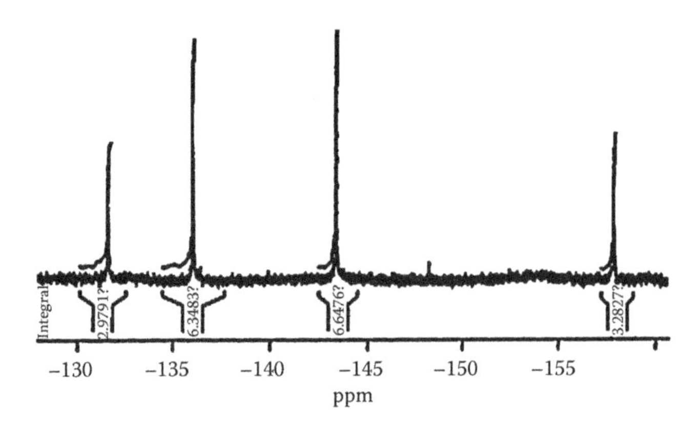

Figure 4.6 ^{19}F NMR spectrum for $C_{60}F_{18}$, showing integration, splitting, and couplings. (From Boltalina, O. V., et al., *Chem. Commun.*, 2549–50, 1996.)

derived from this 2D spectrum allowed for doubtless visualizing of the C_{3v} symmetry composition of the cage carbon atoms.

4.3.4 $C_{60}F_{20}$–$C_{60}F_{36}$ adducts

The next stage of the computational fluorination involves a series of $C_{60}F_{20}$-$C_{60}F_{36}$ adducts. About twenty isomers should be considered to choose the most stable isomers. Such selected adducts are shown in Figure 4.7, and their total energy and geometric parameters are given in Table 4.1. In contrast to the previous series, the current one is not terminated by a high-symmetry adduct. The best isomer, F36, produced computationally has no symmetry whatsoever (C_1).

$C_{60}F_{36}$ is one of the most studied species among other fluorinated fullerenes. However, since the first recording of the substance[28] until the last one,[29] there still has not been a clear vision of the composition and symmetry of the species. A big temptation to see a direct connection between F18 and F36 as a doubling of F18[7] influenced looking for C_3-based structures of F36 both computationally[15–18] and experimentally.[20,28–32] However, all experimental evidence is of a complicated structure and, once interpreted from the high-symmetry position, results in a complicated isomer composition of the produced material that involves isomers of C_3, C_1, and T point symmetry.

As conventionally, this conclusion was based on the exact symmetry dictate of yes or no with respect to the symmetry assignment in view of the symmetry pattern of experimental recordings. However, if we undertake the continuous symmetry analysis of the F36 species in Figure 4.7

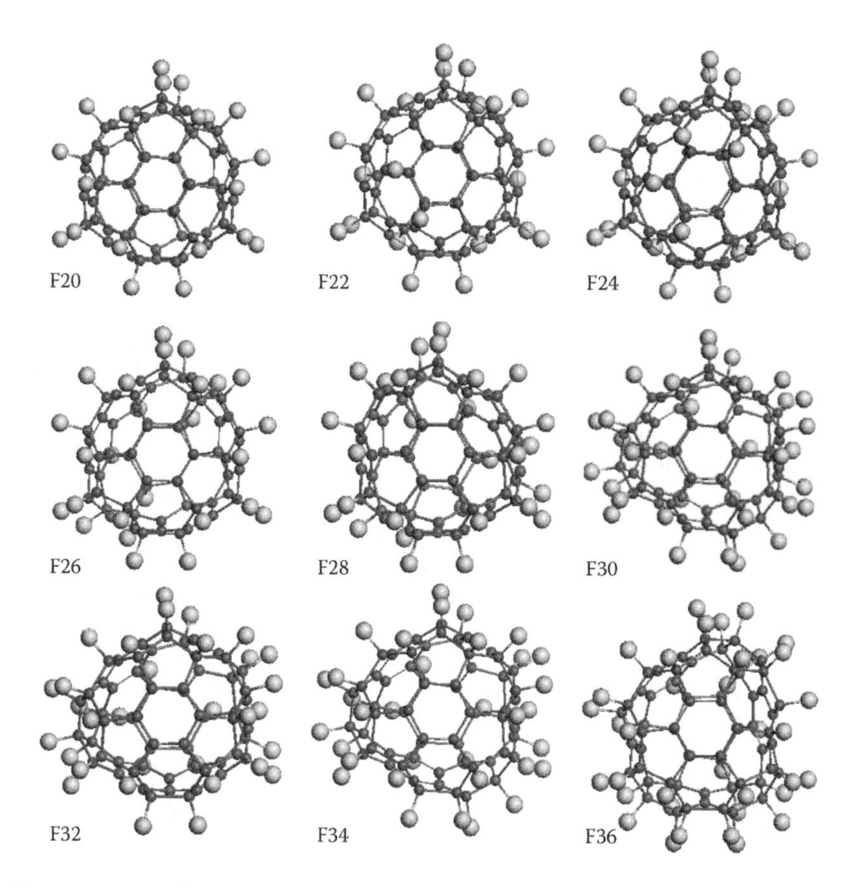

Figure 4.7 Equilibrium structures of $C_{60}F_{20}$ to $C_{60}F_{36}$ fluorinated fullerenes. (From Sheka, E. F., arXiv:0904.4893 (cond-mat.mes-hall).

according to the methodology described in Section 3.2.3, we obtain the following interesting result. As it turns out, the molecule is quite symmetric in reality and involves either 94.3% of point group C_3 symmetry or 95.3% of *T*. The deviation of the continuous symmetry from the exact one in the range of 5.7 to 4.7%, as in the current case, is under bordering conditions (see Section 3.2.5). This does not allow for the experimental discrimination of low- and high-symmetry molecules. This may explain the difficulty of an unambiguous analysis of experimental data for $C_{60}F_{36}$.

As seen from Figure 4.5(b), fluorination of F20 to F36 covers the main stage when the pool of the cage chemical reactivity has been worked out. The remaining N_D values become small. Simultaneously, a considerable decrease of the coupling energy by an absolute value takes place. Both factors evidently point to weakening of the reaction ability, which evidences the reaction termination in a few steps.

4.3.5 $C_{60}F_{38}$–$C_{60}F_{48}$ adducts

Due to the smallness of the N_D value, the next steps of the addition reaction require higher attention and thorough investigation of a number of isomers at each step. As a result of a few roots of the 1,2-addition continuation, a reliable set of energetically stable isomers was obtained. The species are presented in Figure 4.8. Their total energy and geometric parameters are given in Table 4.1. As seen in the figure, the structure of adducts becomes visually more and more symmetric when fluorination proceeds. However, the exact point symmetry of F48 determined quantum chemically is C_1.

Analyzing Figure 4.5(b) for this reaction stage, one should conclude that experimental realization of high fluorinated C_{60} is approaching the end since (1) the chemical reactivity pool is worked out and (2) the coupling energy actively decreases by the absolute value approaching zero. Therefore, F44 to F50 species must be the final ones in the row of fluorinated C_{60} since the coupling energy of F52 and higher species becomes positive. One of the first reports on producing F48[21] states that the substance can be obtained only in due course of long-time fluorination. Just

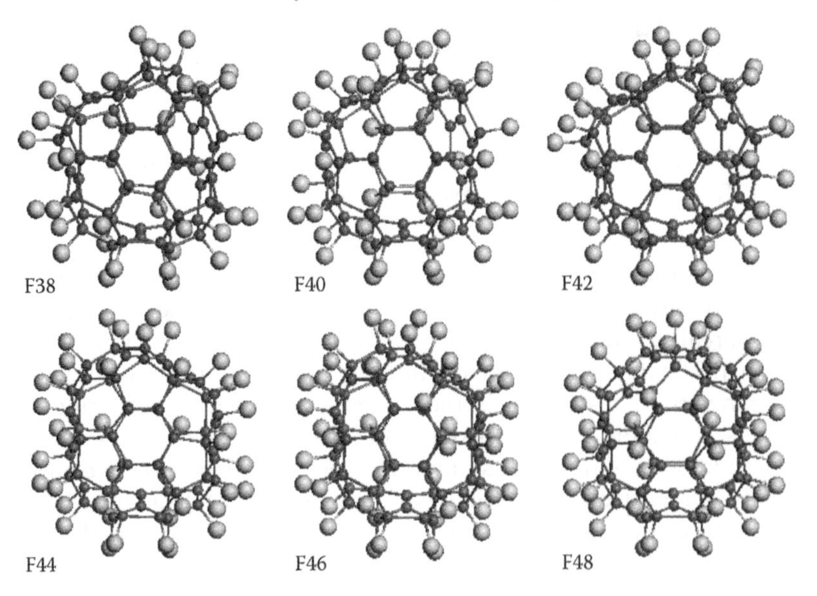

Figure 4.8 Equilibrium structures of $C_{60}F_{38}$ to $C_{60}F_{48}$ fluorinated fullerenes. (From Sheka, E. F., arXiv:0904.4893 (cond-mat.mes-hall).

forthcoming termination of the reaction allows for accumulating the species in a considerable amount. This makes the production of F48 more favorable than either F18 or F36 in spite of much worse conditions for the reaction occurrence concerning N_D values and coupling energy. Indeed, $C_{60}F_{48}$ was found to be the final product of direct fluorination.[33] Furthermore, attempts to achieve perfluorination of C_{60} by applying more harsh experimental conditions resulted in cage disruption, but not in the preparation of $C_{60}F_{60}$.

$C_{60}F_{48}$ is the main target of the stage on which a lot of efforts have been concentrated. The first ^{19}F NMR analysis of the species showed[21] that $C_{60}F_{48}$ demonstrated a clear high-symmetry pattern. Supplemented by the 2D COSY spectrum analysis (Figure 4.9(a)) when only main peaks were taken into account, while incapsulated peaks were not considered, the data were supposedly attributed to eight groups of atoms with six atoms in each. The connectivity obtained from the spectrum allowed suggesting two D_3 enantiomers (Figure 4.9(b)) that fit the experimental data in the best way. All further investigations were aimed at proving this high symmetry of the species, both computationally[14,17] and experimentally.[21,22,31–36] However, in spite of heavy efforts undertaken to clarify the symmetry problem, there are still features that do not fit the high-symmetry suggestion. First, those concern the fact that eight main peaks of the ^{19}F NMR spectrum are located on a significant background consisting of a number of low-intensity peaks, opposite of the perfectly symmetric situation with a zero-point background in the case of $C_{60}F_{18}$. The situation with electron gas diffraction[34] and x-ray measurements[35,36] is not better. At the same time, the continuous symmetry analysis of the F48 molecule shown in Figure 4.8 reveals its 99% D_3-ness. In view of this situation, one may expect a pretty good fitting of the structure to experimental recordings. So far it has been possible to check this suggestion only in regards to x-ray powder diffraction.

Dr. R. Papoular from the Leon Brillouin Laboratory of CEN-Saclay agreed to repeat the Rietveld refining of his high-resolution x-ray powder diffraction structure data, previously performed on the basis of exact D_3 symmetry of the $C_{60}F_{48}$ molecule,[36] by using the F48 molecule. Figure 4.10 presents the refining results. As seen in the figure, the fitting for the C_1 F48 molecule visually is not only not worse than that of the D_3– one, but it is even better in the region of the first most intensive peak. This conclusion is in good agreement with ^{19}F NMR and 2D COSY spectra, the main pattern of which favors D_3 symmetry, but additional features of the spectra, which violate the perfect exact symmetry, evidence some deviations toward lower symmetry, similarly to optical spectra of C_{60}, discussed in Section 3.2.5. Additionally, a close similarity in the behavior of exact and continuous D_3 symmetric $C_{60}F_{48}$ molecules has been supported computationally in regards to the charge distribution presented in Figure 4.11.

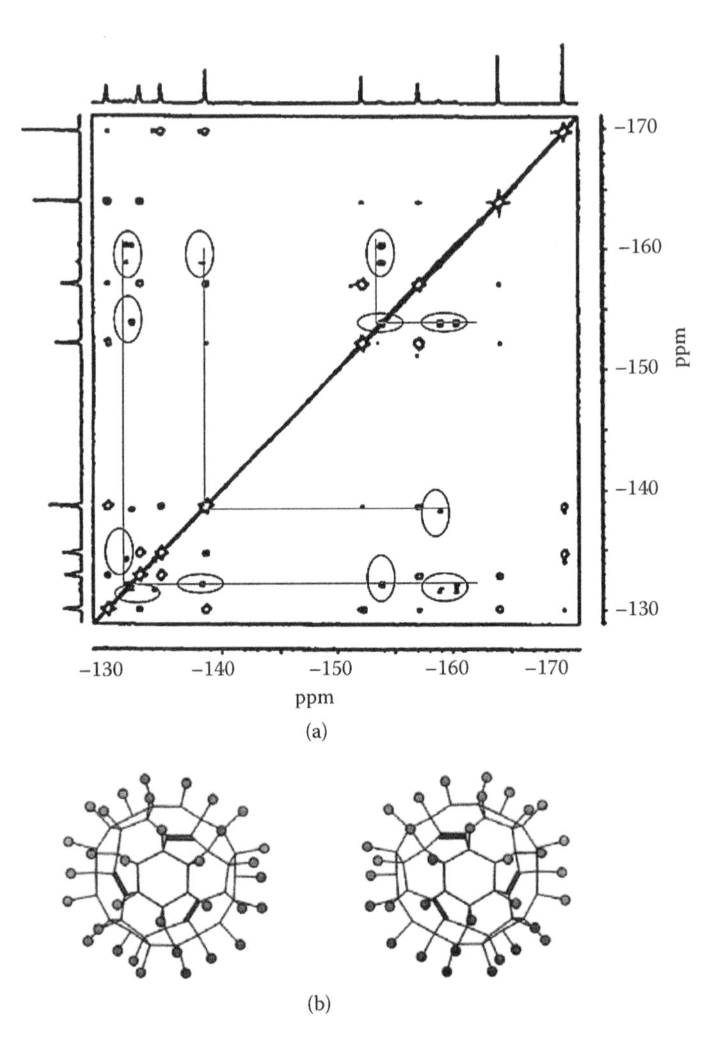

Figure 4.9 ^{19}F NMR spectrum for $C_{60}F_{48}$. (a) Lines between incapsulated peaks indicate connectivity established by COSY. (b) Structure of the D_3 enantiomer pair. (From Gakh, A. A., et al., *J. Am. Chem. Soc.*, 116, 819–20, 1994.)

Comparing the D_3 molecule with F48, it is necessary to stress attention on the following. As shown previously, the F48 molecule has been obtained in due course of successive steps of 1,2-additions that strictly keep long-bond framing of the fullerene cage pentagon. In contrast to this, exact D_3 symmetry of the isomer requires moving a short bond of the fullerene cage to the pentagon framing. This is possible if only the 1,2 manner of addition is replaced by a 1,4 one at the last or one of the preceding steps. A precedent and a possible mechanism for such a spontaneous

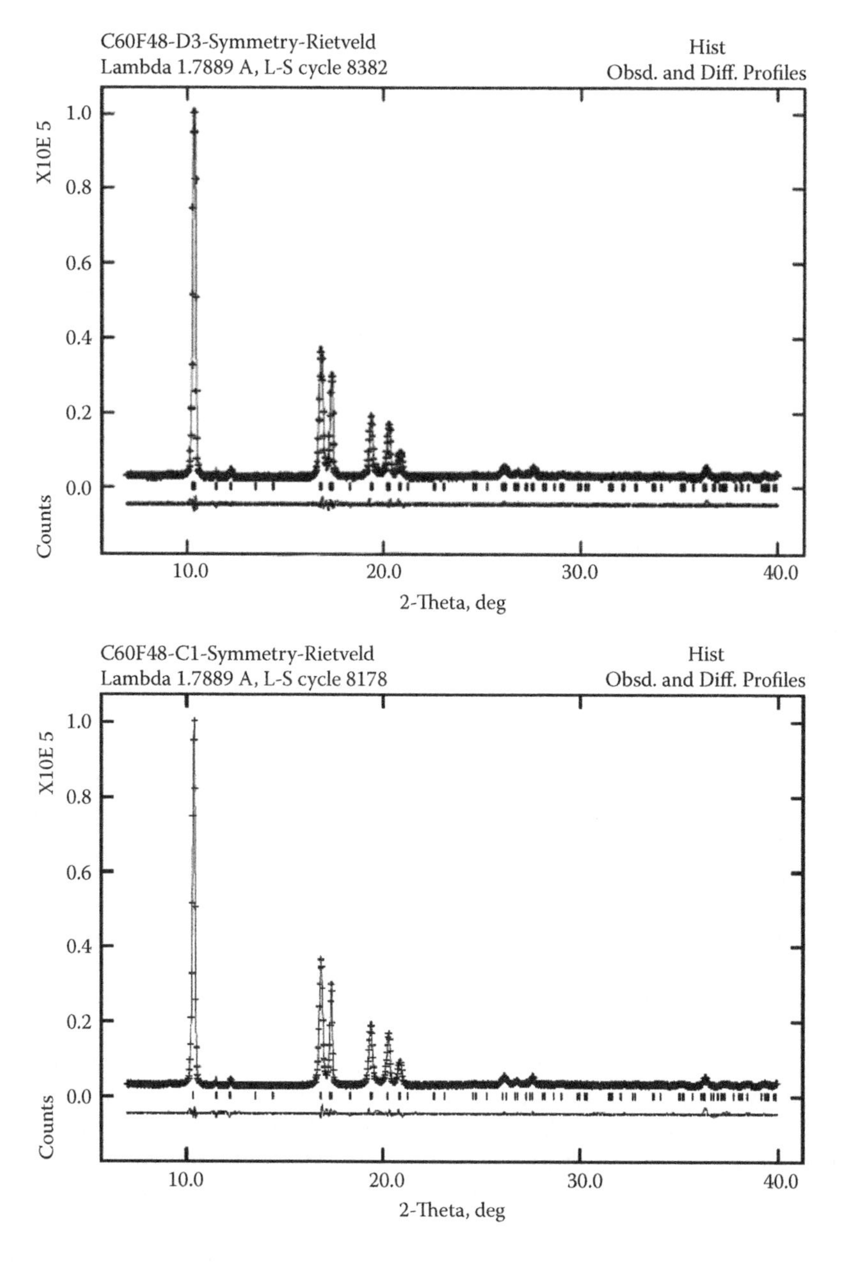

Figure 4.10 Plots of the best GSAS (General Structure Analysis System)-based Rietveld refinements associated with D_3 (top)[36] and C_1 (bottom)[37] molecular models. (Courtesy of Dr. R. Papoular, Leon Brillouin Laboratory, CEN-Saclay.)

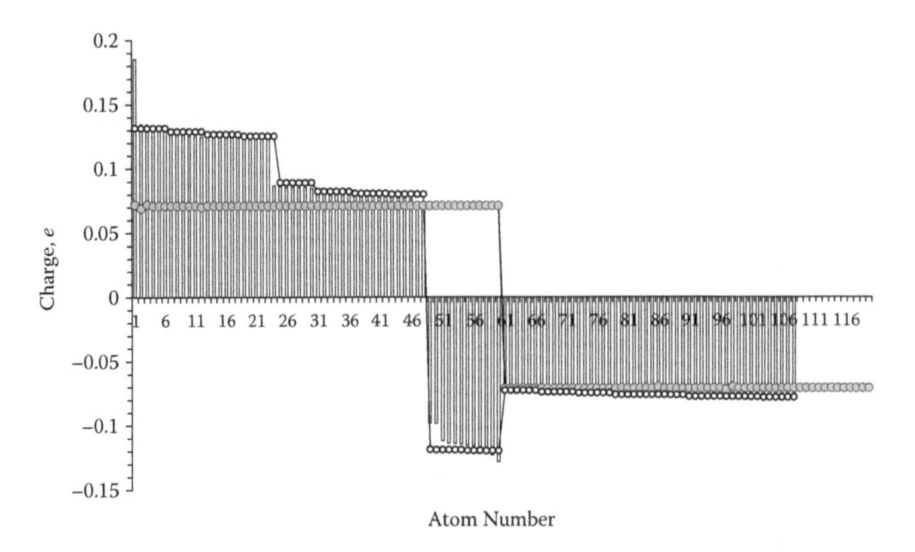

Figure 4.11 Charge distribution over atoms of $C_{60}F_{48}$. Histogram presents C_1 molecule, and curve with empty dots plots data for D_3 molecule. Curve with filled dots is related to $C_{60}F_{60}$. (From Sheka, E. F., arXiv:0904.4893 (cond-mat.mes-hall).)

transformation have not been established so far. On the other hand, geometrical factors given in Table 4.1 show convincingly that a successive fluorination of the cage did not require any serious reconstruction of the latter, which might result in changing the pentagon framing. This means that the main structure factor of the C_{60} cage consisting of the long-single-bond frame of pentagons should be conserved during the whole fluorination process since the fluorination of $C_{60}F_2$ to $C_{60}F_{48}$ follows a single pathway and occurs in regular steps.

At the same time, the total energy of the exact D_3-$C_{60}F_{48}$ isomer is significantly lower (~54 kcal/mol) than that of the F48 molecule. This means that under particular kinetic conditions the formation of a more stable isomer might be energetically profitable. The "fluorine dance," which was discovered ten years ago[38] and was exhibited in a facile migration of fluorine atoms over the fullerene surface at elevated temperature, may obviously provide such particular kinetics. Evidently, the migration itself cannot transform a 1,2-addition into a 1,4-addition. However, to remember the possible presence of the D_{3d} C_{60} isomer (see Section 3.5), it seems reasonable to suppose that the fluorine migration, alongside temperature elevation, may stimulate reisomerization of the C_{60} cage from the C_i or I_h form into a D_{3d} one, which in turn may favor the formation of D_3 symmetric fluoride. Since experimental recordings related to $C_{60}F_{48}$, such as ^{19}F NMR as well as electronic spectra, etc., depend on the chemical prehistory of samples produced, one might think that both exact and continuously D_3 symmetric isomers of the fluoride may coexist, shifting balance to one of them depending on chemical and physical treatments.

4.3.6 $C_{60}F_{50}$–$C_{60}F_{60}$ adducts

Coming back to the continuous chain of successive 1,2-additions, we are facing the final stage of the fluorination involving $C_{60}F_{50}$-$C_{60}F_{60}$ adducts. Since N_D is completely worked out by this stage (Figure 4.5(b)), we cannot use the N_{DA} pointer and have to proceed with further fluorination without it. The problem is facilitated by a comparatively small number of empty places that decrease when fluorination proceeds, so that the isomer study can be performed just routinely by running it over all places one by one. The result of the sorting thus performed is presented in Table 4.1 and Figure 4.12, exhibiting the structure of the species of the lowest total energy. The fluorination process is terminated by the formation of $C_{60}F_{60}$ species of I_h symmetry. All of the above high fluorinated products are thermodynamically stable and could have existed. However, since fluorination is a consequent process, their formation becomes energetically nonprofitable at $k \geq 25$ due to positive partial coupling energy (Figure 4.5(b)), which is why no recording of the species production has been known up until now.

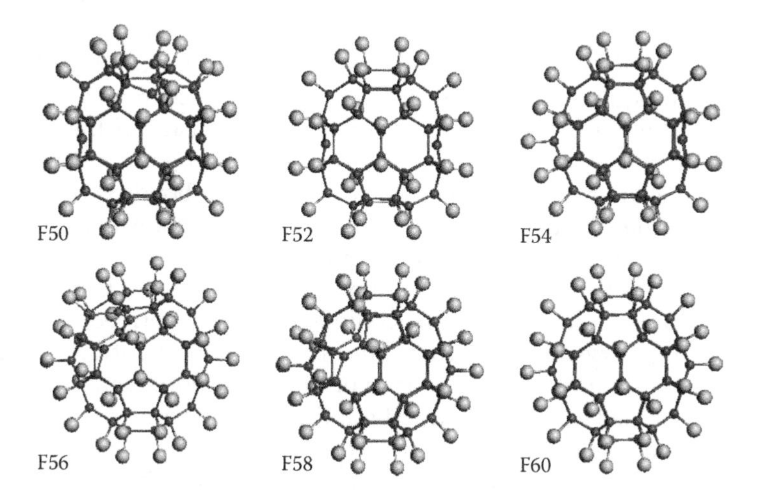

Figure 4.12 Equilibrium structures of $C_{60}F_{50}$ to $C_{60}F_{60}$ fluorinated fullerenes. UBS HF and RHF singlet state coincide. (From Sheka, E. F., arXiv:0904.4893 (cond-mat. mes-hall).)

4.4 Fluorination-induced C_{60} cage structure transformation

Stepwise fluorination is followed by the gradual substitution of *sp*²-configured carbon atoms of pristine fullerene by *sp*³ ones in fluorides. Since both valence angles between the corresponding C-C bonds and the bond lengths are noticeably different in the two cases, the structure of the fullerene cage becomes pronouncedly distorted. Figure 4.13 demonstrates the transformation of the cage structure in due course of fluorination, exemplified by changes within a fixed set of C-C bonds.

The gray-dot curve at each panel in Figure 4.13 presents the bond length distribution for the pristine C_{60}. As seen in the figure, the first steps of fluorination are followed by the elongation of C-C bonds that involve not only newly formed *sp*³ atoms, but some of *sp*² ones as well. This results in growing the total number of effectively unpaired electrons N_D, shown in Figure 4.5(b). The growing continues until $C_{60}F_8$ is formed, after which the former effect is compensated by a gradual decreasing in the N_D value caused by growing the number of *sp*³ atoms. Therefore, the changing touches upon three regions of the bond spectrum: the region of pristine short bonds, the region of long bonds, and the region of new bonds connected with fluorinated carbon atoms.

As seen in Figure 4.13, the pattern is kept throughout the full cycle of fluorination. To illustrate, Figure 4.14 presents the changing in details related to F18. Two shielded zones in Figure 4.14(a) correspond to short and long C-C bonds of the pristine fullerene, and the width of the zones indicates these bonds' dispersion. As was shown in Section 3.5, this dispersion is a consequence of UBS HF solution for all low-energy C_{60} isomers. As seen in the figure, short bonds of the fullerene cage of the fluoride preserve practically the same dispersion, while some bonds are evidently either elongated or squeezed. The dispersion of long bonds is practically doubled. Besides the two zones, a third one appears that is related to the bonds that connect fluorinated carbon atoms.

As seen from Figure 4.14, the bond structure is very complicated. However, in the case of $C_{60}F_{18}$ this complexity can be verified due to an accurate single-crystal x-ray study performed for the fluoride.[26] Figure 4.14(b) presents a three-zone spectrum of F18, while horizontal lines map bonds determined experimentally. The author's list consists of twenty-two bonds attributed to different structural elements of the fullerene skeleton. Some of these bonds are equal in length, which causes a thickening of the corresponding line in the figure. As seen in the figure, computational and experimental data are well consistent, so that the complexity of the fullerene skeleton bond structure in fluorides is really well supported.

As might be naturally expected, the *sp*²-*sp*³ transformation causes a considerable change in valence angles as well. To keep the cage structure

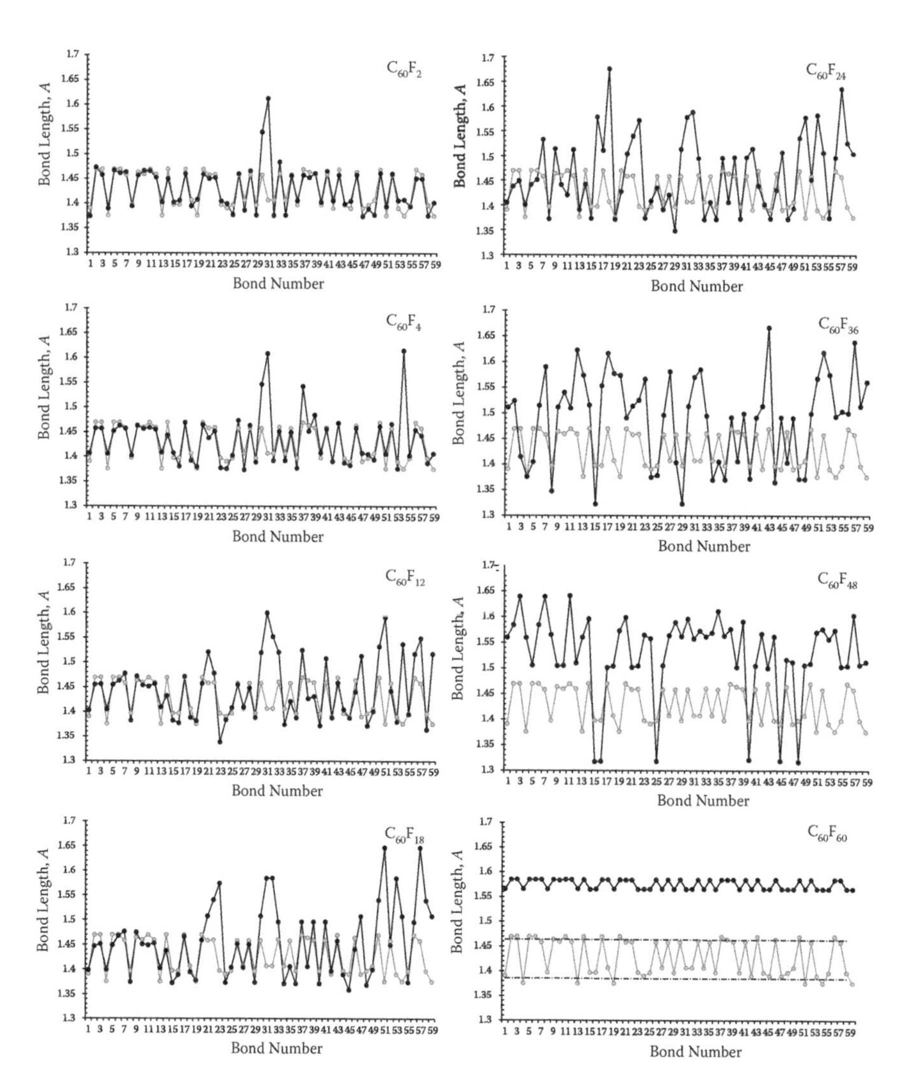

Figure 4.13 *sp*²-*sp*³ transformation of the C_{60} cage structure in due course of successive fluorination. Gray-dot and black-dot diagrams correspond to the bond length distribution related to the C_{60} cage of pristine fullerene and its fluorinated derivatives, respectively. (From Sheka, E. F., arXiv:0904.4893 (cond-mat.mes-hall).)

closed, this effect should be compensated, which, at the level of bonds, causes squeezing of a major part of pristine bonds, both long and short; therewith the squeezing effect grows when fluorination proceeds. This is particularly seen in the right-hand panels of Figure 4.13 related to *k*-high fluorides, where a number of extremely short C-C bonds of 1.325 to 1.320 Å

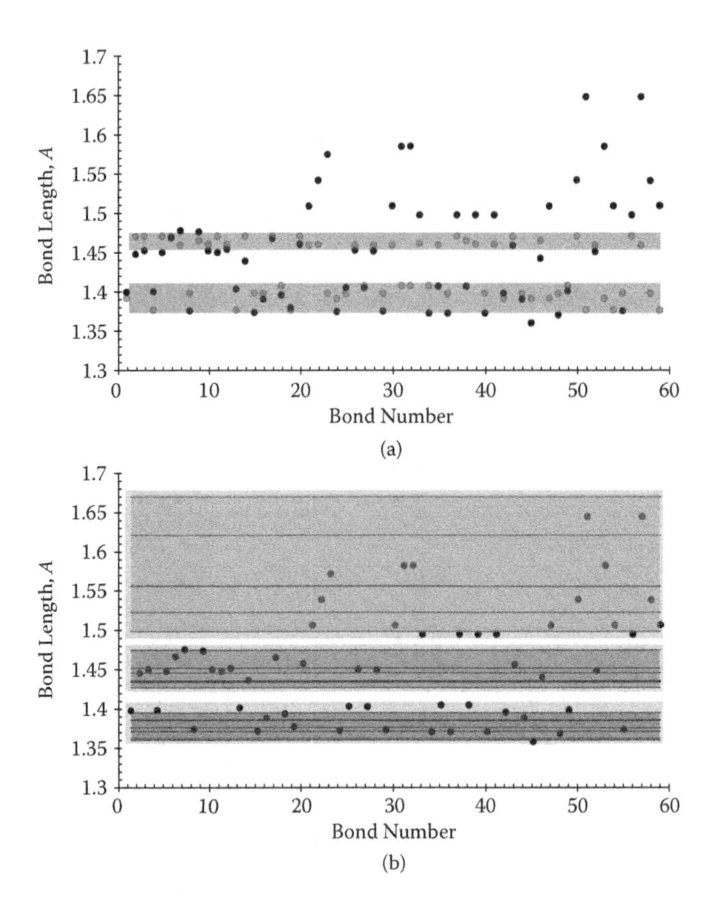

Figure 4.14 (a) Computational C-C bond spectra of F18 (black dots) and C_{60} (C_i isomer) (gray dots). (b) C-C bonds, computed for F18 molecule (black dots) and experimentally determined for $C_{60}F_{48}$ single crystal (lines). (From Neretin I. S., et al., *Angew Chem.*, 112, 3411–14, 2000.)

in lengths are observed. The existence of such bonds is supported experimentally by the single-crystal x-ray data for $C_{60}F_{48}$.[35] These six bonds are the only ones that remain unsaturated in F48 fluoride and carry information about the pristine structure. Once each of them has been saturated when fluorination proceeds from F48 to F60, a new nonstressed cage of I_h symmetry is formed. Important to note is that in spite of the fact that all carbon atoms become sp^3 configured and should be considered fully identical, the short and long bond pattern of the pristine C_{60} is kept for the $C_{60}F_{60}$ cage as well. Moreover, previously short (long) bonds of C_{60} keep their character in the cage. However, their distribution differs from that of

I_h^{RHF} C_{60} (see Figure 3.1) shown in the figure. The presence of fluorine atoms causes a noticeable dispersion. Similar sp^2-sp^3 transformation of the cage was observed under hydrogenation from C_{60} to $C_{60}H_{60}$ (see Chapter 5).

4.5 Concluding remarks

The data presented in the chapter show that the UBS HF approach provides a reliable computational consideration of stepwise fluorination of pristine C_{60} in the manner of a successive computational synthesis of $C_{60}F_{2k}$ species. The consideration is based on the approach's ability to directly select target atoms of the fullerene cage to be subjected to each sequential addition of fluorine. The atomic chemical susceptibility N_{DA} serves as a quantified pointer of the selection. Each addition of fluorine atom causes a remarkable change in the N_{DA} distribution over the C_{60} cage atoms. That is why the fluorination can be considered a series of successive steps, target atoms of which are marked by the largest N_{DA} value of the N_{DA} map related to a preceding member of the series. The reaction stops when the N_D value is fully exhausted. Following this methodology, a complete family of species $C_{60}F_{2k}$ (k = 1, ..., 30) can be easily produced. A good fitting of the computational data to experimental findings convincingly proves a creative role of the suggested synthetic methodology. The methodology forms the grounds for the computational synthesis of adducts based on fullerene-involved addition reactions of any kind. Continuing this methodology application, we will go to C_{60} hydrogenation.

References

1. Sheka, E. F. 2009. Step-wise computational synthesis of fullerene C_{60} derivatives. 1. Fluorinated fullerenes $C_{60}F_{2k}$. arXiv:0904.4893 (cond-mat.mes-hall) (online May 1, 2009).
2. Holloway, J. H., and Hope, E. G. 1995. Fluorination. *The chemistry of fullerenes*, ed. R. Taylor, 109–16. Singapore: World Scientific.
3. Boltalina, O. V. 2000. Fluorination of fullerenes and their derivatives. *J. Fluorine Chem.* 101:273–79.
4. Boltalina, O. V., and Galeva, N. A. 2000. Direct fluorination of fullerenes. *Russ. Chem. Rev.* 69:609–22.
5. Taylor, R. 2001. Fluorinated fullerenes. *Chem. Eur. J.* 7:4074–84.
6. Jaffe, R. L. 2003. Quantum chemistry study of fullerene and carbon nanotube fluorination. *J. Phys. Chem. B* 107:10378–88.
7. Taylor, R. 2004. Why fluorinate fullerenes? *J. Fluorine Chem.* 125:359–68.
8. Balasubramania, K. 1991. Enumeration of isomers of polysubstituted C_{60} and application to NMR. *Chem. Phys. Lett.* 182:257–62.
9. Krusic, P. J., Wasserman, E. P., Keizer, N., Morton, J. R., and Preston, K. F. 1991. Radical reactions of C_{60}. *Science* 254:1183–85.

10. Mitsumoto, R., Araki, T., Ito, E., Ouchi, Y., Seki, K., Kikuchi, K., Achiba, Y., Kurosaki, H., Sonoda, T., Koboyashi, H., Boltalina, O. V., Pavlovich, V. K., Sidorov, L. S., Hattori, Y., and Liu, N. 1998. Electronic structure and chemical bonding of fluorinated fullerenes studied by NEXAFS, UPS, and vacuum-UV absorption spectroscopies. *J. Phys. Chem. A* 102:552–60.
11. Matsuzawa, N., Fukunaga, T., and Dixon, D. A. 1992. Electronic structures of 1,2- and 1,4-C60X2n derivatives with n = 1, 2, 4, 6, 8, 10, 12, 18, 24, and 30. *J. Phys. Chem.* 96:10747–56.
12. Clare, B. W., and Kepert, D. L. 1994. Structure and stabilities of hydrofullerenes. Completion of crown structures at $C_{60}H_{18}$ and $C_{60}H_{24}$. *Mol. Struct. (THEOCHEM)* 303:1–9.
13. Darwish, A. D., Avent, A. G., Taylor, R., and Walton, D. R. M. 1996. Structural characterization of $C_{60}H_{18}$; a C_{3v} symmetry crown. *J. Chem. Soc. Perkin Trans.* 2:2051–54.
14. Clare, B. W., and Kepert, D. L. 1997. An analysis of the 94 possible isomers of $C_{60}F_{48}$ containing a three-fold axis. *Mol. Struct. (THEOCHEM)* 389:97–103.
15. Clare, B. W., and Kepert, D. L. 1999. The structures of $C_{60}F_{36}$ and new possible structures for $C_{60}H_{36}$. *Mol. Struct. (THEOCHEM)* 466:177–86.
16. Clare, B. W., and Kepert, D. L. 2002. Structures, stabilities and isomerism in $C_{60}H_{36}$ and $C_{60}F_{36}$. A comparison of the AM1 Hamiltonian and density functional techniques. *Mol. Struct. (THEOCHEM)* 589–590:195–207.
17. Clare, B. W., and Kepert, D. L. 2002. Structures of $C_{60}H_n$ and $C_{60}F_n$, n=36–60. *Mol. Struct. (THEOCHEM)* 589–590:209–27.
18. Gakh, A. A., and Tuinmann, A. A. 2001. The structure of $C_{60}F_{36}$. *Tetrahedron Lett.* 42:7133–35.
19. Boltalina, O. V., Markov, V. Yu., Taylor, R., and Waugh, M. P. 1996. Preparation and characterization of $C_{60}F_{18}$. *Chem. Commun.* 2549–50.
20. Boltalina, O. V., Street, J. M., and Taylor, R. J. 1998. $C_{60}F_{36}$ consists of two isomers having *T* and C_{3v} symmetry. *Chem. Soc. Perkin Trans.* 2:649–53.
21. Gakh, A. A., Tuinmann, A. A., Adcock, J. L., Sachleben, R. A., and Compton, R. N. 1994. Selective synthesis and structure determination of $C_{60}F_{48}$. *J. Am. Chem. Soc.* 116:819–20.
22. Boltalina, O. V., Sidorov, L. N., Bagryantsev, V. F., Seredenko, V. A., Zapol'skii, A. S., Street, J. M., and Taylor, R. J. 1996. Formation of $C_{60}F_{48}$ and fluorides of higher fullerenes. *Chem. Soc. Perkin Trans.* 2:2275–78.
23. Rogers, K. M., and Fowler, P. W. 1999. A model for pathway of radical addition to fullerenes. *Chem Commun.* 2357–58.
24. Boltalina, O. V., Lukonin, A. Yu., Street, J. M., and Taylor, R. 2000. $C_{60}F_2$ exists! *Chem Commun.* 1601–2.
25. Boltalina, O. V., Darwish, A. D., Street, J. M., Taylor, R., and Wei, X.-W. 2002. Isolation and characterization of $C_{60}F_4$, $C_{60}F_6$, $C_{60}F_8$, $C_{60}F_7CF_3$ and $C_{60}F_2O$, the smallest oxahomofullerene; the mechanism of fluorine addition to fullerenes. *J. Chem. Soc. Perkin Trans.* 2:251–56.
26. Neretin, I. S., Lyssenko, K. A., Antipin, M. Yu., Slovokhotov, Yu. L., Boltalina, O. V., Troshin, P. A., Lukonin, A. Yu., Sidorov, L. N., and Taylor, R. 2000. $C_{60}F_{18}$, a flattened fullerene: Alias a hexa-substituted benzene. *Angew Chem.* 112:3411–14.
27. Boltalina, O. V., Bühl, M., Khong, A., Saunders, M., Street, J. M., and Taylor, R. 1999. The 3He NMR spectra of $C_{60}F_{18}$ and $C_{60}F_{36}$; the parallel between hydrogenation and fluorination. *J. Chem. Soc. Perkin Trans.* 2:1475–79.

28. Boltalina, O. V., Borschevskii, A. Yu., Sidorov, L. N., Street, J. M., and Taylor, R. 1996. Preparation of $C_{60}F_{36}$ and $C_{70}F_{36/38/40}$. *Chem Commun.* 529–30.
29. Bulusheva, L. G., Okotrub, A. V., Shnitov, A. A., Bryzgalov, V. V., Boltalina, O. V., Gol'd, I. V., and Vyalikh, D. V. 2009. Electronic structure of $C_{60}F_{36}$ studied by quantum-chemical modeling of experimental photoemission and x-ray absorption spectra. *J. Chem. Phys.* 130:014704.
30. Avent, A. G., Clare, B. W., Hitchcock, P. B., Kepert, D. L., and Taylor, R. 2002. $C_{60}F_{36}$: There is a third isomer and it has C_1 symmetry. *Chem Commun.* 2370–71.
31. Popov, A. A., Senyavin, V. M., Boltalina, O. V., Seppelt, K., Spandl, J., Feigerle, C. S., and Compton, R. N. 2006. Infrared, Raman, and DFT vibrational spectroscopic studies of $C_{60}F_{36}$ and $C_{60}F_{48}$. *J. Phys. Chem. A* 110:8645–52.
32. Kawasaki, S., Aketa, T., Touhara, H., Okino, F., Boltalina, O. V., Gol'dt, I. V., Troyanov, S. I., and Taylor, R. 1999. Crystal structure of the fluorinated fullerenes $C_{60}F_{36}$ and $C_{60}F_{48}$. *J. Phys. Chem. B* 103:1223–25.
33. Bagryantsev, V. F., Zapol'skii, A. S., Galeva, N. A., Boltalina, O. V., and Sidorov, L. N. 2000. Reactions of fullerenes with difluorine. *Russ. J. Inorg. Chem.* 45:1011–17.
34. Hedberg, L., Hedberg, K., Boltalina, O. V., Galeva, N. A., Zapolskii, A. S., and Bagryantsev, V. F. 2004. Electron-diffraction investigation of the fluorofullerene $C_{60}F_{48}$. *J. Phys. Chem. A* 108:4731–36.
35. Troyanov, S. I., Troshin, P. A., Boltalina, O. V., Ioffe, I. N., Sidorov, L. N., and Kemnitz, E. 2001. Two isomers of $C_{60}F_{48}$: An indented fullerene. *Angew Chem. Int. Ed.* 40:2285–87.
36. Papoular, R. J., Allouchi, H., Dzyabchenko, A. V., Davydov, V. A., Rakhmanina, A. V., Boltalina, O. V., Seppelt, K., and Agafonov, V. 2006. High-resolution x-ray powder diffraction structure determination of $C_{60}F_{48}$. *Fullerenes Nanotubes Carbon Nanostr.* 14:279–85.
37. Papoular, R. J. 2005. Private investigation.
38. Gakh, A. A., and Tuinman, A. A. 2001. 'Fluorine dance' on the fullerene surface. *Tetrahedron Lett.* 42:7137–39.

chapter five

Nanochemistry of fullerene C_{60}

Hydrogenated fullerenes from C_{60} to $C_{60}H_{60}$

The family of hydrogenated fullerene C_{60} presents another example of a large community of fullerene polyderivatives where every member is not subjected to sterical obstacles, so that the family can potentially span the complete series from $C_{60}H_2$ to $C_{60}H_{60}$, at least computationally. This circumstance suggests an attractive possibility of an exhausted comparative examination of two families, $C_{60}H_{2k}$ and $C_{60}F_{2k}$, at the computational level, thus disclosing the ability of the computational synthesis of fullerene derivatives to the greatest extent, as well as revealing the similarities and differences in the families' behavior.

5.1 Grounds of computational methodology

Hydrogenation of C_{60} was one of the first reactions that were considered computationally. The list of available publications is rather long (see a detailed review of papers published by the beginning 1994 in Book and Scuseria,[1] later papers,[2–5] and a collection of works by Clare and Kepert[6–12]). Altogether, the calculations exhibit practically all available computational schemes applied to the species. But the dispersion of the results caused by different techniques was not the reason for so many investigations. The main point concerned structural models of hydrogenates in view of the many-fold isomerism problem described in Chapter 4. Experimental findings that could have assisted in solving the problem have been rather scarce until now, and include the following: (1) polyaddition derivatives are abundant among the products formed, (2) the products are characterized by an even number of hydrogen atoms, and (3) isomerism appearance is limited to a few isomers. Therefore, to be practically useful, simulations had to deal with the construction of one or a few properly selected isomers of polyhydrides. The only clear evidence of how one should reach the goal concerns the pair-hydrogen-atom addition at each successive step. This was common for hydrogenation of both kinds, namely, for molecular addition, that is characteristic for the Birch reduction of C_{60}, and for an atomic one typical for radical reactions.

Therefore, the problems that arose concerning computational hydrogenation of C_{60} were similar to those discussed for fluorination. And the problem solutions looked like those suggested in the case of fluorination. Thus, when constructing structural models, it was accepted that the first and each next pair of hydrogen atoms are accommodated on the fullerene cage, implying a 1,2-addition (suitable for the addition of both the hydrogen molecule and a pair of separated hydrogen atoms), while the 1,4-addition (suitable only for separated atoms) was rejected.[13] The differences in the energy caused by the two additions were attributed to the result that only 1,2-addition does not disrupt the initial polyene structure of the cage, whereas one double (short) bond is moved to a five-membered ring in the case of 1,4-addition.

As for the manner of the successive addition, a contiguous conjectured route[14] to the formation of polyhydrides was accepted. As for the isomer selection, the main suggestion concerned isomers of high symmetry followed by further selection governed by favoring a single one among them with the least total energy. First, T_h symmetry of $C_{60}H_{36}$[15] was proposed, and the high-symmetry selection was supported by structural characterization of $C_{60}H_{18}$,[16] for which a C_{3v} symmetry crown structure was quite reliably established. Afterwards, the idea became common for the majority of calculations related to polyhydrides.

Common to all performed computations is using closed-shell restricted computational schemes. Let us remember that their application to the C_{60} fullerene cage means accepting a complete covalent bonding of the molecule odd electrons. The latter are considered classic π electrons typical for the benzene molecule. However, this approach is valid only for high hydrides with $k \geq 20$ when the effectively unpaired odd electron pool characterized by the molecular chemical susceptibility, N_D, is practically worked out. For lower hydrides, particularly for the lowest ones at $k \leq 10$, real energies of the species are significantly lower than those obtained in the closed-shell approximation that makes the restricted solutions unstable. In contrast to this case, the unrestricted broken symmetry Hartree-Fock (UBS HF) approach provides computations of the species with much lower energy, and thus more stable. Additionally, it reveals the effectively unpaired odd electron pool and offers N_D as a generalized measure of the reaction ability.

The UBS HF approach, similar to the restricted HF one used in Matsuzawa et al.,[13] reveals a preference of 1,2-addition over 1,4-addition.[17] In what follows, this allowed for considering all successive steps of hydrogenation as 1,2-addition stepwise implementations. All hydrogen atoms are bound to the exterior surface of the cage. As for both the contiguous conjectured route and high-symmetry criterion for the isomer selection, this approach does not need any of these assumptions, since changing the atomic chemical susceptibility, N_{DA}, map after each addition creates a definite algorithm of the H_2 addition in accordance with the highest N_{DA} values. Moreover, the two assumptions' validity can readily be checked in due course of the computational synthesis.

5.2 C_{60} hydrogenation as algorithmic process

As was convincingly shown in Chapter 4, the computational synthesis of C_{60} derivatives should base itself on the atomic chemical susceptibility (ACS) (N_{DA}) distribution over the cage atoms. According to Figure 3.9, the first hydride, $C_{60}H_2$, is formed by the addition of two hydrogen atoms to a pair of the cage carbon atoms belonging to group 1. The reaction will proceed around the new cage atoms with the highest N_{DA} values, resulting in the formation of hydride $C_{60}H_4$. A new ACS map reveals the sites for the next addition step, and so on. When the high-rank N_{DA} data are closely packed, a few isomers should be considered, and an additional analysis based on comparing total energies should be performed to choose the isomer with the least energy. Following this stepwise methodology, a complete list of hydrogenated fullerenes $C_{60}H_{2k}$ can be synthesized.[17] The suggested algorithm of computational synthesis of the hydrides family may be directly attributed to the C_{60} hydrogenation in gaseous hydrogen in practice.

When starting hydrogenation of C_{60}, the hydrogen molecule is placed in the vicinity of the selected atoms of group 1 (33 and 22 in this case; see Figure 3.9(d)) and a full optimization of the complex geometry in the singlet state is performed. As occurred, the hydrogen molecule is willingly attached to the cage, forming two C-H bonds, if only the molecule axis is not normal to the chosen C-C bond. In the latter case, the molecule is repelled from the cage. Therefore, the reaction with molecular hydrogen occurs in one stage, in contrast to two-stage molecular fluorination (see Section 4.3.1). Chart 5.1 exhibits the action of the processing algorithm related to C_{60}-to-$C_{60}H_{18}$ hydrogenation. Similarly to fluorides, hydrides are notated for simplicity as HN, where N means the number of hydrogen atoms attached. A structural presentation of hydrides is given in the form of Schlegel diagrams in Figure 5.1, the atom numeration of which is shown in Figure 3.9(d).

The chart presents fragments of the ACS Z → A lists involving the highest-rank N_{DA} data related to the corresponding hydrides. As follows from the chart, the ACS list of H2 points to two pairs of atoms, 5 and 3, and 60 and 57, that are shown by open circles in the $C_{60}H_2$ Schlegel diagram in Figure 5.1. These atoms are connected by a short bond within each pair, thus forming two sites for 1,2-addition of the next hydrogen molecule. Any other atoms differ by the ACS value considerably. As seen at the diagram, these bonds are not contiguous to the starting one, similar to the case of F4 fluoride. The inserted structure of the H2 hydride at the top of Figure 5.1

Chart 5.1 H2 to H18 hydrogenation

H0-C60 atom number	NDA	H2 (22, 33) atom number	NDA	H4 (60, 57) atom number	NDA	H6 (40, 54) atom number	NDA	H8 (55, 38) atom number	NDA
22	**0.27101**	5	0.29052	31	0.30726	**55**	**0.35102**	48	0.34871
33	**0.27102**	**60**	**0.29051**	34	0.30477	24	0.33194	**52**	**0.33261**
	…	3	0.29040	**40**	**0.30247**	31	0.29486	16	0.30018
		57	**0.29039**	**54**	**0.30142**	37	0.29365	12	0.28547
		42	0.26616	43	0.30044	34	0.28754	11	0.28333
		9	0.2660		…		…		…
			…	32	0.25687	23	0.27641	***51***	***0.27866***
				35	0.25605		…		…
					…	***38***	***0.26864***	42	0.27144
							…		…

Isomer analysis by total energy, *kcal/mol*

H6 (40, 54)	**821.16**	**H8 (55, 38)**	**782.45**
H6 (31, 32)	830.31	H8 (24, 23)	788.56
H6 (34, 35)	830.29		
H6 (31, 34)	831.71		

H10 (52, 51) atom number	NDA	H12 (31, 32) atom number	NDA	H14 (24, 23) atom number	NDA	H16 (42, 48) atom number	NDA	H18 (58, 59) atom number	NDA
49	0.36041	**24**	**0.40787**	49	0.37686	**58**	**0.41520**	**5**	**0.26506**
31	**0.35573**	49	0.37738	**42**	**0.35449**	15	0.27992	4	0.26296
42	0.34512	42	0.34928	***48***	***0.34201***	46	0.27809	6	0.26036
48	0.34225	48	0.34036	29	0.29760	27	0.26521	53	0.25752
58	0.33829	29	0.30019	46	0.29412	5	0.26492	27	0.25628
	…		…	47	0.29074		…		…
47	0.28156	47	0.28866		…	***59***	***0.25405***	*3*	***0.25100***
	…		…				…		…
32	***0.27544***	*23*	***0.24824***						
	…		…						
				Isomer analysis by total energy, *kcal/mol*					
H10 (48, 42)	750.79	H12 (49, 47)	712.43	**H14 (24, 23)**	**675.93**	H16 (49,47)	639.72	**H18 (58, 59)**	**600.37**
H10 (52, 51)	**744.05**	**H12 (31, 32)**	**711.31**	H14 (49, 47)	678.98	**H16 (42, 48)**	**637.59**		
		H12 (42, 48)	713.14						

Source: Sheka, E. F., arXiv:0906.2443 (cond-mat.mes-hall).

Note: In parentheses are given the numbers of the C_{60} cage atom to which a pair of hydrogen atoms is attached.

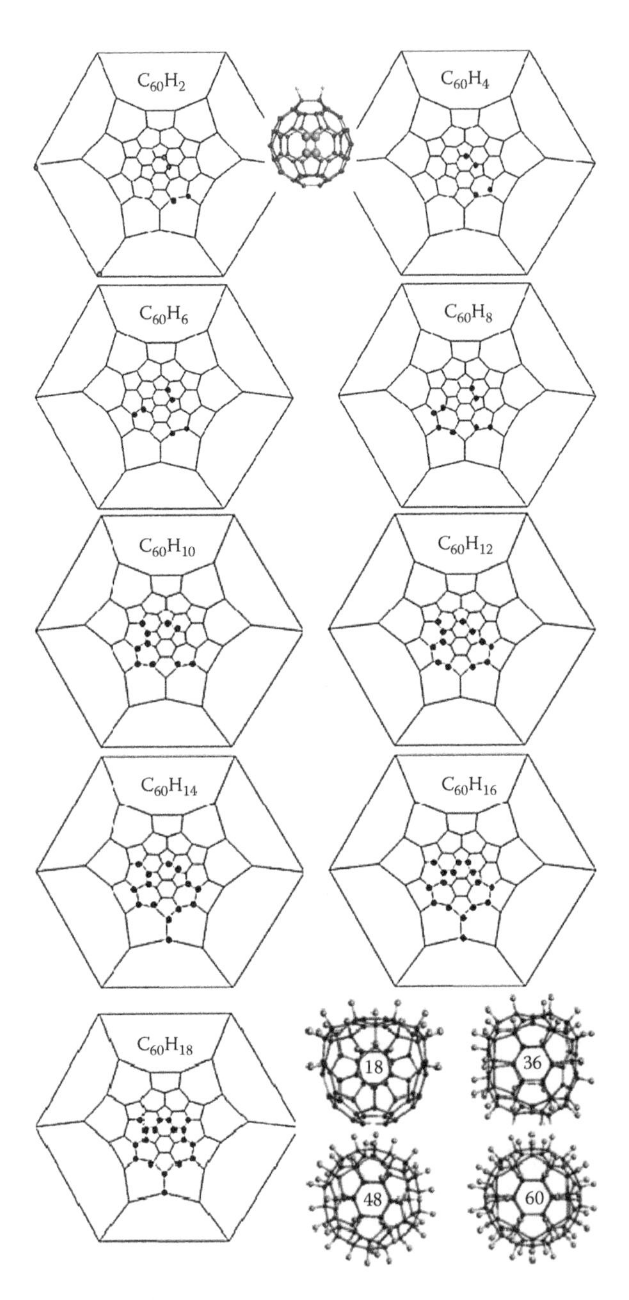

Figure 5.1 Schlegel diagrams of sequential steps of hydrogenation from $C_{60}H_2$ to $C_{60}H_{18}$ and atom structures of $C_{60}H_2$, $C_{60}H_{18}$, $C_{60}H_{36}$, $C_{60}H_{48}$, and $C_{60}H_{60}$. The diagrams are rotated by 30° downward with respect to the diagram in Figure 3.9(d). UBS HF AM1 singlet state. (From Sheka, E. F., arXiv:0906.2443 (cond-mat.mes-hall)).

shows that the two atom pairs marked by light gray are positionally fully equivalent. Thus, the two formed H4 isomers are isoenergetic, so that they generate two equivalent series of successive 1,2-additions. The series started from the 1,2-addition connected with atoms 60 and 57 will be considered below.

The ACS list of H4 highlights at least four high-rank atoms. The first two atoms, 31 and 34, are not bonded via a short bond, and their bond partners (32 and 35, respectively) are located quite far down on the ACS list (dotted lines in the chart indicate the availability of intervals between the data). And only atoms 40 and 54 form a short bond. Facing the situation, one has to perform a set of calculations based on 31 and 32, 34 and 35, and 40 and 54 additions. Obviously, the preference should be done to the isomer with the least energy. As seen from the corresponding part of the chart, the isomer analysis by the total energy definitely favors the H6 (40, 54) isomer. Proceeding with H8 synthesis, one faces the same situation as previously. The high-rank data fragment of the ACS list of H6 is headed by atoms 55 and 24, which are not bond joined. And again, a set of computations concerning 55 and 38, and 24 and 23 additions should be performed as previously, addressing the total energy selection of isomers one gets H8 (55, 38). The same algorithm was used at each next step of synthesis so that a series of hydrides H10 (52, 51), H12 (31, 32), H14 (24, 23), H16 (42, 48), and at last H18 (58, 59) was obtained. When passing from H16 to H18 there was no alternative to the combination (58, 59) according to the ACS list. The obtained H18 hydride is of the crown structure of C_{3v} symmetry, which is in full agreement with experimental findings based on ^{1}H nuclear magnetic resonance (NMR)[16] and ^{3}He NMR[14] studies. Continuing the algorithmic hydrogenation, which is quite similar to fluorination considered in Chapter 4, the higher hydrides from H20 to H60 were obtained. Structures of some of them related to $C_{60}H_{36}$, $C_{60}H_{48}$, and $C_{60}H_{60}$ are given in the bottom part of Figure 5.1.

A series of Schlegel diagrams in Figure 5.1 visualizes the step-by-step chain of 1,2-additions to the C_{60} cage that result in the synthesis of C_{60} to $C_{60}H_{18}$ hydrides. As seen from the figure, the atoms do not follow a contiguous conjectured route. And this feature proceeds when going to high hydrides as well. As said earlier, the addition of a hydrogen molecule at each successive step makes the choice of 1,2-addition quite mandatory. This requirement entails keeping the polyene structure of the fullerene cage unchanged, which is why none of the obtained hydrides possess a short bond moving to a pentagon frame. This conclusion is well supported experimentally by the observation that the Birch reduction of C_{60}, which results in $C_{60}H_{36}$ production, is not followed by alteration of the fullerene skeleton.[5] In turn, the feature excludes the availability of T_h, S_6, and D_{3d} isomers that are characterized by short-bond framing of pentagons among, say, $C_{60}H_{36}$ and $C_{60}H_{48}$ species. H36 and H48, shown in Figure 5.1, look highly symmetrical, while their exact symmetry is C_1. Obviously, the deviation from high symmetry is rather minor, so that the products may show a high-symmetry pattern in various experiments.

Geometric parameters, symmetry, and total energy of the most stable C_{60} hydrides, produced in due course of successive 1,2-additions of hydrogen molecules to the fullerene cage, are given in Table 5.1. The table also involves pairs of the cage atoms that participate in the 1,2-addition at each step. The hydrogenation route is fully similar to that of fluorination (see Chapter 4), with the only difference being in the number of the cage starting atoms: 33 and 22 for hydrogenation and 32 and 31 for fluorination. This finding exhibits the actual identity of atom pairs within group 1 belonging to one of six naphthalene-core fragments of the C_{60} fullerene. In turn, the obtained similarity strongly supports structural parallelism between hydrogenates and fluorinates of C_{60} and C_{70}, pointed out in numerous experimental studies (see Boltalina et al.[14] and references therein).

However, when comparing similar suits from the two families from the viewpoint of their production in practice, the parallelism does not appear to be fully complete. This concerns the production of *k*-high derivatives, first of all. Thus, if, as discussed in Chapter 4, $C_{60}F_{36}$ and $C_{60}F_{48}$ were efficiently produced, only $C_{60}H_{36}$ would be obtained in a significant quantity, whereas no mention of $C_{60}H_{48}$ and higher hydrides is known. A high susceptibility of C_{60} hydrides to oxidation is one of the reasons preventing high hydrides from both stabilizing and efficient production. Another reason is intimately related to the different efficacies of the hydrogenation and fluorination processes.

5.3 *Comparative efficacy of fluorination and hydrogenation reactions*

Leaving aside kinetic parameters, which no doubt are very important for the reaction running in practice, let us concentrate on the static parameters and look at the evolution of the total ΔH and coupling, E_{cpl}, energies of the products, as well as of their molecular chemical susceptibility, N_D, as a function of the pair number, *k*. The quantity E_{cpl} determines the energy that is needed for the addition of every next pair of either hydrogen or fluorine atoms to the fullerene cage. Supposing that the reaction occurs in molecular gases, and reformulating Equation 4.1, the coupling energy is determined as

$$E_{cpl} = \Delta H_{2k} - \Delta H_{2(k-1)} - \Delta H_{mol} \tag{5.1}$$

Here ΔH_{2k} and $\Delta H_{2(k-1)}$ are heats of formation of X2*k* and X2(*k* – 1) (X = H, F) products, while ΔH_{mol} is the heat of formation of either the fluorine or

Table 5.1 Geometric parameters, symmetry, and total energy of hydrogenated fullerenes $C_{60}H_{2k}$

	H2	H4	H6	H8	H10	H12	H14	H16	H18	H20
R(C* – H),[a] Å	1,126	1,126	1,126–1,129	1,126–1,130	1,126–1,130	1,126–1,129	1,125–1,131	1,126–1,130	1,126–1,130	1,126–1,130
R(C* – C*),[b] Å	1,55	1,55, 1,52	1,54–1,50	1,54–1,50	1,54–1,50	1.54–1,50	1,54–1,50	1,54–1,50	1,54–1,50	1,54–1,50
ΔH,[c] kcal/mol	909,57	862,22	821,16	788,56	744,05	711,31	675,93	637,59	600,31	561,95
Symmetry	C_2	C_s	C_1	C_1	C_1	C_1	C_1	Cs	C_{3v}	C_1
Atom pair	22, 33	60, 57	40, 54	55, 38	52, 51	31, 32	24, 23	42, 48	58, 59	5, 3
	H22	H24	H26	H28	H30	H32	H34	H36	H38	H40
R(C* – F),[a] Å	1,126–1,131	1,126–1,133	1,126–1,133	1,126–1,133	1,126–1,135	1,126–1,135	1,127–1,135	1,128–1,135	1,127–1,140	1,127–1,140
R(C* – C*),[b] Å	1,54–1,50	1,58–1,50	1,54–1,50	1,54–1,50	1,57–1,50	1,57–1,50	1,57–1,50	1,57–1,50	1,57–1,50	1,57–1,49
ΔH,[c] kcal/mol	530,03	499,37	460,93	428,56	394,43	369,60	343,06	321,72	301,07	290,98
Symmetry	C_1	C_s	C_s	C_1	C_1	C_1	C_s	C_1	C_1	C_1
Atom pair	6, 26	27, 28	13, 43	10, 17	8, 19	1, 2	12, 14	47, 49	50, 29	20, 18
50, 29	H42	H44	H46	H48	H50	H52	H54	H56	H58	H60
R(C* – F),[a] Å	1,128–1,140	1,129–1,139	1,130–1,141	1,132–1,141	1,133–1,146	1,133–1,146	1,132–1,146	1,138–1,146	1,138–1,146	1,146(1,112)
R(C* – C*),[b] Å	1,57–1,49	1,57–1,50	1,57–1,49	1,56–1,48	1,55–1,48	1,55–1,49	1,55–1,48	1,54–1,49	1,53–1,50	1,53–1,50
ΔH,[c] kcal/mol	273,93	270,67	253,28	255,06	256,98	272,49	287,98	303,45	318,82	334,16
Symmetry	C_1	C_1	C_s	C_1	C_{2v}	C_s	C_{2v}	C_{2h}	C_{2v}	I_h
Atom pair	44, 46	35, 34	36, 37	25, 4	11, 16	53, 56	15, 30	21, 45	39, 41	7, 9

Source: Sheka, E. F., arXiv:0906.2443 (cond-mat.mes-hall).

[a] C* marks the cage atom to which hydrogen is added.
[b] C* – C* marks a pristine short bond of the cage to which a pair of hydrogen atoms is added.
[c] See footnote b to Table 3.2.

hydrogen molecule equal to –22.38 kcal/mol and 0, respectively. For $k = 1$ E_{cpl} constitutes –86.3 and –45.8 kcal/mol for F2 and H2, demonstrating a tight binding of both molecules with the fullerene cage.

Figure 5.2 presents the evolution of ΔH, E_{cpl}, and N_D quantities in due course of hydrogenation[17] and fluorination.[18] As seen from the figure, the two reactions are characterized by significantly different energetic parameters, with an obvious favoring of the fluorination concerning both ΔH and E_{cpl}. Actually, the total energy ΔH of hydrides gradually decreases, which favors polyhydride formation, until k reaches 15, after which the decreasing pace is slowed down and then the energy starts to slightly increase. The coupling energy E_{cpl}, once negative until k reaches 22–25, gradually decreases by absolute value and becomes positive at $k > 25$. The behavior of the total and coupling energies of fluorides is similar with the only difference concerning much bigger values of both amplitude of changing ΔH and absolute magnitude E_{cpl} as well as the absence of the ΔH increasing at $k > 25$.

Molecular chemical susceptibility, N_D, is the other characteristic quantifier. The quantity is shown in the top part of Figure 5.2(b). As seen from the figure, $N_D(k)$ functions are practically identical for both families, gradually decreasing at higher k and approaching zero at $k \sim 20$ to 24. Therefore, decreasing E_{cpl} by the absolute value correlates with decreasing the molecular chemical susceptibility, N_D, or, in other words, working out the pool of effectively unpaired electrons, which results in a considerable lowering of the reaction activity when k changes from 18–20 to 25–26. According to both characteristics, the reaction is terminated at $k > 25$. Important to note is that in spite of obvious obstacles, k-high products might be abandoned among the final products. This is due to the accumulative character of the reaction until the next addition of the atom pair is still energetically favorable. This means that (1) the attachment of a next atom pair will not proceed at positive E_{cpl}, and (2) the accumulation time (and mass yield of the product) will greatly depend on the absolute value E_{cpl}: the less the value, the longer time that is needed. That is why more than four times the difference in the E_{cpl} absolute values for hydrides in favor of fluorides at $k = 18$ and the energy small absolute value results in the termination of the hydrogenation process $C_{60}H_{36}$ product while fluorination still continues and is completed by $C_{60}F_{48}$.

5.4 C_{60} cage structure transformation during hydrogenation

Stepwise hydrogenation is followed by the gradual substitution of sp^2-configured carbon atoms by sp^3 ones. Since both valence angles and the

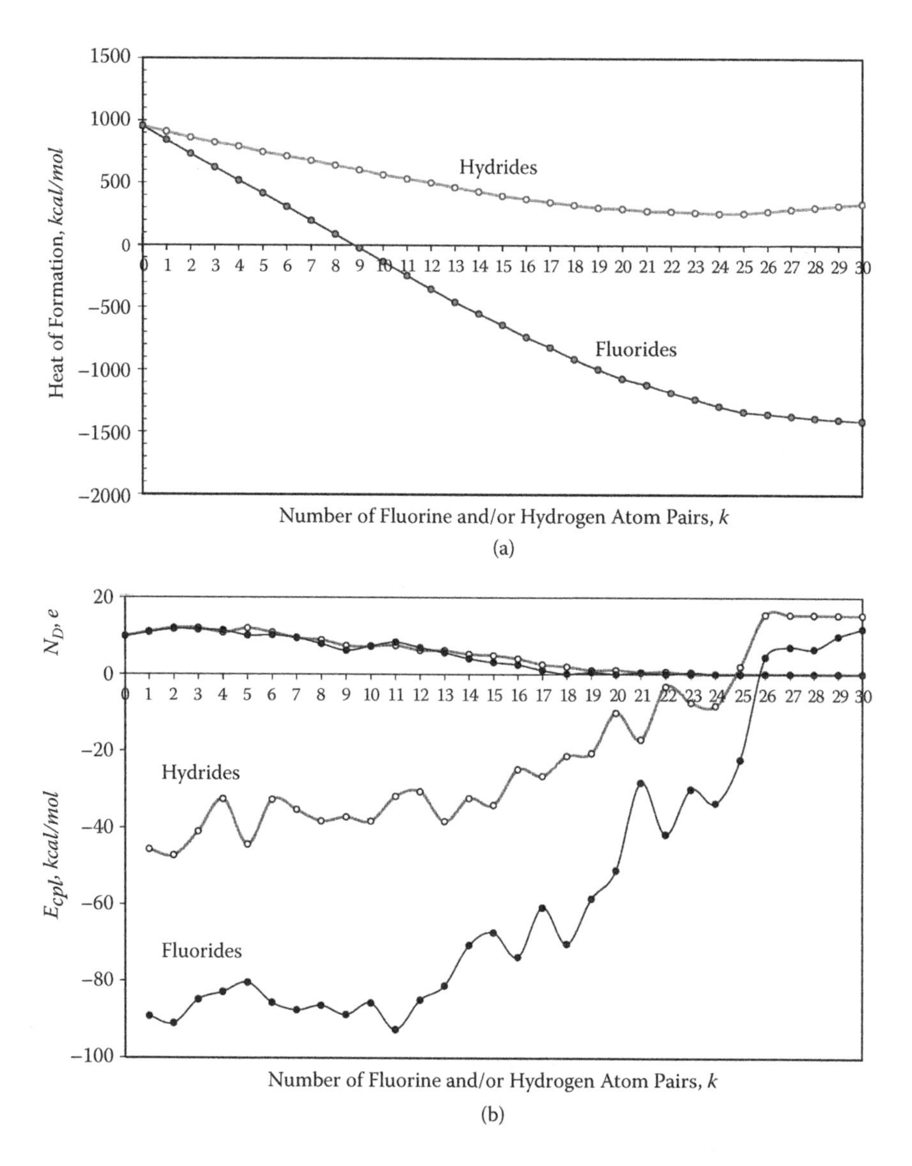

Figure 5.2 Evolution of the total energy (a), molecular chemical susceptibility, N_D, and coupling energy, E_{cpl} (b), at growing the number of atom pairs k for C_{60} hydrides (dotted curves with empty circles) and C_{60} fluorides (solid curves with filled circles). (From Sheka, E. F., arXiv:0906.2443 (cond-mat.mes-hall); Sheka, E. F., arXiv:0904.4893 (cond-mat.mes-hall)).

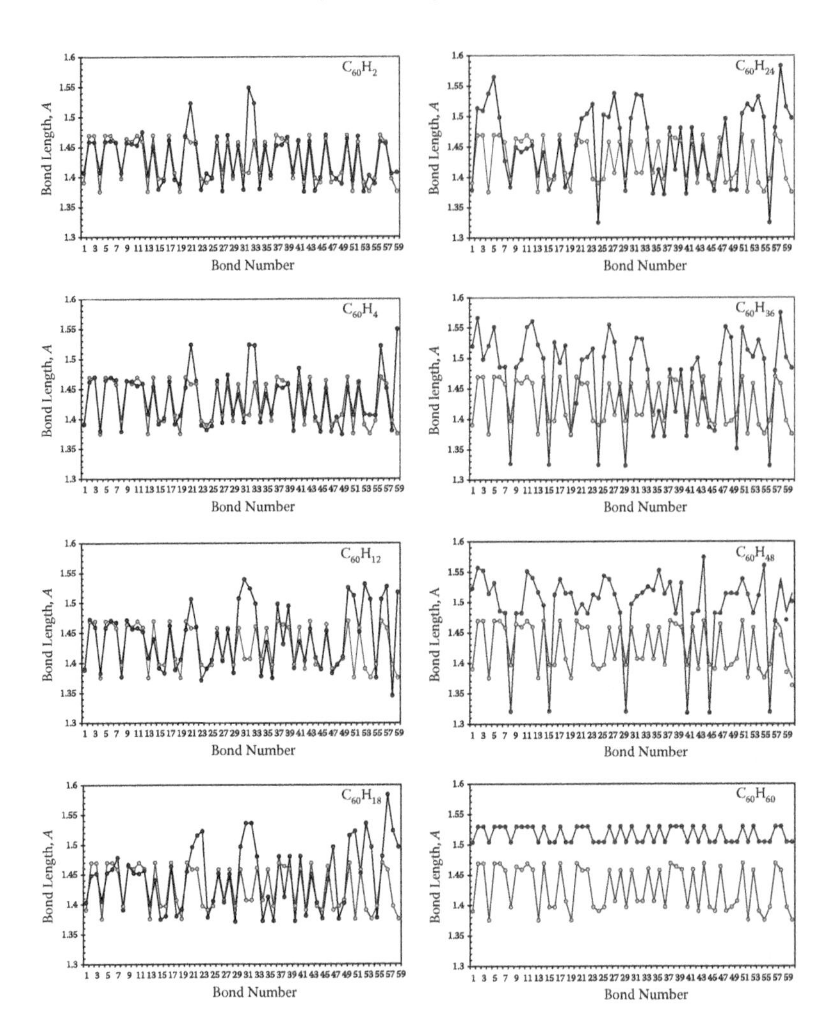

Figure 5.3 *sp^2-sp^3* transformation of the C_{60} cage structure in due course of step-by-step hydrogenation. Gray-dot and black-dot diagrams correspond to the bond length distribution related to the C_{60} cage of pristine fullerene and its hydrides, respectively.

corresponding C-C bond lengths are noticeably different in the two cases, the structure of the fullerene cage becomes pronouncedly distorted. Figure 5.3 demonstrates the transformation of the cage structure in due course of hydrogenation, exemplified by changes within a fixed set of C-C bonds.

The gray dotted curve at each panel in the figure presents the bond length distribution for the pristine molecule. As seen, the first steps of hydrogenation are followed by the elongation of C-C bonds that involve not only newly formed *sp^3* atoms, but also some *sp^2* ones. A noticeable growing of the total number of effectively unpaired electrons N_D ensues from

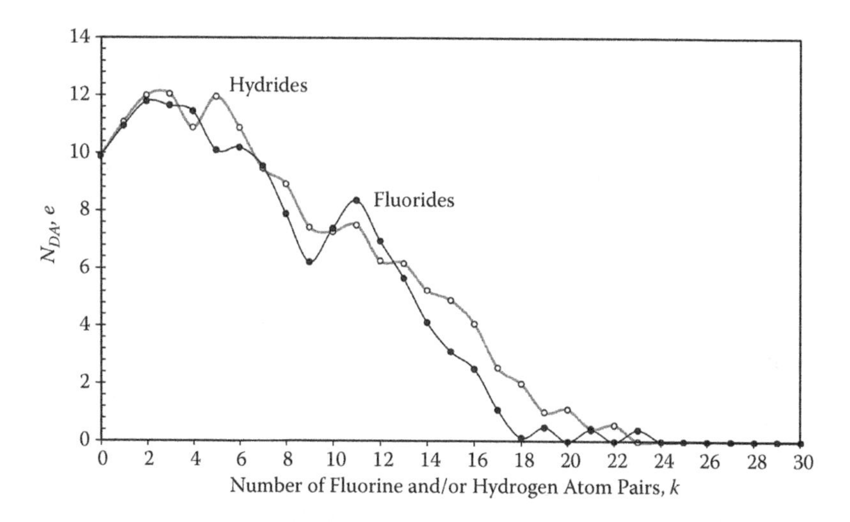

Figure 5.4 Molecular chemical susceptibility, N_D, of $C_{60}X_{2k}$ (k = 1, …, 30) hydrides (dotted curve with empty circles) and fluorides (solid curve with filled circles) vs. k.

this fact (see Figure 2.1). A detailed comparative view of this effect for hydrides and fluorides is presented in Figure 5.4. As seen from the figure, a general similarity in the two families' behavior is quite obvious, which is one more manifestation of the fluorination and hydrogenation parallelism. The N_D growing proceeds until six pairs are attached in both cases, after which the former effect is compensated by a gradual decreasing in the N_D value caused by growing the number of sp^3 atoms. However, the two curves slightly differ in detail, at the same time exhibiting an expected difference in the influence of different attached addends to the fullerene cage.

Comparing the pristine C-C distribution with those belonging to different hydrides in Figure 5.3 makes it possible to trace the fullerene cage structure changes during the progress of the hydrogenation. As might be naturally expected, the sp^2-sp^3 transformation causes the appearance of elongated C-C bonds, the number of which increases when hydrogenation proceeds. To keep the cage structure closed, this effect, as well as changes in valence angles, should be compensated. At the level of bonds, this compensation causes the squeezing of a major part of pristine bonds, both long and short. Empty long bonds become shorter and shorter, achieving 1.395 Å, which is an upper limit, below which a complete covalent bonding of the remaining odd electrons occurs (Section 2.3). This results in the number of effectively unpaired electrons, N_D, approaching zero at k ~ 20 to 24, well before all empty sites will be occupied by hydrogen atoms. The squeezing effect is particularly seen in the right-hand panels of Figure 5.3, related to k-high hydrides where a number of extremely short C-C bonds of

1.325 to 1.320 Å in length are observed. These six bonds are the only ones that remain unsaturated in the $C_{60}H_{48}$ species and carry information about the pristine structure. Once saturated, when hydrogenation proceeds from $C_{60}H_{48}$ to $C_{60}H_{60}$, a new, nonstressed cage of I_h symmetry is formed. Important to note is that in spite of the fact that all carbon atoms become sp^3 configured and should be considered fully identical, the short- and long-bond pattern of the pristine C_{60} is kept for $C_{60}H_{60}$ as well, while discriminating less the two kinds of bonds. Moreover, previously short (long) bonds of C_{60} keep their character in $C_{60}H_{60}$, thus supporting the stability of the fullerene skeleton polyene structure. It is worthwhile to note that the exposition of results for hydrides is fully similar to that for fluorides.

In view of the stability of the polyene structure of the fullerene cage, which is observed at both ends of the hydrogenation and fluorination cycles, it is difficult to suggest its violation somewhere inside the cycle. This means that highly symmetric structures should not be expected for *k*-high species among the family of derivatives obtained within the framework of successive 1,2-additions. Therefore, C_{3v}-$C_{60}H_{18}$ and C_{3v}-$C_{60}F_{18}$ might present the only high-symmetry structure that is allowed for the polyene structure of the C_{60} skeleton. All of the above can be addressed by a similar plotting of sp^2-sp^3 transformation of the cage under fluorination from C_{60} to $C_{60}F_{60}$ (see Figure 4.13), so that a structural parallelism between the two families is deeply rooted in a conservative tendency of the fullerene skeleton not to change its polyene character.

5.5 *Comparison with experiments*

The C_{60} hydrides' chemistry offers two main issues that need a computational explanation. The first one concerns the way in which the hydrogenation reaction is realized and covers such key points as production of a *k*-limited number of hydrides, involving, besides main products $C_{60}H_{18}$ and $C_{60}H_{36}$, only di- and tetrahydroderivatives,[19] the termination of the C_{60} hydrogenation by $C_{60}H_{36}$, and the mass amount of the product. As known, any reaction is subordinated to a complex of static and kinetic conditions. Some arguments that might be useful when solving the problem from the static viewpoint were discussed in Section 5.4. They evidently allow for discriminating hydrogenation and fluorination processes at a semiquantitative level, at least.

The second issue concerns the product structural composition in view of a complicated many-fold isomerism problem. A lot of efforts, both experimental and computational, were undertaken to solve the problem concerning $C_{60}H_{18}$ and $C_{60}H_{36}$. As follows from the experiment, the two hydrides behave quite differently: if the former is mainly observed as a monoisomer, the latter is usually presented by an isomer mixture.[5,14] Moreover, the isomer composition in the latter case depends on the synthetic route, changing in content from one to five components. The monoisomeric form is characteristic for $C_{60}F_{18}$ fluoride as well, while $C_{60}F_{36}$ is produced in a two- to three-isomer form.

The multi-isomer output of a chemical reaction is generally caused by a small difference of the component total energies. Supposing that the output hydrides are formed in practice in due course of successive attachment of hydrogen molecules to the fullerene cage, and that all isomers of each of these hydrides have the same forerun species, one can consider the output isomer family of either $C_{60}H_{18}$ or $C_{60}H_{36}$ within a set of products generated according to the high-rank N_{DA} list of the relevant precursors H18 and H34. Chart 5.2 presents isomer sets of hydrides H18 and H36 generated from H16 and H34 that are the best precursors of the relevant hydrides given in Table 5.1. The chart involves high-rank N_{DA} lists of both precursors with the numbers of atoms to which the relevant N_{DA}s belong, the atom pairs participating in the formation of H18 and H36 isomers, as well as their total energies. The isomer symmetry is shown in parentheses. As seen from the chart, H18 should really be a monoisomer product since the smallest energy difference $\delta(\Delta H)$ related to the best C_{3v} product and to one of the other members of the set is quite large and constitutes 2.3 kcal/mol. The situation for the second hydride is drastically different. The least value $\delta(\Delta H)$ constitutes 0.6 kcal/mol, and the largest does not exceed 1.1kcal/mol, which is quite enough for the competition between the isomers to be real. Evidently, five isomers that open the list should be

Chart 5.2 Comparative energetics of H18 and H36 isomers

H16		H18		H34		H36	
N	NDA	Atom pair	ΔH, kcal/mol	N	NDA	Atom pair	ΔH, kcal/mol
58	0.41520	58, 59 (C3v)	600.307	47	0.21499	47, 49 (C1)	322.688
15	0.27992	15, 30 (C1)	602.604	46	0.21429	44, 46 (C1)	323.743
46	0.27809	46, 47 (C1)	602.612	29	0.21407	29, 50 (C1)	323.270
27	0.26521	27, 28 (C1)	603.657	18	0.21373	18, 20 (C1)	323.723
5	0.26492	3, 5 (C1)	603.673	49	0.20339	21, 45 (Cs)	323.765
				44	0.20286		
				50	0.20197		
				20	0.20179		
				21	0.17842		
				45	0.17785		
				39	0.07026		
				41	0.07014		

Source: Sheka, E. F., arXiv:0906.2443 (cond-mat.mes-hall).

considered main participants of the game. They all are nonsymmetric (C_1), so that experimental plottings related to isomer sets differently produced should have a lot in common that is readily observed in practice.[5]

The situation with fluorides F18 and F36 is different. According to the analysis given in Sections 4.3.3 and 4.3.4, the smallest $\delta(\Delta H)$ values are of 6.2 and 4.2 kcal/mol for F18 and F36, respectively. Both quantities strongly support a monoisomer regime. It can be thought that the two-isomer output of $C_{60}F_{36}$ is connected with particular kinetic conditions due to fullerene atom dancing of fluorine atoms over the fullerene cage, as well as with a possibility of stimulate reisomerization of the cage, discussed in Section 4.3.5. It should be noted that the difference in the energetic parameters of hydrides and fluorides not only governs the different reaction running, exhibited via different conditions for the reaction termination, but also influences the isomer compositions of the reaction outputs.

Now, let us talk a little about the hydrides' symmetry. There is no disagreement between the experimentalist and computational community in attributing C_{3v} exact symmetry to monoisomer $C_{60}H_{18}$. As for $C_{60}H_{36}$, plottings of structure-sensitive experimental techniques such as ^{1}H, ^{13}C, and ^{3}He NMR spectra are rather complicated, not fine structured enough, and this results is an ambiguity of the molecule symmetry determination. That is why references to T_h, T, S_6, D_{3d}, C_3, and C_1 can be found in the available literature (see a comprehensive review[6]). Computationally, numerous attempts were made to select the best candidate for the isomer structure, making the choice between T_h, T, S_6, D_{3d}, C_3, and C_1 representatives by using different computational techniques and by guessing a suitable structure within each symmetric group.

Following algorithmically predetermined sites of successive attachments of a hydrogen molecule to the fullerene cage, a stepwise C_{60} to $C_{60}H_{18}$ hydrogenation results in the formation of C_{3v}-H18 hydride in full accordance with experimental data. A close similarity of this stage of hydride synthesis to that of fluorides suggests that the pattern is connected with a particular algorithmic response of the fullerene skeleton electronic structure to the stepwise attachment in the absence of sterical obstacles for attached units. As has been recently shown, this expectation actually found confirmation when performing a similar computational synthesis of C_{60} to $C_{60}X_{18}$ derivatives, where X presents either a NH or CN entity (see Chapter 6).

Two main results follow from the approach application to $C_{60}H_{36}$:

1. C_{60} to $C_{60}H_{36}$ hydrogenation (involving a C_{60} to $C_{60}H_{18}$ stage) has been achieved by successive 1,2-additions that did not destroy the fullerene skeleton polyene structure. This conclusion can be checked empirically by considering dehydrogenation of the species, as was done in the case of the Birch reduction of C_{60}.[5] Once proved, the

structure type conservation allows for excluding from consideration those H36 isomers, the symmetry of which implies a double-bond framing of pentagons, such as T_h, T, S_6, D_{3d}, C_3, etc. At the same time, the approach logically completes the $C_{60}H_{2k}$ series by the single isomer $C_{60}H_{60}$ of I_h symmetry, as was discussed in Section 5.4.

2. Within the framework of the approach, the best isomer, H36, whose structure is given in Figure 5.1, is only formally nonsymmetric (C_1). Continuous symmetry analysis of the structure, performed in the framework of the methodology discussed in Section 3.2.4, shows that the structure is actually quite high symmetric and can be presented as either 96% C_3-ness or 97% D_3-ness or 84% T-ness. Groups C_3 and D_3, in turn, are subgroups of T_h, S_6, and D_{3d}, point groups. A big contribution of the subgroups in the H36 structure may explain a large variation within T_h, T, S_6, D_{3d}, and C_3 symmetry of $C_{60}H_{36}$ suggested on the basis of experimental data.

5.6 Concluding remarks

The reaction of fullerene C_{60} with molecular hydrogen can be successfully studied in the framework of the unrestricted broken symmetry HF self-consistent field (SCF) semiempirical approach. A complete family of species $C_{60}H_{2k}$ ($k = 1, \ldots, 30$) can be computationally synthesized as successive 1,2-additions of hydrogen molecules to the fullerene cage in a predictable manner. This approach allows for overcoming the difficulty connecting with many-fold isomorphism of the species, as well as for releasing the computational scheme from the contiguous route of sequential additions, and from looking for high-symmetry models for k-high hydride when $k \geq 9$. Added by the least total energy preference, the synthetic methodology makes it possible to obtain a best isomer for any $C_{60}H_{2k}$ ($k = 1, \ldots, 30$) species. The obtained results reveal a structural parallelism between C_{60} hydrides and C_{60} fluorides; this ensues from a conservative tendency to keep the polyene structure of the fullerene skeleton nondestroyed. However, a much weaker interaction of hydrogen molecules with the fullerene cage than that of fluorine molecules causes a significant difference in the species production, limiting the hydride family by $C_{60}H_{36}$, while $C_{60}F_{48}$ is the most abundant among fluorides.

References

1. Book, L. D., and Scuseria, G. E. 1994. Isomers of $C_{60}H_{36}$ and $C_{70}H_{36}$. *J. Phys. Chem.* 98:4283–86.
2. Dunlap, B. I., Brenner, D. W., and Schriver, G. W. 1994. Symmetric isomers of hydrofullerene $C_{60}H_{36}$. *J. Phys. Chem.* 98:1756–57.

3. Bühl, M., Tiel, W., and Schneider, U. 1995. Magnetic properties of $C_{60}H_{36}$ isomers. *J. Am. Chem. Soc.* 117:4623–27.
4. Cahill, P. A., and Rohlfing, C. M. 1996. Theoretical studies of derivatized buckyballs and buckytubes. *Tetrahedron* 52:5247–56.
5. Nossal, J., Saini, R. K., Sadana, A. K., Bettinger, H. F., Alemany, L. B., Scuseria, G. E., Billups, W. E., Saunders, M., Khong, A., and Weisemann, R. 2001. Formation, isolation, spectroscopic properties, and calculated properties of some isomers of $C_{60}H_{36}$. *J. Am. Chem. Soc.* 123:8482–95.
6. Clare, B. W., and Kepert, D. L. 1993. Structures and stabilities of hydrofullerenes $C_{60}H_n$. *Mol. Struct. (THEOCHEM)* 281:45–52.
7. Clare, B. W., and Kepert, D. L. 1994. Structure and stabilities of hydrofullerenes. Completion of crown structures at $C_{60}H_{18}$ and $C_{60}H_{24}$. *Mol. Struct. (THEOCHEM)* 303:1–9.
8. Clare, B. W., and Kepert, D. L. 1994. Structures and stabilities of hydrofullerenes. Completion of a tetrahedral fused quadruple crown structure and a double crown structure of $C_{60}H_{36}$. *Mol. Struct. (THEOCHEM)* 304:181–89.
9. Clare, B. W., and Kepert, D. L. 1994. An analysis of the 63 possible isomers of $C_{60}H_{36}$ containing a three-fold axis. A new structure for $C_{60}H_{20}$. *Mol. Struct. (THEOCHEM)* 315:71–83.
10. Clare, B. W., and Kepert, D. L. 1996. Fullerene hydrides based on skew pentagonal pyramidal arrangements of hydrogen atoms. *Mol. Struct. (THEOCHEM)* 363:179–90.
11. Clare, B. W., and Kepert, D. L. 2002. Structures, stabilities and isomerism in $C_{60}H_{36}$ and $C_{60}F_{36}$. A comparison of the AM1 Hamiltonian and density functional techniques. *Mol. Struct. (THEOCHEM)* 589–590:195–207.
12. Clare, B. W., and Kepert, D. L. 2002. Structures of $C_{60}H_n$ and $C_{60}F_n$, n = 36–60. *Mol. Struct. (THEOCHEM)* 589–590:209–27.
13. Matsuzawa, N., Fukunaga, T., and Dixon, D. A. 1992. Electronic structures of 1,2- and 1,4-$C_{60}X_{2n}$ derivatives with n = 1, 2, 4, 6, 8, 10, 12, 18, 24, and 30. *J. Phys. Chem.* 96:10747–56.
14. Boltalina, O. V., Bühl, M., Khong, A., Saunders, M., Street, J. M., and Taylor, R. 1999. The ^{3}He NMR spectra of $C_{60}F_{18}$ and $C_{60}F_{36}$; the parallel between hydrogenation and fluorination. *J. Chem. Soc. Perkin Trans.* 2:1475–79.
15. Austin, S. J., Batten, R. C., Fowler, P. W., Redmond, D. B., and Taylor, R. 1993. A prediction of the structure of $C_{60}H_{36}$. *J. Chem. Soc. Perkin Trans.* 2:1383–86.
16. Darwish, A. D., Avent, A. G., Taylor, R., and Walton, D. R. M. 1996. Structural characterization of $C_{60}H_{18}$; A C_{3v} symmetry crown. *J. Chem. Soc. Perkin Trans.* 2:2051–54.
17. Sheka, E. F. 2009. Stepwise computational synthesis of fullerene C_{60} derivatives. 2. Hydrogenated fullerenes from C_{60} to $C_{60}H_{60}$. arXiv:0906.2443 (cond-mat.mes-hall) (online June 13, 2009).
18. Sheka, E. F. 2009. Step-wise computational synthesis of fullerene C_{60} derivatives. 1. Fluorinated fullerenes $C_{60}F_{2k}$. arXiv:0904.4893 (cond-mat.mes-hall) (online May 1, 2009).
19. Avent, A. G., Darwish, A. D., Heimbach, D. K., Kroto, H. W., Meidine, M. F., Parsons, J. P., Remars, C., Roers, R., Ohashi, O., Taylor, R., and Walton, D. R. M. 1994. Formation of hydrides of fullerene-C_{60} and fullerene-C_{70}. *J. Chem. Soc. Perkin Trans.* 2:15–22.

chapter six

Nanochemistry of fullerene C_{60}

Cyano- and azo-polyderivatives

6.1 Introduction

The idea to write this chapter came to the author when thinking about how general arguments are favoring electronic and structural parallelism of C_{60} fluorides and hydrides, described in the previous chapter. Actually, besides their individual intriguing peculiarities, two features are common for both fluorinated and hydrogenated fullerenes C_{60}. The former is evidently connected with the absence of steric limitations, which is a result of a monoatomic structure of the addends and their small size. The latter concerns the 1,2 fashion of the atom additions to the fullerene cage. A natural question arises: How common are the two criteria for the structural parallelism to be observed in other cases when the requirements are met? The author had to divert her attention from the continuation of writing the book to perform a number of computations to get an answer to the question. From the computational viewpoint, a crown-like C_{3v} structure for the eighteen-fold members of the family of polyderivatives seemed to be the best manifestation of the serial event. It was hard to overcome the temptation to look for other eighteen-fold derivatives that might be expected to fit the two requirements mentioned above and to look at their structure. CN and NH units as possible addends looked like the best candidates. Actually, their addition to the C_{60} cage is not accompanied by strong steric limitations, on one hand, and must obey the 1,2-addition algorithm, on the other. Just performed calculations have shown that the structural parallelism for both sets of derivatives with fluorinated and hydrogenated fullerene C_{60} families indeed takes place. But when the author turned her attention to chlorine, it turned out that the empirical reality definitely showed noparallelism of chlorides and, say, fluorides. Upon first glance, the strictness of the above two criteria seems not be violated drastically. Is there any extra reason to provide parallelism of C_{60} polyderivatives? Let us look at what computations tell us.

6.2 Grounds of computational methodology

As in the previous two chapters, the algorithmic approach, based on precalculated atomic chemical susceptibility (ACS), is explored to computationally synthesize two families of C_{60} polyderivatives related to $C_{60}(CN)_n$ and $C_{60}(NH)_m$ compounds. The stepwise successive synthesis is limited to n = 18 and m = 9 since linear CN units are added to individual carbon atoms of the fullerene cage, while every NH unit blokes one C-C bond due to interaction of the nitrogen atoms with two carbons. The limitation is subordinated to checking if the numbers n = 18 and m = 9 might be "magic" if, suppose, that the C_{3v} crown structure of polyderivatives in this case is the brightest manifestation of the parallelism.

6.3 Polyhydrocyanides $C_{60}H(CN)_{2n-1}$ and polycyanides $C_{60}(CN)_{2n}$

Cyano- and hydrocyano C_{60} fullerenes invented by Wudl et al.[1,2] have attracted great attention as components of proton exchange membrane fuel cells[3,4] due to their high proton conductivity. The inventors have elaborated a multistep route of producing polyhydrocyanofullerenes from C_{60} to $C_{60}H(CN)_5$ that can evidently be expanded over higher members of the family, and present the series

$$C_{60} \rightarrow C_{60}H(CN)_1 \rightarrow C_{60}(CN)_2 \rightarrow C_{60}H(CN)_3 \rightarrow C_{60}(CN)_4 \rightarrow \ldots \rightarrow C_{60}H(CN)_{2n-1} \rightarrow C_{60}(CN)_{2n} \tag{6.1}$$

This production route has been supported by other chemists (see, for example, Konarev and coworkers[5,6]) demonstrating the best way to produce hydrocyano $C_{60}H(CN)_{2n-1}$ and cyano $C_{60}(CN)_{2n}$ polyderivatives and their ions. Different molecular and anionic representatives of the two series covering members with n = 1, 2, and 3 have been obtained.

Let us let lay the foundation of the production algorithm of Equation 6.1 for the stepwise computational synthesis of the species. Although in practice alkali metal cyanides are the main reagents when obtaining anion salts of fullerene C_{60}, quenched afterwards by different acids to get the products of the series,[1] it is evident that in computational experiment the products' formation can be considered a result of intermolecular interaction between the partners of the C_{60}-HCN dyad. As it turns out, the final product depends on the starting distance between C and H atoms of the molecule with the cage atoms C_c. Particularly, it is critical with respect to the hydrogen atom. If HCN is aligned along one of the C-C bonds joining two carbon atoms (C_c) of fullerene belonging to group 1, with the highest ACS (see Figure 3.9), the molecule is added to the fullerene cage associatively (product I) (see Figure 6.1) when the C_c-H distance exceeds 1.7 Å. When the distance is less than 1.4 Å, the molecule is attached to

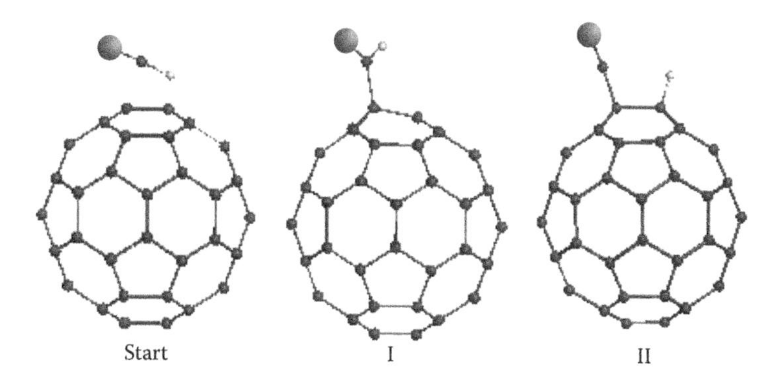

Figure 6.1 The first meeting of the HCN molecule with fullerene C_{60}. A general starting composition and equilibrium structures of products I and II (see text). Small white and big dark gray balls mark hydrogen and nitrogen atoms, respectively. UBS HF AM1 singlet state. (From Sheka, E. F., arXiv:1007.4089 (cond-mat. mes-hall), 2010.)

the fullerene cage dissociatively (product II). The energy gain constitutes 48.5 kcal/mol in favor of product II, so that the formation of the first product of Equation 6.1 is energetically profitable. In turn, the interval of 1.4 to 1.7 Å marks the space where the barrier of the HCN molecule dissociation in the vicinity of fullerene C_{60} is positioned.

Starting from product II, keeping its equilibrium structure and orientation of the hydrogen atom, let us substitute the hydrogen atom with the second CN unit and continue the structure optimization. Thus, the equilibrium structure of bis-cyanofullerene $C_{60}(CN)_2$ is obtained. Its ACS map is analyzed for two new carbon atoms to be selected for the next step of the HCN molecule dissociative addition. As it turns out, similarly to the case of fluorides and hydrides, these two atoms are located in the equatorial space of the fullerene cage. The addition of the HCN molecule to these atoms results in $C_{60}H(CN)_3$ formation. Substituting the hydrogen atom of the species by a CN unit, one obtains tetra-cyanofullerene $C_{60}(CN)_4$, the ACS map of which points to new carbon atoms for the next HCN addition, resulting in $C_{60}H(CH)_5$ production, and so forth. It should be noted that in contrast to the fluorides and hydrides case, the difference between the highest-rank N_{DA} and remaining data in the N_{DA} list is well pronounced, so that there is no necessity in isomer analysis by energy in this case. Following this algorithm, the final product, $C_{60}(CN)_{18}$, of this series was obtained.

Figure 6.2 presents odd members of Equation 6.1 related to the obtained hydrocyano[C_{60}] fullerenes. The structures are drawn in a fixed projection related to the fullerene cage to make possible a viewing of successive change in the structure of the molecules as a whole and the fullerene cage in particular. The even members of the series related to polycyanofullerenes are shown in Figure 6.3. The molecules are presented in a particular

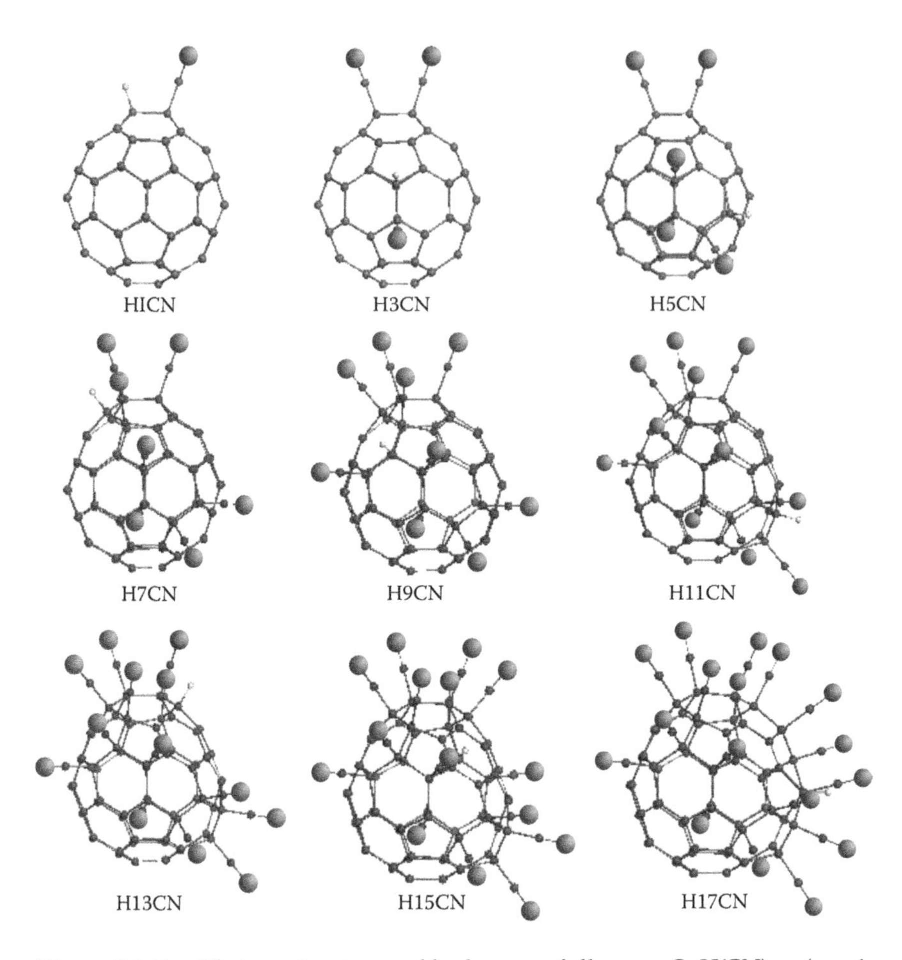

Figure 6.2 Equilibrium structures of hydrocyanofullerenes $C_{60}H(CN)_{2n-1}$ (n = 1, ..., 9). To simplify the product notation, a concise marking H(2n – 1)CN is used. For atoms' marking, see caption to Figure 6.1. (From Sheka, E. F., arXiv:1007.4089 (cond-mat.mes-hall), 2010.)

projection—let us call it a central-hexagon one—to exhibit the evolution of the molecule as a whole and the fullerene cage structure toward that of 18CN of the exact C_{3v} symmetry. The molecule is of a crown structure similar to those of $C_{60}F_{18}$ and $C_{60}H_{18}$. The two last panels in the bottom right corner compare crown structures of 18CN and 18F molecules from Chapter 4. As seen from both figures, no vivid steric complications accompany the successive cyanation of the fullerene cage performed as a successive chain of 1,2-additions. The carbon skeleton preserves its closed shape and responds to the deformation caused by the cyanation, just minimizing angular and bond strengths in a way that is similar to that of and characteristic for fluorination and hydrogenation.

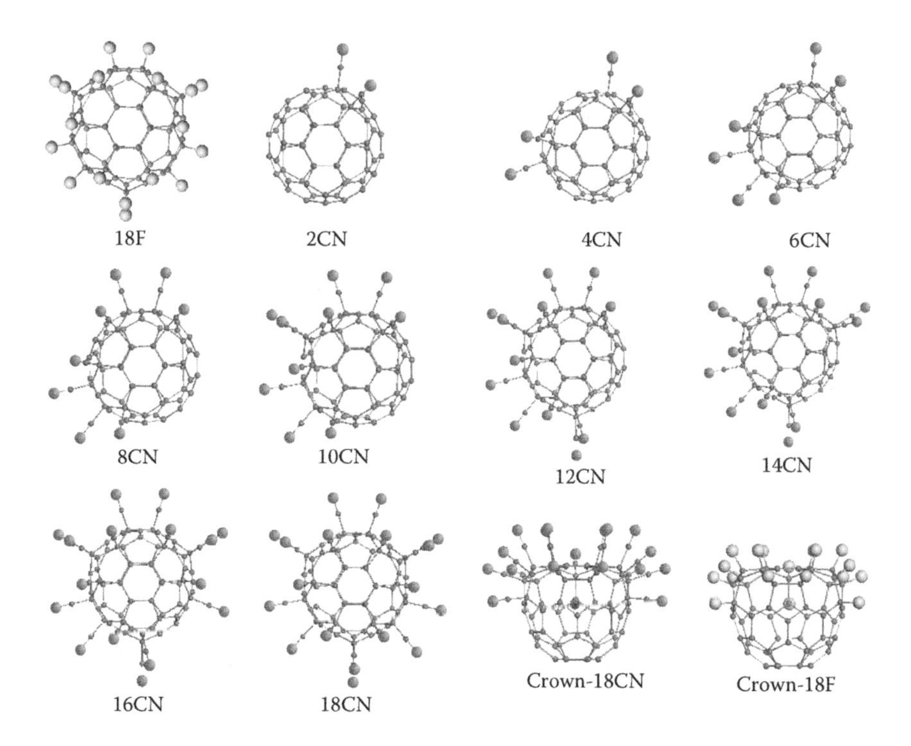

Figure 6.3 Equilibrium structures of cyanofullerenes $C_{60}(CN)_{2n}$ (n = 1, ..., 9) in the central-hexagon projection. To simplify the product notation, a concise marking 2nCN is used. For atoms' marking, see caption to Figure 6.1. (From Sheka, E. F., arXiv:1007.4089 (cond-mat.mes-hall), 2010.)

A successive change in the fullerene skeleton structure under cyanation is presented in Figure 6.4 for a fixed set of C-C bonds. The main changing corresponds to the formation of a long C-C bond between cyanated carbon atoms and neighboring atoms of a noncyanated skeleton. The bond length, starting at 1.55 Å for bis-cyanofullerene, gradually rises and achieves 1.67 Å for high members of the series. So, a large increase in the bond length is similar to that observed for C_{60} fluorides (see Figure 4.13) and is a measure of the skeleton deformation caused by sp^2-sp^3 transformation of the carbon atom configuration.

The product symmetries and total energies are listed in Table 6.1. According to two different stage productions of the series, the coupling energy, E_{cpl}, that is needed to produce a new species at each successive step should be estimated separately. Thus, the production of hydrocyanofullerenes is described by E_{cpl}, determined as

$$E_{cpl}\left[H(CN)_{2n-1}\right] = \Delta H\left[H(CN)_{2n-1}\right] - \Delta H\left[(CN)_{2n-2}\right] - \Delta H(HCN) \quad (6.2)$$

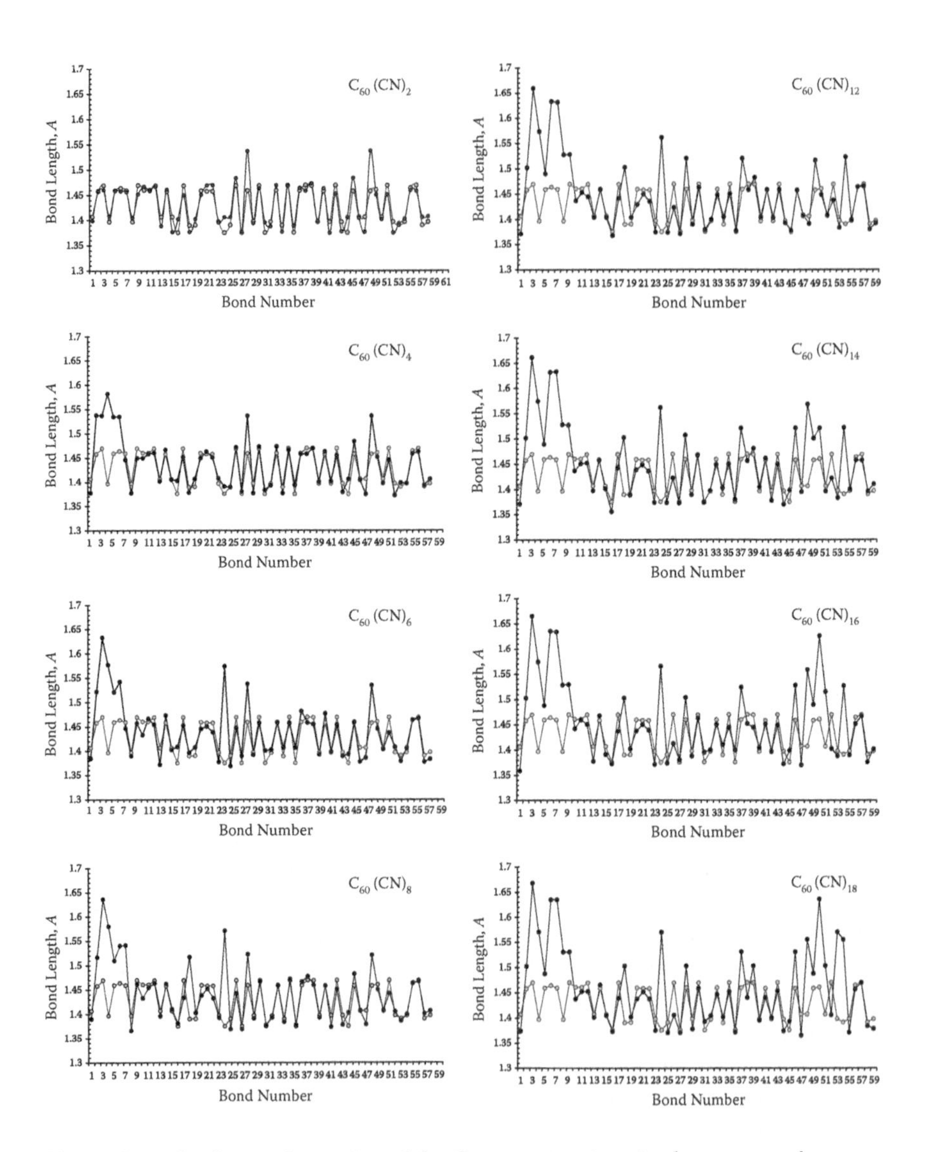

Figure 6.4 *sp*²-*sp*³ transformation of the C_{60} cage structure in due course of successive cyanation. Gray-dot and black-dot diagrams correspond to the bond length distribution related to the C_{60} cage of pristine fullerene and its cyanated derivatives, respectively. (From Sheka, E. F., arXiv:1007.4089 (cond-mat.mes-hall).)

Table 6.1 Total energy and symmetry of polycyanides of fullerenes C_{60}

	Polycyanides $C_{60}(CN)_{2n}$								
	2CN	4CN	6CN	8CN	10CN	12CN	14CN	16CN	18CN
ΔH,[a] kcal/mol	996.58	1,042.44	1,087.45	1,146.94	1,102.22	1,247.62	1,298.09	1,349.46	1,400.48
E_{cpl}, kcal/mol	–67.39	–60.76	–65.40	–64.02	–66.41	–64.20	–67.39	–60.23	–90.36
Symmetry	C_{2v}	C_s	C_1	C_s	C_s	C_1	C_s	C_1	C_{3v}
	Polyhydrocyanides $C_{60}H(CN)_{2n-1}$								
	H1CN	H3CN	H5CN	H7CN	H9CN	H11CN	H13CN	H15CN	H17CN
ΔH,[a] kcal/mol	951.70	990.93	1,040.59	1,098.68	1,148.24	1,199.55	1,246.63	1,297.47	1,383.57
E_{cpl}, kcal/mol	–34.65	–36.63	–32.84	–34.41	–29.69	–33.42	–31.98	–31.61	–24.65
Symmetry	C_1	C_s	C_s	C_1	C_s	C_1	C_1	C_1	C_1

Source: Sheka, E. F., arXiv:1007.4089 (cond-mat.mes-hall).

[a] See footnote b to Table 3.2.

where the right-hand-part components present the total energies (heats of formation) of hydrocyanofullerene $C_{60}H(CH)_{2n-1}$, cyanofullerene $C_{60}(CN)_{2n-2}$, and the HCN molecule, respectively. In turn, the production of cyanofullerenes is described by the coupling energy, E_{cpl}:

$$E_{cpl}\left[(CN)_{2n}\right] = \Delta H\left[(CN)_{2n}\right] - \Delta H\left[H(CN)_{2n-1}\right] - \Delta H(CN) \qquad (6.3)$$

Here the right-hand-part components present the total energies of cyanofullerene $C_{60}(CN)_{2n}$, hydrocyanofullerene $C_{60}H(CH)_{2n-1}$, and the CN molecular unit, respectively. The corresponding energy coupling data are listed in Table 6.1. As seen from the table, the serial production is energetically favorable. C_s symmetry for compound H1CN and C_{2v} symmetry for compound 2CN were assigned on the basis of ^{1}H nuclear magnetic resonance (NMR), ^{13}C NMR, UV-vis, and infrared (IR) spectra.[2] The latter finding is fully consistent with calculations. As for the former, the calculations attribute the H1CN structure to C_1 symmetry. However, a detailed continuous symmetry analysis in the framework of the methodology discussed in Section 3.2.3 gives 100% C_s-ness for the C_1 H1CN structure, thus removing the inconsistency.

6.4 Polyazoderivatives $C_{60}(NH)_m$

Fullerene C_{60} derivatives, which involve nitrogen atoms within the attached units, present the largest class of chemicals produced on the fullerene basis (see Hirsch and Brettreich[8] and Troshin et al.[9]). The product structure critically depends on the addend atomic composition, particularly in the case of polyadditions. Polyamines, which are produced by using large molecular addends, will be considered in Chapter 7. In this section, we will consider a family of azoderivatives of fullerene C_{60}[10] that are rooted in a large class of fulleroaziridines.[8,9] A small, two-atom composition of the addend (NH) makes it interesting from the viewpoint of expected structural parallelism. On the other hand, since attaching an NH unit to the fullerene cage occurs bidentatly through the formation of two C-N bonds, the 1,2 fashion of additions is supported as well. The computational synthesis under discussion covers a series of polyfulleroaziridines $C_{60}(NH)_m$ for m = 1, 2, ..., 9. $C_{60}(NH)_9$ aziridine can be compared with the corresponding eighteen-added units fluoride, hydride, and cyanide.

A full set of synthesized products is shown in Figure 6.5. As previously, the first product, C_{60} $(NH)_1$, is obtained when attaching the nitrogen atom to a couple of skeleton carbon atoms of group 1 (Figure 3.9) joined by a short bond. The N_{DA} map of the $C_{60}(NH)_1$ product points to two equivalent pairs of carbon atoms situated in the equatorial region of the cage, which result in the

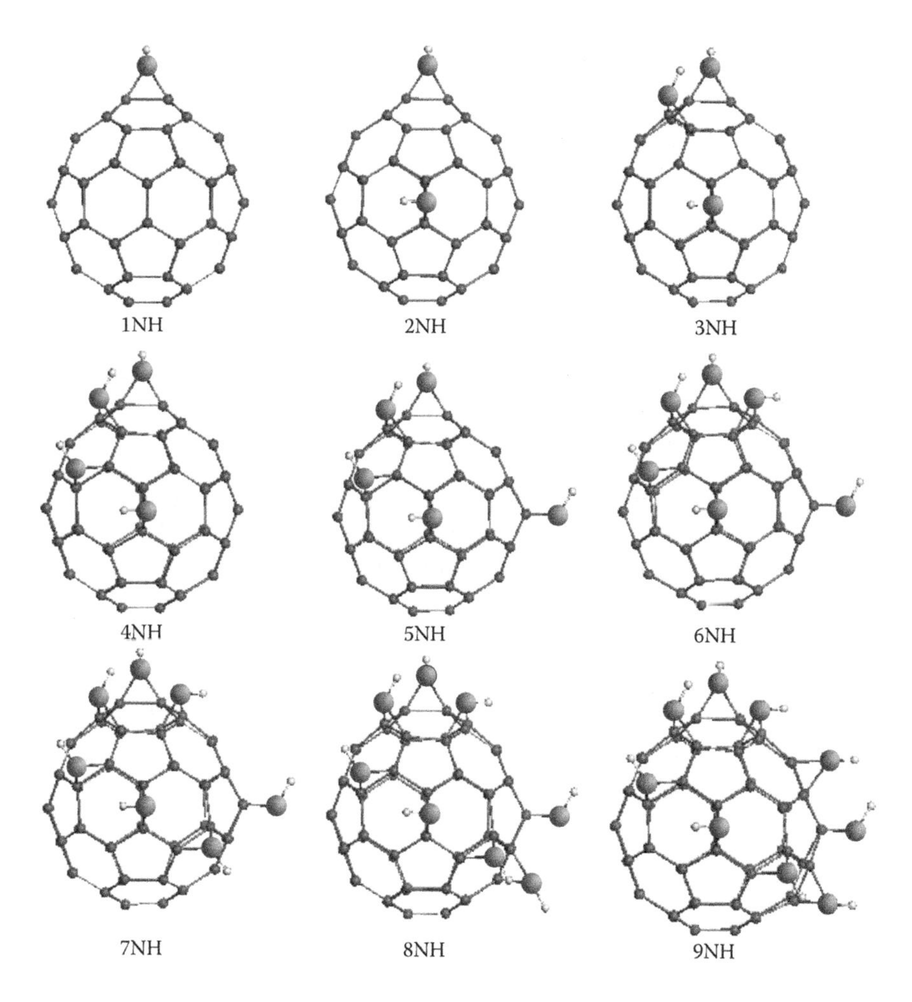

Figure 6.5 Equilibrium structures of polyaziridines $C_{60}(NH)_m$ (m = 1, …, 9). To simplify the product notation, the concise marking mNH is used. For atoms' marking, see the caption to Figure 6.1. UBS HF AM1 singlet state. (From Sheka, E. F., arXiv:1007.4089 (cond-mat.mes-hall), 2010.)

production of the second aziridine, $C_{60}(NH)_2$. Its ACS map exhibits a couple of carbon atoms in the direct vicinity of the first addition, thus causing the formation of aziridine $C_{60}(NH)_3$. Following this algorithm, the final product of the series, $C_{60}(NH)_9$, was produced. Similarly to cyanides, the high-rank N_{DA} data are well separated from the others in the list, which is why the presented synthesis does not require an additional isomer analysis by energy.

The structure projection in Figure 6.5 is similar to that in Figure 6.2, making it possible to compare both a general pattern and the difference in details under the successive additions in the two cases. As seen in the figure, the succession of additions in both cases is different, in spite of the final structures appearing to be similar. Another view on the aziridine's structure is given in Figure 6.6. As previously, the central-hexagon projection was chosen to trace the origin and development of the C_{3v} symmetry pattern, if it occurs. As seen from the figure, the nitrogen-carbon carcass of the $C_{60}(NH)_9$ aziridine has an evident characteristic C_{3v} pattern. However, due to the large dispersion of the N-H bond space orientations at practically the same energy, which causes a pronounced scattering in the hydrogen atom positions, the exact symmetry of the molecule is C_1. However, the structure analysis from the viewpoint of continuous symmetry shows that the structure is 99.7% C_{3v}, so we can convincingly talk about a deep structural parallelism of this aziridine, with eighteen-fold attached fluoride, hydride, and cyanide.

The addition of NH units to the fullerene cage, as of all other addends considered earlier, causes a noticeable deformation of the latter. Figure 6.7 presents a successive deformation of the cage in terms of selected C-C bonds throughout the aziridination. The product symmetries and total energies are listed in Table 6.2. The evolution of the total energy of the products and

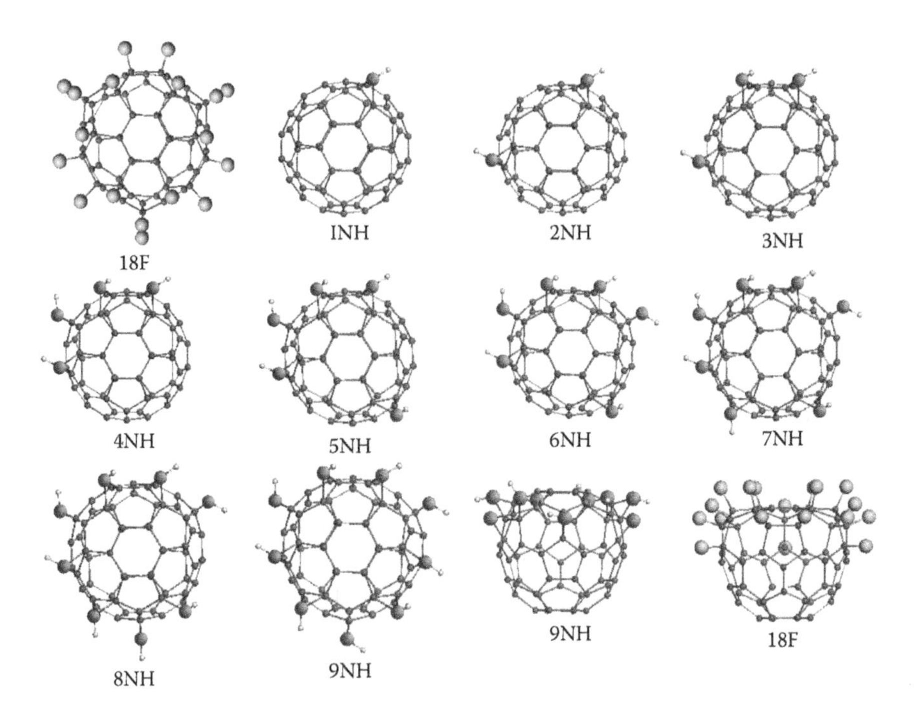

Figure 6.6 Equilibrium structures of polyaziridines $C_{60}H(CN)_m$ (m = 1, ..., 9) in the central-hexagon projection. (From Sheka, E. F., arXiv:1007.4089 (cond-mat.mes-hall), 2010.)

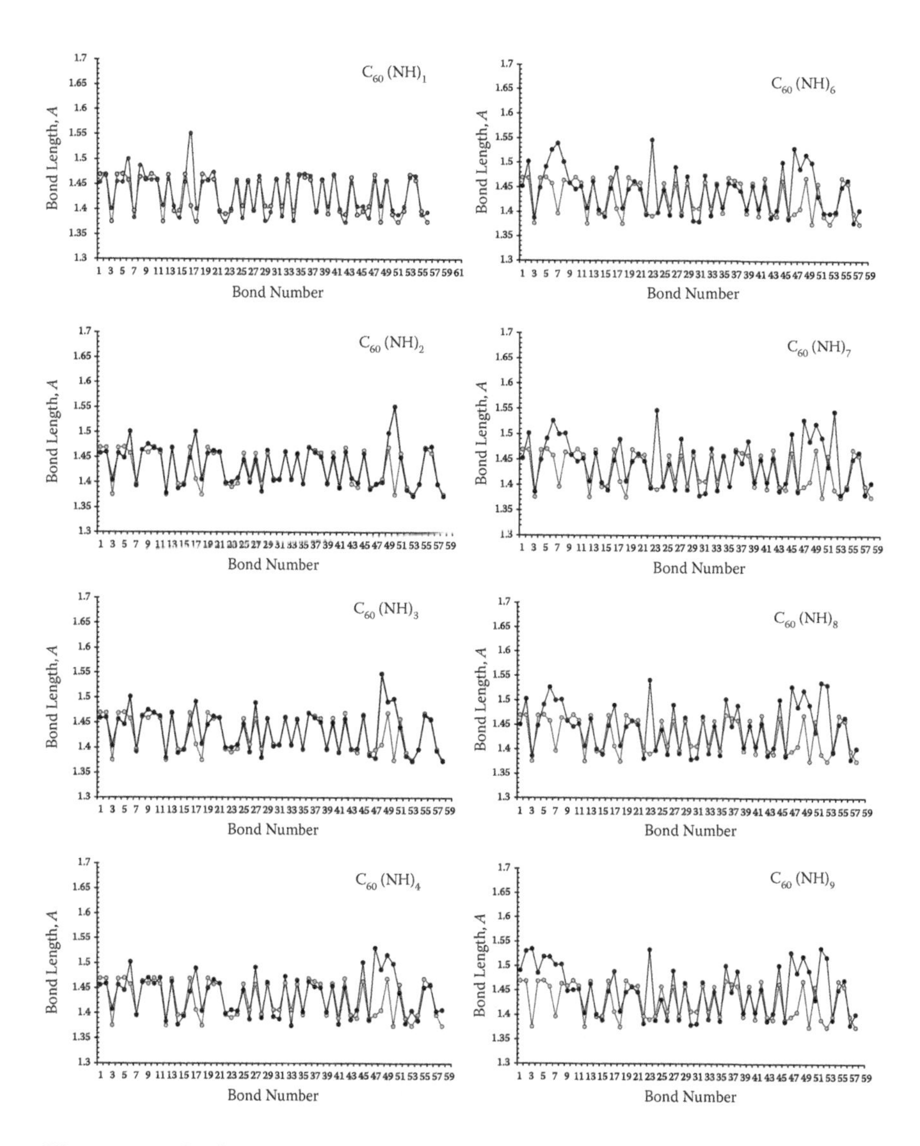

Figure 6.7 sp^2-sp^3 transformation of the C_{60} cage structure in due course of successive aziridination. Gray-dot and black-dot diagrams correspond to the bond length distribution related to the C_{60} cage of pristine fullerene and its fulleroaziridines, respectively. (From Sheka, E. F., arXiv:1007.4089 (cond-mat.mes-hall).)

Table 6.2 Total energy and symmetry of parent polyazidines of fullerenes C_{60}

	Polyaziridines C_{60} $(NH)_m$								
	1NH	2NH	3NH	4NH	5NH	6NH	7NH	8NH	9NH
ΔH,[a] kcal/mol	965.24	974.24	990.05	992.57	1,006.21	1,011.58	1,030.02	1,038.36	1,051.78
Symmetry	C_1	C_1	C_1	C_1	C_1	C_1	C_1	C_1	C_1

Source: Sheka, E. F., arXiv:1007.4089 (cond-mat.mes-hall).

[a] See footnote b to Table 3.2.

the coupling energy needed for every next addition of the NH unit is presented in Figure 6.8. The coupling energy is determined as

$$E_{coupl}[(NH)_m] = \Delta H\left[(NH)_m\right] - \Delta H\left[(NH)_{m-1}\right] - \Delta H(NH) \qquad (6.4)$$

As seen in the figure, the total energy gradually grows when aziridination proceeds. In spite of this fact, the coupling energy, determined according to Equation 6.4, remains negative and quite large in value. This means that the serial production is energetically favorable. The evolution of the total number of effectively unpaired electrons in the family, shown in the top panel of Figure 6.8(b), demonstrates high chemical reactivity of all members of the series.

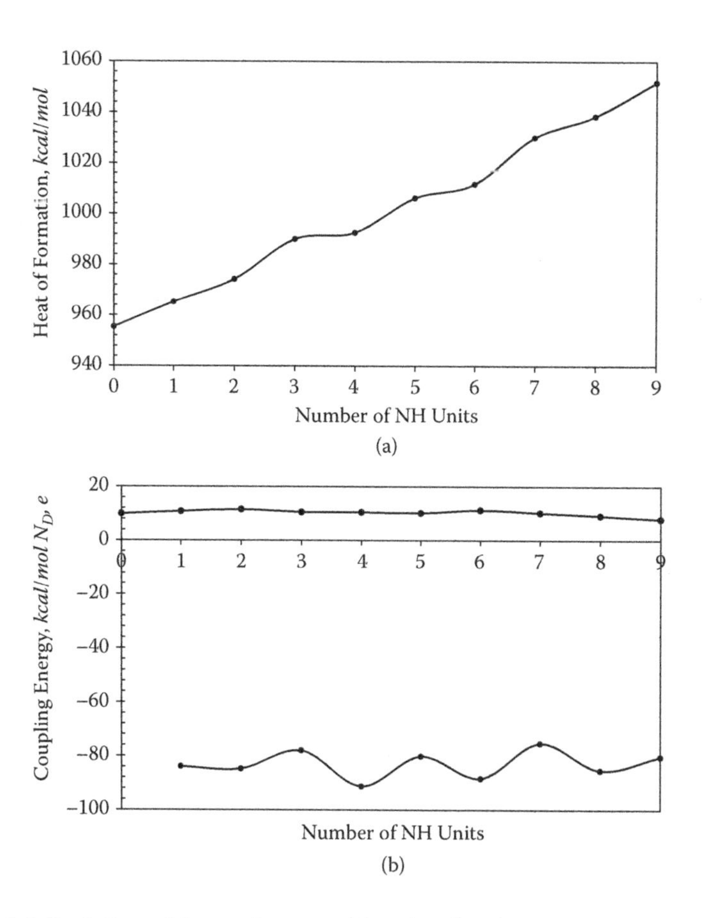

Figure 6.8 Evolution of the total energy (a) molecular chemical; susceptibility N_D and coupling energy, E_{opl} (b), at growing the number of CN units attached to the C_{60} skeleton. (From Sheka, E. F., arXiv:1007.4089 (cond-mat.mes-hall).)

So far there has been only one report[10] on the production and characterization of the first member of the series attributing its symmetry to the C_{2v} point group. Such occurred that this very substance (noted as molecule II) we discussed in Section 3.2.3 in regards to continuous symmetry of fullerene derivatives and their electronic spectra and had to make a conclusion about high continuous symmetry of the C_1 molecule consisting of ~94% I_h symmetry, this was supported by a high-symmetry pattern of its optical spectra close to the I_h one. Obviously, a high I_h grade of continuous symmetry implies that symmetric patterns related not only to I_h itself but also to its subgroups might be expected. In this view, attributing a high-symmetry pattern of ^{13}C NMR spectra of the species to the C_{2v} group[10] does not seem strange and inconsistent with computational predictions. The continuous symmetry contribution of C_{2v} to the 9NH molecule structure constitutes 98.7%.

6.5 *Concluding remarks: A little about C_{60} chlorination*

Our suggestions concerning a deep similarity in the structural patterns of the fullerene C_{60} derivatives when the relevant additions occur without steric limitations and in the 1,2 fashion seem to be true. Additional support of the statement can be obtained when comparing structures of the eigtheen-fold C_{3v} crown members of all four families in Figure 6.9. When

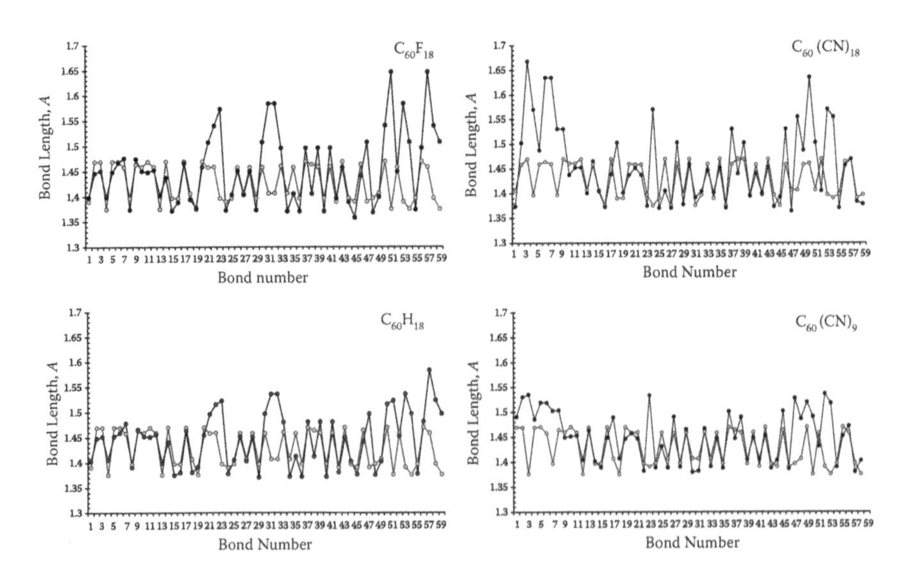

Figure 6.9 A comparative view of the sp^2-sp^3 transformation of the C_{60} cage structure of eighteen-fold derivatives under fluorination, hydrogenation, cyanation, and aziridination.

looking at the picture, one should keep in mind that the bond sets are the same only in the case of fluorination and hydrogenation, while other two sets are grouped in a different manner due to changing the atom numeration in the pristine fullerene molecule in the two latter cases. Nevertheless, the presented picture reveals quite clearly both a big similarity in the general behavior of the bond distribution and the different extent of the cage deformation under different additions. Obviously, fluorination and cyanation cause the largest effect, while hydrogenation and aziridination disturb the cage to a much weaker but comparable extent.

Evidently the absence of steric limitations and the 1,2 fashion of successive additions are necessary to support a deep parallelism of the relevant derivatives. However, the question arises whether they are enough to cover all kinds of possible addition reactions that meet the requirements. Thus, we cannot pass by C_{60} chlorination that occurs quite differently in practice in comparison with fluorination and hydrogenation.[11] Evidently, steric limitations are more severe in this case. However, they are not strong enough to explain the drastic difference. We are not going to discuss in depth the chlorination chemistry of fullerene C_{60} and will restrict ourselves to only a short excursion into first-step computational synthesis of C_{60} chlorides within the framework of the approach based on ACS's guiding role.

As it turns out, a stepwise synthesis of chlorides is subordinated to a sequence of 1,2-additions, as in all cases considered earlier. Thus, a predicted preference of first-step 1,4-additions of chlorine atom to the C_{60} cage over 1,2-additions,[12] which is rather small, is inverted when the distribution of effectively unpaired electrons over cage atoms is taken into account. However, the chlorination proceeds absolutely in a different way in regards to, say, fluorination. This difference is originated at the third step of the reaction, when the third chlorine atom is attached to the cage after the first two were situated at two carbon atoms belonging to group 1. In contrast to the third fluorine (hydrogen, cyan) and the second NH units, which were to be settled in the equatorial space of the fullerene cage, the third chlorine atom takes a seat in the nearest vicinity of the first two atoms, revealing a tendency to a contiguous sequence of addition steps. It should be noted that the preference of the highest-rank N_{DA} data over the remaining ones is the strongest in the chlorination case in regards to the four processes considered earlier.

The contiguous addition occurs until chloride $C_{60}Cl_6$ is formed. Figure 6.10 presents $C_{60}F_6$ (Figure 6.10(a)) and $C_{60}Cl_6$ (Figure 6.10(b)) structures to be compared. A noncontiguous and contiguous sequence of the addition steps is clearly seen. The contiguity of chlorine addition is terminated at the sixth step, and subsequent 1,2-additions occur quite far from the first six atoms (see Figure 6.10), leading to an evidently symmetric structure of chloride $C_{60}Cl_{18}$, shown in Figure 6.11. In spite of an absolutely different pattern of the structure in comparison with that of $C_{60}F_{18}$, the added

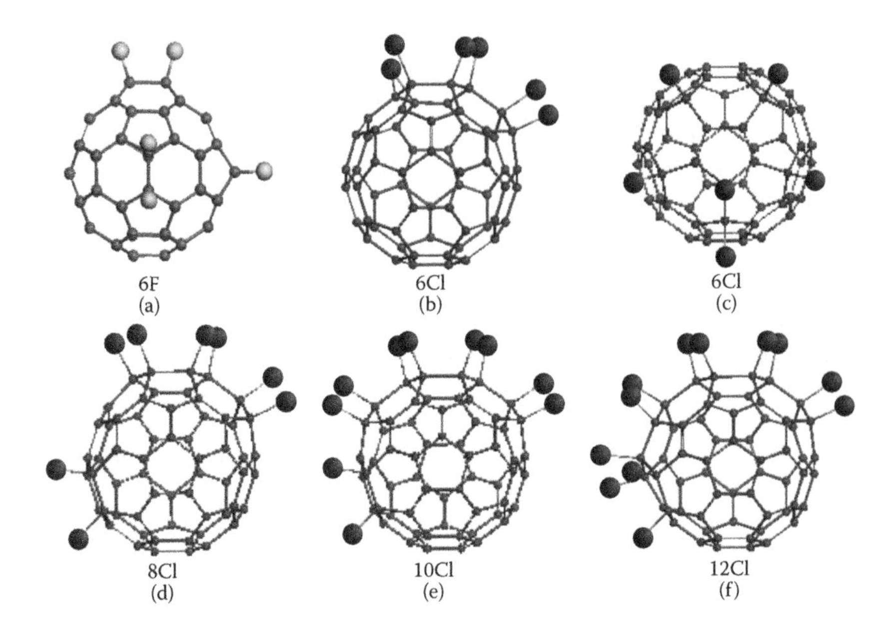

Figure 6.10 Equilibrium structures of fluoride $C_{60}F_6$ (a) and chlorides $C_{60}Cl_6$ (A) (b), $C_{60}Cl_8$ (d), $C_{60}Cl_{10}$ (e), and $C_{60}Cl_{12}$ (f). UBS HF AM1 singlet state. Chloride $C_{60}Cl_6$ (B) suggested on the basis of experimental data[13,14] (c). Light gray and black balls mark fluorine and chlorine atoms, respectively. (From Sheka, E. F., arXiv:1007.4089 (cond-mat.mes-hall), 2010.)

atoms concentrated in the upper part of the molecules form a crown-like flower bowl in the first case and a flower bowl with arms in the second. The addition of the twentieth chlorine atom causes a destruction of the cage that proceeds when the number of the attached atoms increases. The chemical bond rupture greatly raises the cage radicalization, which in turn strengthens the chemical reactivity of the chlorides, stimulating their further chlorination.

In 1991 was shown that fullerene C_{60} easily reacted with both gaseous and liquid chlorine and originated a mixture of species with either twelve or twenty-four chlorine atoms in general per one molecule of fullerene.[11,13] This appears to be connected with the fullerene cage destruction due to massive chlorination, mentioned above. The first individual chloride, $C_{60}Cl_6$, was obtained two years later,[14] and until now it has been the most studied representative of the family.[15,16] The chloride structure was attributed to C_s symmetry, which is consistent with that of the $C_{60}Cl_6$ species given in Figure 6.10(b) (chloride A). However, the structure of the $C_{60}Cl_6$ molecule presented in Figure 6.10(c) (chloride B), which was proposed on the basis of numerous synthetic data where the chloride was used as

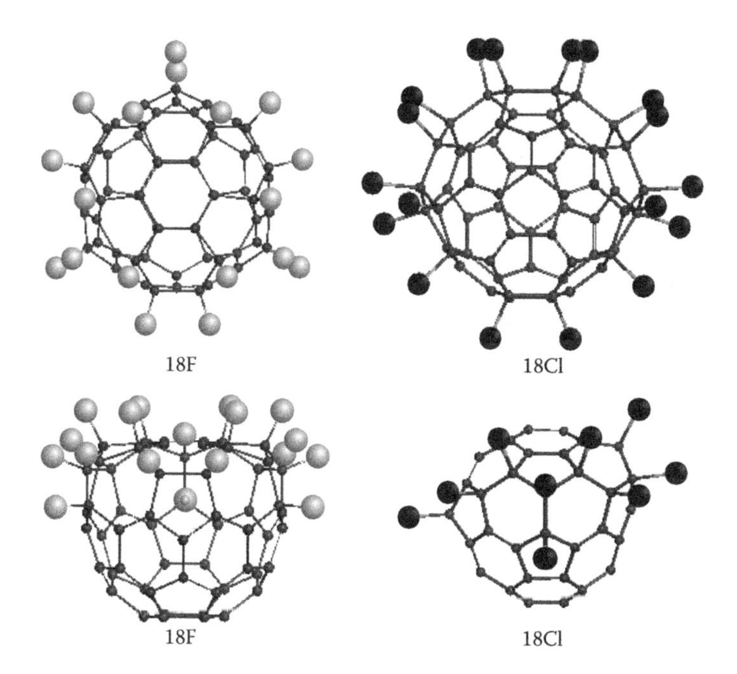

Figure 6.11 Two projections of the equilibrium structures of fluoride $C_{60}F_{18}$ and chloride $C_{60}Cl_{18}$. UBS HF AM1 singlet state. For atoms' marking, see caption to Figure 6.10. (From Sheka, E. F., arXiv:1007.4089 (cond-mat.mes-hall).)

an initial chemical reactant, differs from the calculated one quite drastically. Since the energy of molecule B is much lower than that of molecule A, and since chlorine atoms can easily move over the fullerene surface,[17] one may suggest that the initially formed A structure is transformed into the B one due to chlorine atom dancing, simultaneously causing a reisomerization of the fullerene cage to suit new requirements concerning the pentagon framing, similarly to the case of $C_{60}F_{48}$ fluoride discussed in Section 4.3.5.

Besides $C_{60}Cl_6$, other chlorides, such as $C_{60}Cl_8$, $C_{60}Cl_{12}$, $C_{60}Cl_{28}$, and $C_{60}Cl_{30}$, were produced under severe experimental conditions.[9] As seen from the list, it differs drastically from those of fluorides and hydrides and does not involve 18-, 24-, 36-, and 48-fold species, thus supporting a cardinal difference of the chlorination process. A contiguous-addition initial stage of the chlorination seems to be responsible for the difference. Therefore, to specify the structure evolution of a derivative family, three criteria involving steric limitations, the 1,2 manner of the additions, and a sequence of the additions on the first stage of the reaction should be taken into account.

References

1. Keshavarz, K. M., Knight, B., Srdanov, G., and Wudl, F. 1995. Cyanodihydrofullerenes and dicyanodihydrofullerene: The first polar solid based on C_{60}. *J. Am. Chem. Soc.* 117:11371–72.
2. Jousselme, B., Sonmez, G., and Wudl, F. 2006. Acidity and electronegativity enhancement of C_{60} derivatives through cyano groups. *J. Mater. Chem.* 16:3478–82.
3. Wang, H. B., DeSousa, R., Gasa, J., Tasaki, K., Stucky, G., Jousselme, B., and Wudl, F. 2007. Fabrication of new fullerene composite membranes and their application in proton exchange membrane fuel cells. *J. Membrane Sci.* 289:277–83.
4. Tasaki, K., Venkatesan, A., Wang, H., Jousselme, B., Stucky, G., and Wudl, F. 2008. Hydrogen cyano fullerene derivatives as acid for proton conducting membranes. *J. Electrochem. Soc. B* 155:1077–84.
5. Konarev, D. V., Khasanov, S. S., Otsuka, A., Yoshida, Y., and Saito, G. 2003. First ionic multi-component complex of fullerene $C_{60}(CN)_2$ with Co(II)tetraphenylporphyrin and bis(benzene)chromium. *Synth. Metals* 133–134:707–9.
6. Troshin, P. A., Khakina, E. A., Peregudov, A. S., Konarev, D. V., Soulimenkov, I. V., Peregudov, S. M., and Lyubovskaya, R. N. 2010. [C60(CN)5]-: A remarkably stable [60]fullerene anion. *Eur. J. Org. Chem.* 3265–68.
7. Sheka, E. F. 2010. Stepwise computational synthesis of fullerene C_{60} derivatives. 3. Cyano- and parent azo-polyderivatives. arXiv:1007.4089v1. (cond-mat.mes-hall).
8. Hirsch, A., and Brettreich, M. 2004. *Fullerenes chemistry and reactions*. London: Wiley-VCH.
9. Troshin, P. A., Troshina, O. A., Lyubovskaya, R. N., and Razumov, V. F. 2009. *Functional derivatives of fullerenes. Synthesis and applications to organic electronics and biomedicine* [in Russian]. Ivanovo: Ivanovo State University.
10. Averdung, J., Luftmann, H., and Mattay, J. 1995. Synthesis of 1,2-(2,3-dihydro-1H-azirino)-[60]fullerene, the parent fulleroaziridine. *Tetrahedron Lett.* 36:2957–58.
11. Tebbe, F. N., Becker, J. Y., Chase, D. B., Firment, L. E., Holler, E. R., Malone, B. S., Krusic, P. J., and Wasserman, E. 1991. Multiple, reversible chlorination of C_{60}. *J. Am. Chem. Soc. 113*:9900–1.
12. Matsuzawa, N., Fukunaga, T., and Dixon, D. A. 1992. Electronic structures of 1,2- and 1,4-$C_{60}X_{2n}$ derivatives with n = 1, 2, 4, 6, 8, 10, 12, 18, 24, and 30. *J. Phys. Chem.* 96:10747–56.
13. Troshin, P. A., Popkov, O., and Lubovskaya, R. N. 2003. Some new aspects of chlorination of fullerenes. *Fullerenes Nanotubes Carbon Nanostr.* 11:163–85.
14. Kuvychko, I. V., Streletskii, A. V., Popov, A. A., Kotsiris, S. G., Drewello, T., Strauss, S. H., and Boltalina, O. V. 2005. Seven-minute synthesis of pure C_s-$C_{60}Cl_6$ from [60]fullerene and iodine monochloride: First IR, Raman, and mass spectra of 99 mol % $C_{60}Cl_6$. *Chem. Eur. J.* 11:5426–36.
15. Olah, G. A., Bucsi, I., Lambert, C., Aniszfeld, R., Trivedi, N. J., Sensharma, D. K., and Prakash, G. K. S. 1991. Chlorination and bromination of fullerenes. Nucleophilic methoxylation of polychlorofullerenes and their aluminum trichloride catalyzed Friedel-Crafts reaction with aromatics to polyarylfullerenes. *J. Am. Chem. Soc. 113*:9385–87.

16. Birkett, P. R., Avent, A. G., Darwish, A. D., Kroto, H. W., Taylor, R., and Walton, D. R. M. 1993. Preparation and ^{13}C NMR spectroscopic characterisation of $C_{60}Cl_6$. *Chem. Soc. Chem. Commun.* 1230–32.
17. Troshin, P. A., Lyubovskaya R. N., Ioffe, I. N., Shustova, N. B., Kemnitz, E., and Troyanov, S. I. 2005. Synthesis and structure of the highly chlorinated [60]fullerene $C_{60}Cl_{30}$ with a drum-shaped carbon cage. *Angew. Chem. Int. Ed.* 44:234–37.

chapter seven

Nanochemistry of fullerene C_{60}

Donor–acceptor reactions of fullerene C_{60} with amines

7.1 Introduction

Four and half decades ago, in a search for a luminescent probe to test intermolecular interaction on the zeolite-organic molecule interface, the author studied luminescence of naphthalene and its two derivatives, β-methylnaphthalene and β-naphthol, adsorbed in zeolite pore.[1] On the basis of a time-dependent change in the luminescence spectra, it has been found for the first time that in the adsorbed phase a chemical reaction proceeds that is completed during storing by oxidation of the adsorbed molecules. The time required to achieve a stable spectrum was estimated in days. The study of the effect on this reaction of either gas atmosphere composition, in which the samples were stored, or temperature did not reveal noticeable correlations and the reaction mechanism remained obscure. And it was not until recently that it became clear that the effect of yet another parameter, light, remained unexplored in the study. Already then, forty years ago, it was found that the reaction proceeded only in those samples, in fresh portions of which broad bands of luminescence and absorption were observed in the visible light region. The bands were attributed to optical transitions related to the charge transfer complex formed during the adsorption. At the same time, no measures to shadow the samples from visible light were taken. The motivation to return to this puzzle happened to be the features of reactions of fullerenes with amines of a different nature,[2–4] regularly described in the literature. Thus, synthesizers are well aware that in most cases, the solutions of mixtures of primary reagents are colored, and the color changes with time until a new stable color is obtained.[2,5] As a rule, the product taken from the reactor after the stable color is achieved turns out to be an amine derivative of fullerene. From private talks with some chemists, it is known that putting the vessel in which the reaction occurs in the refrigerator substantially slows down the reaction, but no systematic studies have been made yet. As traditionally noted, slowing down has been attributed to a decrease of temperature, though less than a 10% temperature decrease would hardly slow down the rate of color change a few times.

Color changeability of reaction solutions underlies the assumption of a two-stage (at the least) character of the reaction involving a charge transfer.[6] Indeed, a large value of the energy of electron affinity (EA) of a fullerene molecule at relatively low ionization potentials (IPs) of amines trivially provides a high contribution of donor–acceptor (D–A) interaction to the total intermolecular interaction (IMI). A general consideration of this factor role is presented in Section 1.2. In this chapter, we address the feature in regards to reactions of fullerene C_{60} with a particular set of amines—an interaction that allows describing the typical zones of contact of the fullerene with different classes of amines, on the one hand, and disclosing different types of D–A-affected IMI potentials, on the other.

7.2 *About intermolecular interaction and donor–acceptor chemical reactions*

To make clear a complex consequence of donor–acceptor interaction in the fullerene-based binary systems, we will briefly recall the main issues discussed in Section 1.2. The first concerns the necessity to examine the configuration interaction of the states of neutral molecules and their ions for the IMI terms of both ground and excited states of the system to be constructed.[7] Generally, the IMI term of a D–A system, $E_{int}(r, R)$, is composed of two terms, $E_{int}(A^+B^-)$ and $E_{int}(A^0B^0)$, and is a complicated function of intra- and intermolecular coordinates. The term $E_{int}(A^+B^-)$ describes the interaction of ions leading to their bonding in the point R^{+-}. Similarly, the term $E_{int}(A^0B^0)$ binds neutral molecules in the point R^{00}. The second states that at short distances, $E_{int}(A^+B^-)$ is always below $E_{int}(A^0B^0)$. However, at longer distances the situation may change. The term $E_{int}(A^0B^0)$ tends on infinity to its asymptotic limit $E_{inf}(A^0B^0)$ equal to zero. The asymptotic limit of the term $E_{int}(A^+B^-)$ is equal to $I_A - \varepsilon_B$, where I_A determines the IP of molecule A, and $E_{inf}(A^+B^-)$ and ε_B is the EA of molecule B. If $I_A - \varepsilon_B > 0$, the terms $E_{int}(A^+B^-)$ and $E_{int}(A^0B^0)$ intersect. The third deals with the configuration interaction between the states of neutral molecules and their ions provides avoiding the intersection, so that in the vicinity of R_{scn} the terms split, forming two branches of combined IMI terms: lower branches, which describe the ground state of the system, and upper ones, which are related to excited states.

A two-well shape of IMI terms of the ground state is the main issue concerning a D–A system pointing to a possible existence of two stable structural configurations formed by the system constituents.[8,9] The configurations involve charge transfer complexes $[A^0B^0]$ ($A + B$ later on) at comparatively large intermolecular distances, R^{00}, and new chemical products $[A^+B^-]$ (AB) at intermolecular distances, R^{+-}, compared to the lengths of chemical bonds. It should be noted that actually each IMI term represents a surface in a multidimensional space; therefore, the existence of several minima, mainly in the region R^{+-}, cannot be excluded.

A practical realization of both $A + B$ and AB products depends on a particular shape of the IMI term, four possible types of which are shown in Figures 1.2 and 1.3. The IMI term of type 1 (Figure 1.2(b)) implies that both products are energetically stable. The AB product formed at R^{+-} has a clear ionic origin, so that its structure and electron properties are determined mostly by interaction of molecular ions. In contrast to this, the neutral molecules of the system are responsible for the properties of the $A + B$ product in the vicinity of R^{00}. Evidently, the reaction starts with the formation of the $A + B$ product. Passing to AB is possible when overcoming a barrier. There are many empirical ways to overcome the barrier. This topic will be considered in detail in Chapter 8, and particularly exemplified by the dimerization of fullerene C_{60}. Here it is worthwhile to mention only a photoexcitation that willingly promotes the transformation from the $A + B$ stage to an AB one. Actually, excitation within the B_2 band in Figure 1.2(b) causes ionization of the neutral molecules. A Coulomb interaction between the ions afterwards will facilitate passing through the intersection point to the minimum at R^{+-}. Photosensitized reactions of this kind are well known.[6] In the case of IMI terms of type 2 (Figure 1.2(c)), a single bound state, mainly ionic by nature, is responsible for the formation of the final product, AB. Reactions of this type are characteristic for the interaction of fullerene with alkali metal atoms.[8–10]

In the two above cases, both coupling energies are negative and $|E^{+-}_{cpl}| > E^{00}_{cpl}|$, so that the minimum at R^{+-} dominates, and D–A interaction obviously plays a governing role, leading to the formation of the AB adduct. In contrast to this case, the IMI terms shown in Figure 1.3 correspond to the case when $E^{+-}_{cpl} >$ while $E^{00}_{cpl} < 0$. Changing the coupling energy sign indicates weakening the D–A interaction and decreasing its contribution to the total IMI term. At the same time when the inequality $E^{+-}_{cpl} > |E^{00}_{cpl}|$ is not too strong, the IMI term. At the same time when the inequality $E^{+-}_{cpl} > |E^{00}_{cpl}|$ is not too strong, the IMI term of the ground state can still be two-well, but with a definite preference for the R^{00} minimum over the R^{+-} one (Figure 1.3(a)). Under these conditions, the formation of the AB product from neutral molecules is energetically nonprofitable. However, if the molecules are preliminary ionized (photoionized from the $A + B$ state in particular), the ions may form a stable bound product AB. Reactions of this kind are often called *hidden photochemical reactions.* Just this very reaction happened for naphthalene and its derivatives adsorbed inside a zeolite pore.[1,11] In contrast to the situation that is governed by the IMI term of type 1, in the case of the IMI term of type 3, the bigger the inequality, $E^{+-}_{cpl} > |E^{00}_{cpl}|$, the less probable is the formation of the AB product from an energetically stable $A + B$ complex. In addition, when the R^{+-} minimum is rather shallow or is absent, the IMI term of type 4 takes a one-well shape constructed predominantly of the term $E_{int}(A^0B^0)$ (Figure 1.3(b)). This is the limiting case when the D–A interaction contribution is the weakest.

C_{60}-based binary systems offer a large variety of the IMI terms covering all four possible types. Thus, the IMI term of the dyad C_{60}-Li (as well as other alkali metals) belongs to type 2,[8,9] while the term of the dyad C_{60}-H_2O is evidently ofF type 4. As for C_{60}-amine dyads, IMI terms of types 1 and 3 dominate. Let us consider this case as exemplified by a particular set of amine components.

7.3 Donor–acceptor reactions for fullerene dyads with different types of intermolecular interaction terms

7.3.1 Methodology of a D–A dyad consideration

The multimode character of the IMI term of the ground state for binary molecular systems involving fullerene C_{60} makes it obvious that a particular computational procedure should be elaborated to distinguish dyads in regards to the IMI term character, thus highlighting the way chemical reactions occurred between the dyad constituents. The following sequence of computational actions has been suggested.[8,9]

- Computations, opened by quantum chemical (QCh) calculations of free molecules A and B as well as their ions involve structure optimization when seeking the energy minimum. A broken symmetry unrestricted version of a selected computational tool should be used due to fullerene constituents of the dyad. However, the second reactant sometimes possesses a radical character as well. A comparative study of the structure of molecules and ions is necessary to clear up how much the molecule changes under chemoionization and if there is a possibility of the dissociation of any chemical bond under this condition.
- The dyad calculations start by selecting sets of initial configurations that differ by the positions of molecules A and B in the neighborhood of minima R^{+-} and R^{00}. Distance R^{+-} should slightly exceed the chemical bond length, the formation of which is expected for the molecule pair. Distance R^{00} is assumed to start from a sum of van der Waals radii of atoms that form the shortest intermolecular contacts.
- A full QCh calculation cycle is performed in the ground state for each initial configuration of the D–A dyad, and a possible way of the *AB* product formation is predicted.
- The barrier energy, E_{barr} (see Figure 1.2(b)) is additionally determined with the QCh reaction-coordinate approach.

Table 7.1 Characteristics of electron states of molecules constituting C_{60}-based dyads[a]

Calculated quantities	Fullerene C_{60}	DMMA C_3NH_7	TDAE $C_{10}N_4H_{24}$	TAE $C_2N_4H_8$	COANP $C_{13}H_{19}N_3O_2$
		UBS HF AM1 singlet state			
Heat of formation, ΔH, kcal/mol	955.56	40.15	53.31	12.91	9.42
Ionization potential, I, eV	9.86	7.98	8.69	7.51	9.32
Electron affinity, ε, eV	2.66	−1.94	−0.96	−1.75	0.94
Dipole moment, *Db*	0.01	1.29	0.04	0.02	7.95
Squared spin, (S**2)	4.92	0.75	0.09	0.43	0.0
Number of effectively unpaired electrons, N_D, e	9.84	1.50	0.18	0.86	0
Symmetry	C_i	C_s	C_s	D_2	C_1
ΔE^{RU}, kcal/mol[b]	17.14	9.81	0.05	0.23	0.00

[a] Equilibrium molecule structures are shown in Figure 7.1.

[b] ΔE^{RU} describes lowering the molecule total energy when going from RHF to UHF calculations (see Equation 1.1) and points to the molecule radicalization if not zero.

Let us look at how this methodology works in regards to a few selected C_{60}-amine dyads.

7.3.2 Molecular partners

Calculation results for selected molecules are summarized in Table 7.1. Leaving aside fullerene C_{60}, let us focus attention on the amines' properties.

7.3.2.1 Dimethylenemethylamine (DMMA)

This molecule exhibits well-seen radical properties provided by two methylene radicals. Radicalization of the C-C bonds of methylene units is significant since ΔE^{RU} is of more than 20% of the molecule heat of formation and $<S^{**}2>$ is nonzero, which provides $N_D = 1.5\ e$. For the DMMA (A)–C_{60} (B) dyad, $I_A < I_B$ and $\varepsilon_A < \varepsilon_B$, so that DMMA is an electron donor, while fullerene accepts the electron.

7.3.2.2 Tetrakis(dimethylamino)ethylene (TDAE)

This molecule is a rather strong covalent compound with a small, but not zero, energy ΔE^{RU} and squared spin. The latter causes a noticeable radicalization of the molecule characterized by $N_D = 0.18\ e$. $I_A < I_B$ and $\varepsilon_A < \varepsilon_B$ in this case, so that TDAE in the dyad TDAE (A)–C_{60} (B) can serve as an electron donor, while fullerene plays the acceptor role.

7.3.2.3 Tetrakisaminoethylene (TAE)

Substitution of amino units by dimethylamines somewhat weakens the covalent bonding of the ethylene bridge compared to that of TDAE, which results in increasing ΔE^{RU} and $N_D = 0.83\ e$. Otherwise, the molecule behaves similarly to TDAE and is a proper donor partner for fullerene C_{60}.

7.3.2.4 2-Cyclooctylamine-5-nitropyridine (COANP)

This molecule is an example of a complete covalent pairing of all atoms with zero energy ΔE^{RU} and $N_D = 0$, but offers suitable characteristics to donate an electron to fullerene C_{60}.

7.3.3 Molecular ions

According to the partner role of the molecules in the studied dyads, Table 7.2 presents basic QCh characteristics of the relevant single-charged ions in the ground state. The ion structures are compared with those of the neutral molecules in terms of valence bond lengths in Figure 7.1. As seen from the table, the negative ion formation lowers the molecular heat of formation, while the positive ion formation results in a considerable increase of the latter. Simultaneously, IPs and EAs of all positive ions increase, while the quantities decrease for the negative ions. The deviation of the

Table 7.2 Characteristics of electron states of single-charged ions

Calculated quantities	$(C_{60})^-$	$(C_{60})^+$	$(DMMA)^+$	$(TDAE)^+$	$(TAE)^+$	$(COANP)^+$
			UBS HF AM1 doublet state			
Heat of formation, ΔH, kcal/mol	878.79	1,166.68	200.99	193.57	157.17	199.61
Ionization potential, I, eV	4.23	12.89	15.17	11.63	12.43	14.03
Electron affinity, ε, eV	−0.40	6.83	5.94	3.99	4.59	5.45
Squared spin, $(S^{**}2)$	5.44	5.36	0.86	0.76	0.76	1.07
Symmetry	C_i	C_i	C_1	C_1	C_2	C_1

Source: Sheka, E. F., *Int. J. Quant. Chem.*, 100, 388–406, 2004.

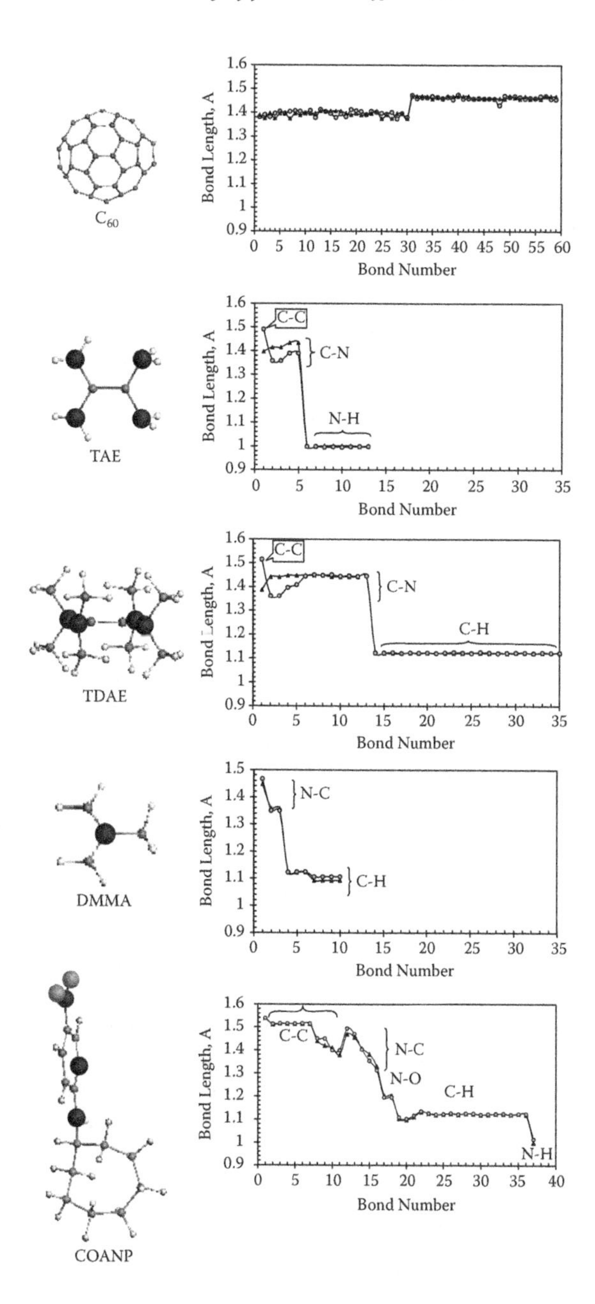

Figure 7.1 Left column: Equilibrium structures of dyad partners. Right column: Structure of neutral molecules (triangles) and molecular ions (circles) in terms of valence bonds. Carbon atoms are shown by small dark balls. Dark gray and light gray balls mark nitrogen and oxygen atoms, respectively. Hydrogen atoms are shown by small white balls.

<*S***2> value from the exact value of 0.75 characterizes a spin-mixing of the considered doublet states.

As follows from the data, the molecules form two groups. The first group combines C_{60}, DMMA, and COANP, the ionization of which does not cause lengthening of the valence bond by more than 0.02 Å. It should be noted that this is equally related to both negative and positive C_{60} ions. This means that one should not expect a facilitated dissociation of any chemical bonds of the molecules under ionization. As for TAE and TDAE belonging to the second group, the ionization results in a considerable lengthening of the C-C bond of the ethylene bridge by 0.10 and 0.12 Å, respectively. Consequently, it can be said that the molecular ion formation is accompanied by a significant shift of the atom equilibrium positions along this internal coordinate. According to the schemes in Figure 1.4, the shift is escorted by the ion vibrational excitation during vertical diabatic transition. As a result, the ion dissociation along the C-C bond is greatly facilitated.

7.4 C_{60}–based dyads

Following item 2 of the methodology recommendations given in Section 7.3.1, we will consider two starting configurations of the dyads under interest. The first one (I below) is related to $R_{st} \sim R^{00}$, while the second (II) corresponds to $R_{st} \sim R^{+-}$. Parameter R_{st} describes the shortest distance between a selected atom of the amines and a reference atom of the fullerene molecule. When R_{st} is large, final results do not critically depend on which atom of the fullerene cage is taken as a reference. In contrast, in the case of short starting distances, the choice of the reference atom is quite crucial, as was shown in the three preceding chapters. Obviously, the reference atoms should be chosen among cage atoms of group 1 with the highest ACS. A full QCh calculation cycle has resulted in obtaining equilibrium structures of the dyad products as well as a set of electronic characteristics of the latter. For the discussed dyads, these data are collected in Figure 7.2 and Table 7.3. The obtained data evidently show that R_{st} plays the decisive role. Let us comment on the obtained data for each studied dyad separately.

7.4.1 Dyad C_{60}–DMMA

R_{st} corresponds to the C-C_f distance between a carbon atom of the methylene unit of DMMA and the reference atom of the C_{60} fullerene cage. At $R_{st} \geq 3.5$ Å (the lower limit is between 2.5 and 3.5 Å), C_{60} and DMMA molecules are largely separated by the final distance, R_{fin}, of 5.88 Å and, as seen from Table 7.3, form a weak charge transfer complex: the coupling energy is small; the molecules preserve their individuality to a great extent; the

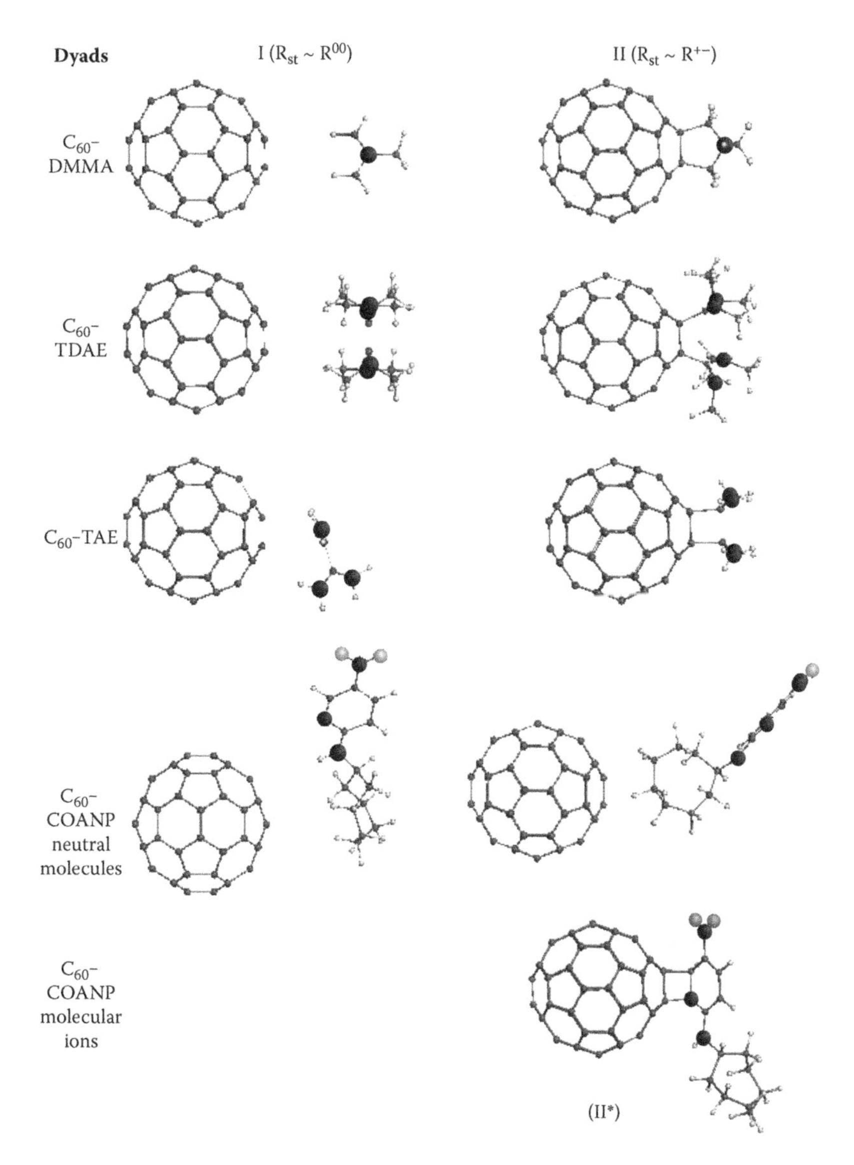

Figure 7.2 Equilibrium structures of C_{60}-based dyads with amines at two starting distances. UBS HF AM1 singlet state. See atom's marking in caption to Figure 7.1.

transferred charge is zero; and the partner compositions of highest occupied (HOMO) and lowest occupied (LUMO) molecular orbitals show a complete cross-partitioning, providing the appearance of a traditional charge transfer absorption band of B_2 type in Figure 1.2(b).

At $R_{st} \leq 2.5$ Å the partners form a chemical compound that is basic for a large series of fulleropyrrolidines (FPs), firstly synthesized in 1998.[2–4,12]

Table 7.3 Characteristics of electron states of D–A dyads C_{60} + DMMA, C_{60} + TDAE, C_{60} + TAE, and C_{60} + COANP[a]

	C_{60}-DMMA		C_{60}-TDAE		C_{60}-TAE		C_{60}-COANP	
	I	II	I	II	I	II	I, II Neutral molecules	II* Ions
	UBS HF AM1 version							
UBS HF AM1 version	R_{st} = 3.54 Å	R_{st} = 2.52 Å	R_{st} > 2 Å	R_{st} = 1.7 Å	R_{st} > 2 Å	R_{st} = 1.7 Å	R_{st} = 1.65–3.53 Å	R_{st} = 1.63 Å
Heat of formation, ΔH, kcal/mol	995.50	927.67	1,008.66	1,017.08	<968.36>[b]	945.34	964.75	973.9
Coupling energy, E_{cpl} kcal/mol	–0.21	–68.04	–0.21	–75.03	<–0.11>[b]	–83.39	–0.23	+8.92
ΔE^{RU}, kcal/mol	27.06	20.73	17.31	22.90	17.40	21.76	17.26	20.28
Squared spin, (S**2)	5.69	5.50	5.02	6.54	5.16	5.48	4.94	5.43
Ionization potential, I, eV	7.98	9.64	8.69	7.96	8.20	9.57	9.32	8.91
Electron affinity, ε, eV	2.64	2.44	2.67	2.33	1.84	2.44	2.70	2.54
Dipole moment, Db	1.38	2.52	0.05	4.84	1.44	3.65	7.63	8.24
Symmetry	C_1	C_1	C_1	C_1	C_1	C_2	C_1	C_1
Partner charge, C_{60}/XXX,[c] a.u.	0.0/0.0	–0.208/0.208	0.00/0.00	–0.03/0.03	0.0/0.0	–0.136/0.134	0.00/0.00	–0.02/+0.02
Partner composition of HOMO, C_{60}/XXX,[c] %	0.0/100	99.0/1.0	0.0/100	0.0/100	0.0/100	7.2/92.8	0.0/100	3.6/96.4
Partner composition of LUMO, C_{60}/XXX,[c] %	100/0.0	100/0.0	100/0.0	100/0.0	100/0.0	100/0.0	100/0.0	100/0.0

[a] Equilibrated structures of dyads are shown in Figure 7.2.
[b] Data in < > correspond to R_{fin} = 3.7 Å. TAE structure is very similar to that of the free molecule.
[c] Notation *XXX* means amine partners, namely, DMMA, TDAE, TAE, and COANP.

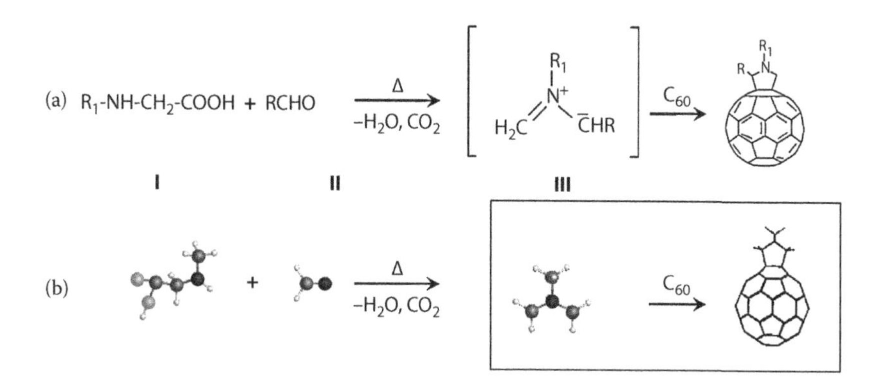

Figure 7.3 (a) Scheme of 1,3 bipolar addition of azomethyne alides to C_{60}. (b) Equilibrium structures of *N*-methylglycin (I), formaldegide (II), and dimethylenemethylamine (DMMA) (III). (From Prato, M., and Maggini, M., *Acc. Chem. Res.*, 31, 519–26, 1998.)

1,3 dipolar cycloaddition of azomethine ylides to C_{60} lays the foundation of the reaction, which is schematically shown in Figure 7.3(a). *N*-Methyl-3,4-fulleropyrrolidine (*N*-MFP) (or *N*-methyl-pyrrolo[3,4]C_{60}[13]) is produced according to the reaction shown in Figure 7.3(b). The reaction barrier is located over 2.52 Å. Analyzing data collected in Table 7.3, one can see that the fulleroid is characterized by a considerable coupling energy and a pronounced ΔE^{RU} caused by odd electrons of fullerene. The coupled ions' charge constitutes about 14% of that of free ions and provides a considerable increase of the dipole moment. As for the dyad orbitals, LUMO is provided by fullerene only. At the same time, HOMO is partner mixed, and consequently, a B_1 type band (see Figure 1.2(b)) should be expected in the absorption spectrum, the excitation of which is accompanied by a considerable charge transfer in addition to the charge available in the ground state. Since $E^{+-}_{cpl} > |E^{00}_{cpl}|$, the two-well IMI term is of type 1 with a deep minimum at R^{+-}.

7.4.2 Dyad C_{60}–TDAE

When analyzing data of Table 7.3, it becomes obvious that binary system C_{60}-TDAE has much in common with C_{60}-DMMA. Thus, similar to the latter, the C_{60}-TDAE dyad has two regions of the starting parameter, R_{st}, which marks the distance between carbon atoms of the ethylene bridge in the TDAE molecule and reference carbon atoms of fullerene. At $R_{st} \geq 2.0$ Å, a traditional weak charge transfer complex is formed, with the only difference that the region of the charge transfer complex existence is shifted to shorter distances, which allows for localizing the barrier location in the region of 2.0 to 1.7 Å. The relevant coupling energy is small, which is consistent with a shallow minimum on the IMI term at R^{00}. However, just this very minimum is responsible for the formation of the C_{60} – TDAE molecular crystal with the experimental R_{fin} value within the range of 3.8 to 4.0 Å.[14] A presence of a B_2 type charge transfer band in the absorption spectrum of the crystal is well supported by the partner composition of HOMO and LUMO, shown in Table 7.3.

At $R_{st} \leq 2.0$ Å, two dimethylamine radicals are willingly added to two reference carbon atoms of the fullerene cage. The radical formation is promoted by the TDAE dissociation along the C-C bond. A possibility of this dissociation was discussed in Section 7.3. Thus formed, TDAE-fulleroid is similar to DMMA-fulleroid (*N*-MFP) in many respects. It is characterized by a large coupling energy (deep minimum), which is determined in accordance with the following expression:

$$E_{cpl} = \Delta H_{compl} - \Delta H_{C60} - \Delta H_{TDAE} - E_{diss}(R_{eq}) \quad (7.1)$$

Here $E_{diss}(R_{eq})$ is the dissociation energy of TDAE equal to 83.24 kcal/mol at the C-C distance $R_{eq} = 2.8$ Å, at which the atoms occurred after TDAE-fulleroid was formed. ΔE^{RU} is similar to *N*-MFP. In contrast to the latter, TDAE-fulleroid practically loses its ionic character. As for the partner composition of HOMO and LUMO, it is of a cross-partitioning character, typical of traditional charge transfer complexes with a charge transfer band of the B_2 type. That is why charge transfer will accompany the compound photoexcitation. Similar to the C_{60}-DMMA dyad, C_{60}-TDAE is well described by the two-well IMI term of type 1, with a deep minimum at R^{+-}.

7.4.3 Dyad C_{60}–TAE

Results of the complex testing have shown that, as in previous cases, the complex configuration depends on the starting intermolecular distance. The latter is presented by the shortest distance between carbon atoms of the C-C bond of the TAE molecule and one of the reference carbon atoms of the fullerene C_{60}. At $R_{st} \geq 2.0$ Å, the distance in the final complex is enlarged up to 3.7 to 19 Å, depending on the TAE orientation with respect to the fullerene. The heat of formation is changed from the value given in Table 7.3 to that which exceeds the former by ~20 kcal/mol. This change is connected with the changes in molecular structure, the charge distribution over atoms, and the molecule heat of formation of TAE, which is attributed to considerable polarization effects that resulted from the inductive IMI.[7]

Variations in the complex heat of formation, as well as the TAE structure, affect the other complex characteristics only slightly. As seen from the table, both ΔE^{RU} and squared spin are mainly determined by the C_{60} molecule. EA is also originated from the molecule value, while IP has clearly shown a TAE origin. The dipole moment of the complex is rather significant, which evidences a charge polarization between its components, discussed earlier. The transferred charge is equal to zero, while the partner composition of HOMO and LUMO explains the complete cross-partitioning of these orbitals over the complex components. The latter provides appearing a charge transfer band of a B_2 type in the complex absorption spectrum.

At $R_{st} \leq 2$ Å, the only equilibrated configuration has been formed. It is related to a di(diamino)fulleroid (TAE-fulleroid), which is formed due to dissociation of the TAE molecule along its C-C bond, similarly to C_{60}-TDAE, discussed above. The coupling energy is determined by Equation 7.1, where $E_{diss}(R_{eq})$ at $R_{eq} = 1.66$ Å is equal to 32.02 kcal/mol. The coupled ions' charge constitutes about 14% of that of free ions and provides a considerable increase of the dipole moment. As for fulleroid orbitals, LUMO is provided by fullerene only, while HOMO is partner mixed and favors TAE. Consequently, photoexcitation of the fulleroid within the B_1 band is accompanied by a considerable charge transfer in addition to the charge available in the ground state. Similar to C_{60}-DMM and C_{60}-TDAE, the C_{60}-TAE dyad is well described by the two-well IMI term of the first type, with a deep minimum at R^{+-}.

7.4.4 Dyad C_{60}–COANP

If for configurations in the vicinity of R^{00} at large intermolecular distances a setting of initial configuration is quite trivial, then in the region of small intermolecular distances close to R^{+-}, structural features of the COANP molecule determine parameter R_{st} as the distance connecting the nitrogen atom of the pyridine ring of amine with the reference atom of the fullerene molecule. Three initial configurations of the binary system C_{60}-COANP have been chosen for calculations: $R_{st} = 1.4$, 2.2, and 3.5 Å. In all cases, not only short final distances, R_{fin}, compared with chemical bond length are obtained, but shortened intermolecular contacts are absent and R_{fin} values constitute 4.5, 4.6, and 3.8 Å. In spite of the clear difference in the mutual orientation of the partners, the complex energetic parameters are practically identical. This feature, related to a very weak dependence of the coupling energy of the charge transfer complexes on mutual orientation of molecular partners, well explains the stability of molecular crystals belonging to an extended class of intermolecular configurations of the C_{60}-X type,[14,15] in spite of the practically free rotation of the fullerene in these crystals at ambient temperature.

Therefore, the binary system C_{60}-COANP, consisting of neutral molecules, forms a typical weak charge transfer complex: the charge transfer does not occur in the ground state; there is a complete cross-partitioning of HOMO and LUMO so that a charge transfer band of B_2 type is observed in the absorption spectrum of the complex.[16] From this follows that the minimum R^{00} plays the main role on the IMI term, which should be attributed to the term of either type 3, shown in Figure 1.3(a), with a shallow minimum or type 4, in Figure 1.3(b).

To distinguish the two possibilities, it is necessary to look at whether the dyad of molecular ions C_{60}^--COANP$^+$ can form a stable chemical compound. According to the data discussed in Section 7.3, the equilibrated structures of the COANP molecule and its ion are practically coincident, so that a stable molecular ion $[COANP]^+$ is formed at a vertical diabatic transition without any indication to any bond dissociation. Results of the calculations performed for the ionic dyad are listed in Table 7.3. At R_{st} = 1.63 Å the two partners are stacked together, forming a COANP monoderivative of fullerene II*, shown in Figure 7.2. The coupling energy is positive, which undoubtedly points to the IMI term of type 3, shown in Figure 1.3(a). All other characteristics of the II* adduct are in the range typical for other C_{60}-based amine monoderivatives. A peculiarity of the IMI term tells us that the COANP monoderivative can be obtained in due course of a photochemical reaction only when photoexcitation transforms a pair of neutral molecules into a pair of molecular ions. A Coulomb interaction between the latter stimulates the formation of a new molecule, similarly to harpoon reactions in molecular beams.[17]

7.5 Concluding remarks about donor–acceptor chemical reactions of fullerene C_{60}

The donor–acceptor contribution to the intermolecular interaction of fullerene C_{60} with amines is obviously significant. Two-well IMI potentials of types 1 and 3 are characteristic for the dyads. If weakly coupled van der Waals charge transfer complexes $A + B$ are usually formed at large intermolecular distances in both cases, the tightly coupled monoadduct *AB* formation at short distances is not always evident and is dependent on a particular type of potential. If the latter is of type 1, the fulleroid formation is energetically favorable with respect to both free neutral molecules and molecular ion states. If we deal with a potential of type 3, the fulleroid formation is not energetically favorable with respect to free neutral molecules. As a result, a fulleroid can be readily formed by combining neutral molecules in the first case (this is the case of C_{60}-DMMA, C_{60}-TAE, and C_{60}-TDAE dyads), while interaction of the C_{60} neutral molecule with COANP does not result in the monoadduct formation. At the same time, experimentally, a precipitation of some nonsoluble species is observed when combining solutions of neutral molecules in both cases. The reaction yield is low and its rate is small. Quantum chemical calculations show that fulleroids in the latter case can actually be formed if molecular ions, not neutral molecules, are considered. This finding suggests fulleroid formation in due course of a hidden photochemical reaction. After obtaining a conventional solution of two neutral species, which is usually deeply colored due to charge transfer complex formation between the molecules at large distances, molecular ions are formed by light absorption. When ions appear, the Coulomb interaction provides their attraction until the fulleroid is formed.

The features exhibited when discussing the C_{60}-COANP dyad have thus highlighted reasons for time-dependent color changeability of fullerene solutions, which is common for numerous chemical reactions with participation of fullerene C_{60}. It is possible to conclude that in all these cases, the primary coloring of solutions is determined by wide adsorption bands in the visible range of the spectrum corresponding to the formation of charge transfer complexes between the components of the relevant dyads. Absorption of light (including *uncontrolled* absorption of sunlight or the artificial illumination of the laboratory room) leads to the formation of ions A^+ and B^-, the interaction of which is accompanied by the formation of a new chemical product, *AB*. Therefore, the formation of the final product goes through a stage of a hidden photosynthetic reaction. When a dyad is subordinated to the IMI potential of type 1, photoexcitation promotes the *AB* product formation in this case, as well just overcoming the reaction barrier when starting the reaction from the component solution.

Not only monoadducts *AB*, but polyadducts A_nB (*B* is C_{60}) production is typical for fullerene derivatives. Moreover, just polyadducts formation is more energetically preferable in many cases. A stepwise consideration seems to be the best approach for the relevant reaction description. It concerns not only the consideration of the *k*th successive additions of the corresponding addend *A* to the fullerene cage of a preceding adduct $A_{k-1}B$, as was discussed in Chapters 4 to 6, but also a thorough analysis of the IMI potential of a dyad A_nB-A at each step. As shown in Table 7.3, *AB* monoproducts are characterized by active donor–acceptor parameters. Similar behavior is characteristic for C_{60}-based polyadducts, which was directly shown for such addends as pirrolidone and styrene.[18] Consequently, each next *k*th step of the addend attaching should be preceded by the analysis of the type of the IMI potential of the $A_{k-1}B$-A dyad to make sure neutral molecules $A_{k-1}B$ and *A* or their molecular ions are considered when forming the A_kB product.

References

1. Eremenko, A. M., and Sheka, E. F. 1967. Peculiarity of electronic states of naphthalene adsorbed in zeolites [in Russian]. *Teor. Eksp. Khim.* 3:390–96.
2. Wilson, S. R., and Schuster, D. I. 1997. Overview of recent advances in the organic functionalization of fullerenes. *Proc. Electrochem. Soc.* PV 97–14:237–39.

3. Maggini, M., Scorrano, G., and Prato, M. 1993. Addition of azomethine ylides to C_{60}: Synthesis, characterization, and functionalization of fullerene pyrrolidines. *J. Am. Chem. Soc.* 115:9798–99.
4. Prato, M., and Maggini, M. 1998. Fulleropyrrolidines: A family of full-fledged fullerene derivatives. *Acc. Chem. Res.* 31:519–26.
5. Janaki, J., Premila, V., Gopalan, P., Sastry, V. S., and Sundar, C. S. 2000. Thermal stability of a fullerene-amine adduct. *Thermochem. Acta* 356:109–16.
6. Lobach, A. S., Goldshleger, N. F., Kaplunov, L. G., and Kulikov, A. V. 1995. Near-IR and ESR studies of the radical anions of C_{60} and C_{70} in the system fullerene-primary amine. *Chem. Phys. Lett.* 243:22–28.
7. Kaplan, I. G. 1986. *Theory of intermolecular interaction*. Amsterdam: Elsevier.
8. Sheka, E. F. 2004. Intermolecular interaction in C_{60}-based donor acceptor complexes. *Int. J. Quant. Chem.* 100:388–406.
9. Sheka, E. F. 2004. Multi-mode ground state interaction terms in C_{60}-based electron donor-acceptor complexes. *Centr. Eur. J. Phys.* 2:1–29.
10. Shinohara, H. 2000. Endohedral metallofullerenes. *Rep. Progr. Phys.* 63:843–92.
11. Sheka, E. F. 2006. Photosynthetic reactions in fullerene based donor-acceptor complexes. *J. Struct. Chem.* 47:600–7.
12. Yurovskaya, M. A., and Trushkov, I. V. 2002. Reactions of cycloaddition to buckminsterfullrene C_{60}. Achievements and trends. *Russ. Chem. Bull. Int. Ed.* 51:1186–93.
13. Liu, Y., Zhang, D., Hu, H., and Liu, C. 2001, Theoretical study of the electronic spectra and second-order non-linear optical properties of *N*-methyl-2-(2′-thiophen)-pyrrolo [3,4] C_{60}. *J. Mol. Spectr. (THEOCHEM)* 545:97–103.
14. Narymbetov, B., Omerzu, A., Kabanov, V. V., Tokumoto, N., Kobayashi, H., and Mihailovich, D. 2000. Origin of ferromagnetic exchange interactions in a fullerene-organic compounds. *Nature* 407:883–85.
15. Konarev, D. V., and Lyubovskaya, R. N. 1999. Donor-acceptor complexes and ion-radical salts based on fullerenes. *Russ. Chem. Rev.* 68:19–38.
16. Kamanina, N. V., and Sheka, E. F. 2004. Limitters of laser emission and diffraction elements based on COANP-fullerene system: Non-linear optical properties and quantum-chemical modeling. *Opt. Spectr.* 96:599–612.
17. Herschbach, D. R. 1966. Molecular beams *Adv. Chem. Phys.* 10:319–403.
18. Sheka, E. F., and Yevlampieva, N. P. 2006. Adducts $A_nC_{60}H_n$: Electro-optical properties and quantum chemical calculation data. *Fuller. Nanotubes Carb. Nanostr.* 14:343–48.

chapter eight

Nanochemistry of fullerene C_{60}

C_{60} dimerization and oligomerization

8.1 Introduction

Now we are passing to a particular kind of chemical reaction involving multiple fullerene molecules, namely, to reactions of their dimerization or oligomerization. The peculiarity of these reactions derives from the fact that if conventional reactions are carried out mainly in chemical laboratories, polymerization of fullerenes occurs mainly in due course of physical experiments—enough to look at the production of dimers by means of photochemical,[1] thermal and high-pressure,[2–4] plasma,[5] electron-induced,[6] and electric-field-stimulated[6,7] techniques. Besides that, the reactions occur not in chemical flasks at the molecular level, but inside physical objects, such as solid films, freely sustained or deposited on some surface, as well as in a crystalline state. This "physical fragrance" of the reaction undoubtedly sets fullerene dimers or oligomers apart from other polymers. Obviously, there are a few reasons for such exclusiveness, the most important of which is a vacuum or chemical intermolecular interaction (IMI) between the molecules and the molecule packing that greatly influences the polymer energetics. We introduce the jargon terms *vacuum* and *chemical* IMI to point out that the interaction concerns molecules that constitute individual dimers or oligomers only. This kind of event was considered in previous chapters. The influence of packing the molecules themselves as well as dimers and oligomers is traditionally considered from other standpoints, while basing it on vacuum quantum chemical data (see, for example, Dzyabchenko et al.[8]).

We will follow our previous approach and will restrict ourselves to the consideration of chemical IMI only. In the case of fullerenes, the IMI, as we know, is considerably aggravated with a donor–acceptor (D–A) interaction. This is particularly important for fullerene homomolecular systems since the fullerene C_{60}, as well as other members of the fullerenic family, is simultaneously both a good electron donor, due to rather low ionization potential (IP), and a good electron acceptor, due to extremely high electron affinity (EA). This particular bimodal feature of the D–A ability of participants is the main feature of D–A-assisted reactions to be considered below.

8.2 Ground-state term of the C_{60}–C_{60} dyad

Experimentally it is known that dimerization is energetically favorable with respect to monomer molecules, so that the IMI potential for C_{60}-C_{60} dyads should be of type 1 (see Section 1.2 and Figure 1.2), which is schematically redrawn in Figure 8.1. The reaction can be considered as moving the two molecules toward each other, being spaced initially at large intermolecular distance R, then equilibrated and coupled as an $A + B$ complex in the R^{00} minimum, and afterwards achieving a minimum at R^{+-} to form a tightly bound adduct AB. The last stage implies overcoming a barrier, which is followed by the transition from the (A^0B^0) to the (A^+B^-) branch of terms, after which a Coulomb interaction between molecular ions completes the formation of the final AB adduct at the R^{+-} minimum. As we saw in Section 7.3.3, neither ionization nor positive charging of C_{60} causes lengthening of the molecule valence bonds by more than 0.02 Å. Therefore, the ion formation is not accompanied by a noticeable shift of the atom equilibrium positions along any internal coordinate and does not cause any significant vibrational

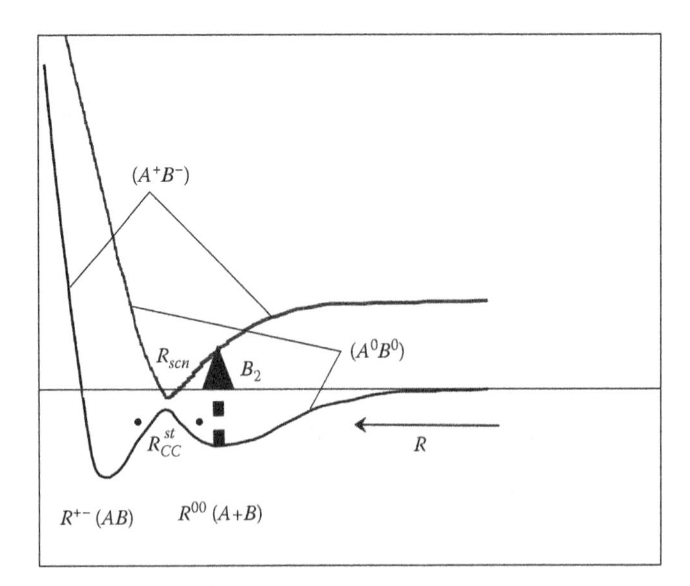

Figure 8.1 Scheme of terms corresponding to the IMI potential of type 1. A detailed presentation of the terms is given in Figure 1.2(b). (A^0B^0) and (A^+B^-) match branches of the terms related to the IMI between neutral molecules and their ions, respectively. Two black spots mark starting points R^{st}_{CC}.

excitation during the relevant transition that could have caused the molecule decomposition under ionization. That is why the fullerene dimerization or oligomerization should occur as a direct addition reaction between nondecomposed molecules.

Equilibrium configurations $A + B$ and AB are the main goal of computations in this case. Passing from $A + B$ to AB or, in other words, overcoming the barrier may be substituted by two independent calculations performed at starting intermolecular distances, R^{st}_{CC}, situated on the right- and left-hand sides from the barrier maximum position, respectively, as shown in Figure 8.1.

On the basis of discussions in the preceding chapters, it can be said with confidence that computational covalent chemistry of the C_{60} molecule is well quantitatively guided by the atomic chemical susceptibility (ACS) of its atoms. Among the latter, the most active atoms collected in group 1 are the first targets involved in the initial stages of any addition reaction. Consequently, the initial composition of a pair of the C_{60} molecules (Figure 8.2(a)) is quite evident. Two selected starting configurations correspond to R^{st}_{CC} equal to 3.07 (Figure 8.2(b)) and 1.7 Å (Figure 8.2(c), which are determined by distances between 1-1′ and 2-2′ target atoms of group 1.

When R^{st}_{CC} = 3.07 Å, optimization of the initial structure leads to a weakly bound pair of molecules spaced by R^{fin}_{CC} = 4.48 Å (Figure 8.2(b)). Monomer molecules preserve their initial configurations and, as seen from Table 8.1, the formed complex is a classical charge transfer one. The fragment composition of the highest occupied (HOMO) and lowest occupied (LUMO) molecular orbitals is cross-partitioned, showing that a charge transfer occurs when the complex is photoexcited. The distance between molecules is bigger that that in the pristine C_{60} crystal. However, remember that the calculations are related to vacuum, so that the spacing shortening when going from a free molecular pair to crystalline packing caused by a collective effect is obviously expected; this is actually typical for molecular crystals.[10]

When R^{st}_{CC} = 1.7 Å, a bound dumbbell-like dimer is formed (Figure 8.2(c)). Two monomers within the dimer are contacted via a [2 + 2] cycloaddition of sixty-six bonds that form a cyclobutane ring. The latest structure study of the crystalline photopolymers[11] has convincingly exhibited the most abundant component related to [2 + 2] cycloadduct dimers indeed.

The main electronic characteristics of the $(C_{60})_2$ adduct are presented in Table 8.1. Since the listed data have been obtained in the UBS HF approximation, opposite of other calculations performed in different closed-shell restricted approximations, the comparison will be done for the differential coupling energy and structural parameters only. The available E_{cpl} data are given in footnote c of Table 8.1. As seen, the available E_{cpl} values fill quite a large interval from +13.77 to −49 kcal/mol. The largest by absolute value

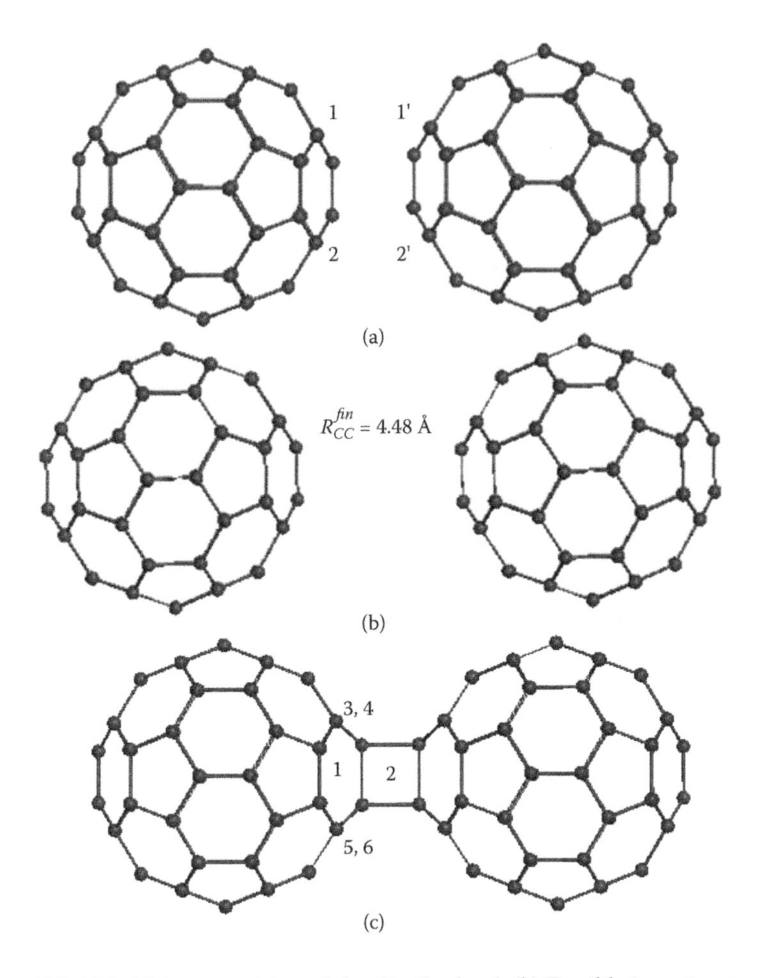

Figure 8.2 (a) Initial composition of the C_{60}-C_{60} dyad. (b) Equilibrium structure of the $C_{60}+C_{60}$ charge transfer complex; $R_{CC}^{st} = 3.07$ Å and $R_{CC}^{fin} = 4.48$ Å. (c) Equilibrium structure of the $(C_{60})_2$ dimer; $R_{CC}^{st} = 1.7$ Å and $R_{CC}^{fin} = 1.55$ Å. UBS HF AM1 singlet state. All distances correspond to the spacing between 1-1′ and 2-2′ atoms. (From Sheka, E. F., *Chem. Phys. Lett.*, 438, 119–26, 2007.)

data are obtained by semiempirical techniques, and they seem to be the most realistic since dimer $(C_{60})_2$ is a rather stable species.[2] A large negative E_{cpl} value undoubtedly evidences that the $(C_{60})_2$ dimer is actually a typical AB adduct attributed to the R^{+-} minimum on the IMI ground-state term. Therefore, calculations have shown that the ground-state IMI term of the C_{60}-C_{60} binary system is a two-well one of type 1, with two minima wells different by the depth.

The calculations have revealed as well that the dimer $(C_{60})_2$ formation is accompanied by a considerable change of charge distribution over

Table 8.1 Electronic characteristics of the C_{60}-C_{60} dyad

Computed quantities (AM1) Singlet, UBS HF AM1	Monomer C_{60}	R^{st}_{CC}, nm 3.58 Å	1.71 Å
Heat of formation,[a] ΔH, kcal/mol	955.56	1,910.60	1,868.49
Coupling energy,[b] E_{cpl}, kcal/mol	—	−0.52	−42.63[c]
Ionization potential,[d] I, eV	9.86 (8.74[1])	9.87	9.87
Electron affinity,[d] ε, eV	2.66 (2.69[2])	2.66	2.62
Dipole moment, *Db*	0.01	0.001	0.001
Squared spin, (S**2)[a]	4.92	9.87	10.96
Total number of effectively nonpaired electrons, N_D	9.84	19.75	21.93
Gained charge to Mol 1	—	0.0	0.0
Transferred charge from Mol 2	—	0.0	0.0
Symmetry	C_i	C_i	C_{2h}
HOMO, fragment compositions, η	—	$\eta_{Mol1} = 0\%$	$\eta_{Mol1} = 61.8\%$
		$\eta_{Mol2} = 100\%$	$\eta_{Mol2} = 38.1\%$
LUMO, fragment compositions, η	—	$\eta_{Mol1} = 100\%$	$\eta_{Mol1} = 83.9\%$
		$\eta_{Mol2} = 0\%$	$\eta_{Mol2} = 15.8\%$

Source: Sheka, E. F., *Chem. Phys. Lett.*, 438, 119–26, 2007.

[a] See footnote b to Table 3.2.

[b] Coupling energy is determined by $E_{cpl} = \Delta H_{\text{dim}} - 2\Delta H_{mon}$.

[c] Available data on the coupling energy are presented in the following format: value in kcal/mol (method), reference: −48.2 (MNDO), Strout, D. L., et al., *Chem. Phys. Lett.*, 214, 576, 1993; 13.77 (TBMD), Menon, K. R., et al., *Phys. Rev. B*, 49, 13966, 1994; −33.0 (AM1), −36.9 (PM3), and −24.2 (AM1/LDFT), Matzuzawa, N., *J. Phys. Chem.*, 98, 2555, 1994; −48.9 (MNDO), −29.8 (3-21GLDA), −10.3 (3-21HF), and +5.05 (3-21BLYP), Scuseria, G. E., *Chem. Phys. Lett.*, 257, 583, 1996; −6.9 (DF-TB) and −27,6 (SCF-LDA), Poregaz, D., et al., *Appl. Phys. A*, 64, 321, 1997; −45.9 (AM1), Stafstrom, S., and Fagerstrom, J., *Appl. Phys. A*, 64, 307, 1997; −36.7 (PM3) and +6.4 (TB), Lee, K. H., et al. *J. Phys. Chem. B*, 104, 7038, 2000; −10 (DCM AM1) and −23 (DCM PM3), Cabrera-Trujillo, J. M., and Robles, J., *Phys. Rev. B*, 64, 165408, 2001; −14.46 (BP86-D/SVP), Bihlmeier, A., et al., *Chemphyschem*, 6, 2625, 2005.

[d] Here IP and EA correspond to the energies of HOMO and LUMO, respectively, just inverted by sign. Experimental data for the relevant orbitals are taken from Weaver et al.[12] (1) and Wang et al.[13] (2).

monomer atoms, compared with that of pristine C_{60} (Figure 8.3). However, the change is quite localized and concerns mainly four atoms directly involved in the cycloaddition, as well as four pairs of atoms attached to the cycle. The changes are fully identical for both molecules, and in spite of pronounced charge redistribution, the molecules remain charge neutral. The case differs from that of other fullerene-based D–A systems, described in Chapter 7, where *AB* adduct formation is accompanied by a significant charge transfer. The matter seems to be connected with the fact that in spite of the undoubtedless leadership of D–A in IMI related to the C_{60}-C_{60}

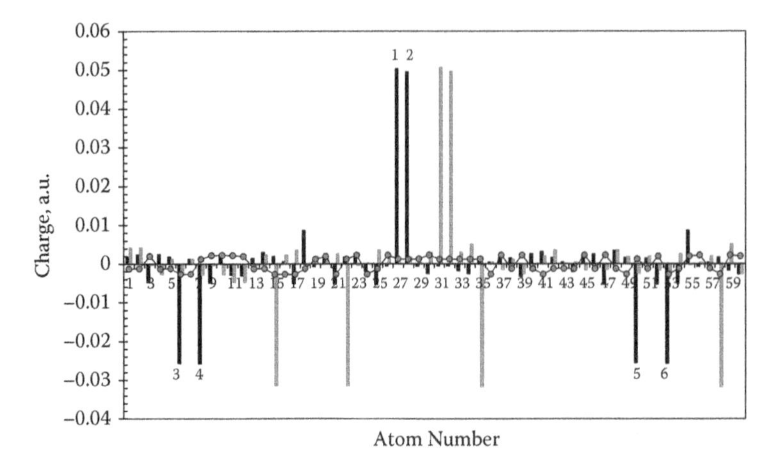

Figure 8.3 Charges of monomer atoms: molecule 1 (black) and molecule 2 (gray). Digits mark atoms denoted in Figure 8.2(c). Curve with half-filled circles presents charge distribution over atoms of pristine C_{60} molecule. (From Sheka, E. F., *Chem. Phys. Lett.*, 438, 119–26, 2007.)

dyad, the location of both donor and acceptor cannot be attributed to a particular monomer molecule, in contrast to heterogeneous binary systems, where the partner D–A character is directly pointed by chemicals.

As seen from Table 8.1, the dimer IP and EA practically coincide with those of monomers. It is better to say that the values are close to those of the IP of the donor and the EA of the acceptor molecule, respectively. Generally, such a correlation is typical for D–A fullerene-based binary systems with different donors. The same can be said about energies of molecular orbitals located below HOMO and above LUMO within the interval of ~1.5 eV in width. As well known, these energies do not reproduce exactly positions of the ground and excited electron states of molecular species, but they put a frame for the energy interval where these states are located. That is why a comparison of HOMO and LUMO energies with those of the monomer allows concluding that permitted electron transitions of the dimer spectrum should occupy the same energy interval as those of a free monomer, although slightly blue shifted. These very features are observed experimentally.[14] The same obviously concerns the lowest electron transition. It is important to stress that experimentally recorded absorption spectra of both species related to this transition are quite similar. Particularly, it should be noted that the dimer spectrum is as weak as that of the monomer. This behavior shows that the dimer spectrum is caused by the excitation of its monomer component, and that this monomer continuous symmetry is high. Actually, for C_1 symmetric formally, the monomer molecule is 99.99% I_h-ness, which is the same for the C_i pristine monomer (see Section 3.2.4), and which explains the optical spectrum peculiarity. Similar peculiarities are observed for Raman spectra of C_{60} oligomers,[15] where the spectra complication caused by going from monomer to oligomer (via dimer, trimer, and other stages) does not follow increasing the number of allowed frequencies to be observed according to the symmetry lowering, but just reflects increasing the number of cycloaddition units.

Structural experiments performed by now offer a large pool of data to be compared with calculated ones. The data related to the [2 + 2] cycloaddition region are presented in Table 8.2. As seen from the table, all calculations show that the initial double C-C bond 1 (hinge bond) is elongated so that its value exceeds the upper extreme that is characteristic for a single C-C bond. Pristine single bonds (2 and 3) that are adjacent to the carbon atom involved in the cycloaddition are elongated as well, and take values characteristic for a single C-C bond. Lengths of two newly formed intermolecular bonds 4 (pivot bonds) are either close to the upper limit of single C-C bonds (all semiempirical calculations) or exceed the limit by 20 to 30% (all different density functional theory (DFT) techniques). A comparison with the only experimental x-ray examination of a solvated $(C_{60})_2$ crystal[17] shows that the data are well consistent with all calculated ones for bonds 1 to 3, while they fit better the DFT data for bonds 4. As for the distance between the molecule centers in the dimer, the UBS HF calculated value of 9.08 Å[9] is in good agreement with the available experimental data: 9.1 Å,[17] 9 Å,[18] and 8.8 (4) Å.[19]

The barrier profile shown in Figure 8.4 for the transition from charger transfer complex C_{60} + C_{60} to dimer completes the IMI term description.[9] The barrier is identical to that obtained in Stafstrom and Fagerstrom[20] within the framework of the restricted Hartree-Fock (RHF) approach concerning both the barrier energy, E_{bar}, of 34.26 kcal/mol (1.49 eV) (34.44 kcal/mol in Stafstrom and Fagerstrom[20]) and critical distance, R_{CC}^{fin} = 2.12 Å (2.12 Å in Stafstrom and Fagerstrom[20]), of the maximum position.

8.3 Dimerization mechanisms

Experimental observations show that the IMI interaction in the C_{60} pair is described by a two-well term of type 1. At ambient conditions, the dimerization does not occur spontaneously, which points to the reaction barrier. According to experimental findings, the barrier (activation energy) is 1.25 eV (28.7 kcal/mol)[21] in the case of photostimulated dimerization and 1.40 eV (32.14 kcal/mol)[22] when high pressure is applied. The barrier can be overcome by different ways, which explains a large variety of the technological schemes in use.

Table 8.2 Lengths of C-C bonds forming cyclobutane ring in C_{60} dimer, Å

Molecule	Reference [reference]	C-C bond number 1 (hinge)	2	3	4 (pivot)
C_{60}	AM1_UBS HF [9]	1.391 ± 0.032	1.464 ± 0.013	1.464 ± 0.013	—
	Exp. [16]	1.398 (10)	1.455 (6)	1.455 (6)	—
$(C_{60})_2$	AM1_UBS HF [9]	1.596–1.597	1.515–1.516	1.515–1.516	1.548
	MNDO[a]	1.616	1.537	1.537	1.561
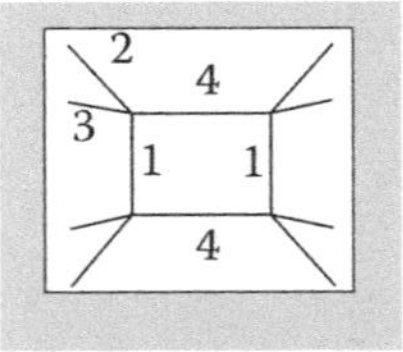	DCM AM1_RHF[b]	1.605	1.515	1.515	1.552
	AM1_RHF[c]	1.603	1.515	1.515	1.546
	PM3_RHF[c]	1.598	1.516	1.516	1.550
	AM1_RHF[d]	1.603	—	—	1.546
	TB[e]	1.615	1.519	1.519	1.629
	LDA-VWN/STO[f]	1.582	1.508	1.508	1.586
	BLYP/3-21G[a]	1.606	1.525	1.525	1.594
	Exp. [17]	1.581 (7)	1.530 (8)	1.530 (8)	1.575 (7)

[a] Scuseria, G. E., *Chem. Phys. Lett.*, 257, 583, 1996.
[b] Cabrera-Trujillo, J. M., and Robles, J., *Phys. Rev.* B 64, 165408, 2001.
[c] Matzuzawa, N., Alta, M. Dixon, D. A. and Fitzgeralg, G., *J. Phys. Chem.* 98, 2555, 1994.
[d] Osawa, S., Sakai, M. and Osawa, E., *J. Phys. Chem.* A 101, 1378, 1997.
[e] Lee, K. H., Eun, H. M., Park, S. S., Suh, Y. S., Jung, K.-W., Lee, S. M., Lee, Y. H., and Osawa, E., *J. Phys. Chem.* B 104, 7038, 2000.
[f] Onoe, J., Nakayama, T., Aono, M., and Hara, T., *J. Appl. Phys.*, 96, 443, 2004.

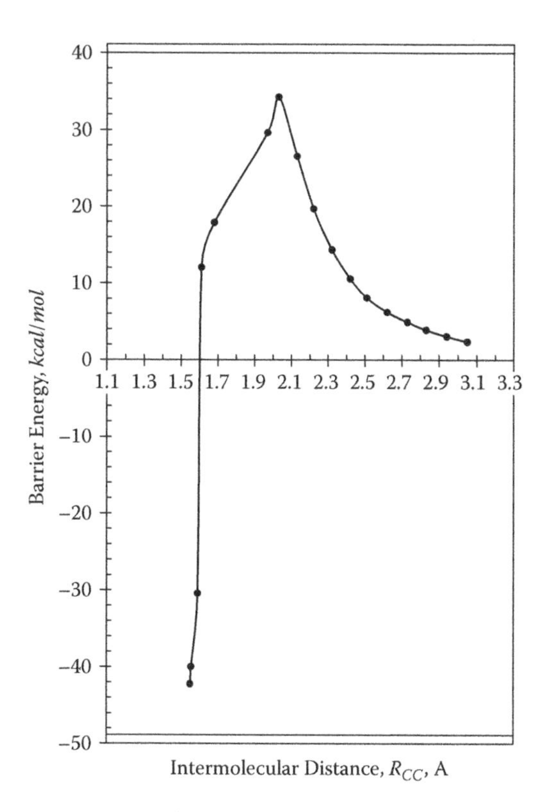

Figure 8.4 Profile of the barrier of C_{60} dimerization. (From Sheka, E. F., *Chem. Phys. Lett.*, 438, 119–26, 2007.)

8.3.1 Photoexcitation technology[12]

A convenient triplet state photochemical mechanism, where a monomer in the excited triplet state $^3M^*$ reacts with a monomer in the ground-state M to yield the dimer D, i.e., $^3M^* + M \rightarrow D$, is usually accepted. Not disputing this clearly evident process, another mechanism can be suggested. As shown in the previous section, two C_{60} molecules separated by ~4.5 Å form a charge transfer complex. The complex formation is evidenced experimentally by the presence of a wide, structureless absorption band in the visible spectral region of solid solution of the C_{60} molecules in a toluene matrix,[23]as well by changing color when aggregating in water solutions.[24] In solid C_{60} photo-stimulated charge transfer between adjacent molecules causes the appearance of charge transfer excitons, the spectral region of which coincides with that of Frenkel excitons.[25,26] The important role of charge transfer excitons for the dimerization was first discussed in Suzuki et al.[27] Not excitons themselves, but a multiphonon exciton-phonon interaction that causes the lattice relaxation was considered to be able to provide better conditions for the molecules' coupling. However, another view might be suggested on the exciton role. The related absorption bands, both in molecular solutions and in solids, are provided by phototransitions of B_2 type (Figure 8.1) in due course of which the ground state of weakly interacting neutral molecules is transformed into the charge transfer state of molecular ions due to excitation of charge transfer excitons in a perfect crystal or local ones in the case of a solidified solution of fullerenes in molecular matrices. By this means, photoexcitation of a C_{60} pair by visible-UV light in sets of B_2 bands is followed by the formation of an ion pair located above the barrier. A Coulomb interaction between the ions easily provides their passing through the R_{scn} region toward the R^{+-} minimum where dimers are formed.

8.3.2 *Thermal and high-pressure technologies*[2–4]

These include tools that assist in directly overcoming the barrier. The barrier of 1.25 to 1.40eV can be overcome by high-pressure application, which means shortening the intermolecular distance in a forced manner. The mechanism universality is clearly exhibited in producing not only dimers and other linear oligomers, but also complicatedly packed carpet polymers of C_{60} (see, for example, one of the first observations of the events[28]).

8.3.3 *Plasma*[5] *and electron beam processing*[6]

These are quite well explained by thermal effects caused by applied tools.

8.3.4 *Field-stimulated formation and decomposition of dimers*[6,7]

If mechanisms governing the processes described above are quite transparent, a discovered ability to manipulate with the dimer formation or decomposition by application of a static electric field seems at first glance quite astonishing. If previous processes can be explained even without taking into account a double-well profile of the IMI term related to the ground state of the molecular pair, as was done previously,[1–6] it is difficult to suggest the field mechanism explanation without considering D–A interaction.

An exhausted study of C_{60} dimerization under a static electric field was performed by M. Aono and his team when studying an STM image of the molecules deposited onto different substrates (see Nakaya et al.[7] and references therein). As occurred, the dimerization proceeds when 3 V sample-to-tip bias is applied. Exclusion of the tight bonding of the molecule with the substrate by multilayer deposition has revealed a peculiar feature of the reaction: the molecule dimerization takes place at the sample negative bias only. Another one is related to the decomposition of the previously formed dimers at the sample negative bias when the bias polarity is inverted.

A suggested explanation of the effect as a weakly bound molecular assemblage response to the electron flux of ~1×10^8 s^{-1} has to be declined due to the fact that a direct exposition of the object to an even more dense flux of ~10^9 s^{-1} of energetic electrons does not cause any distinguishable effect.[6] Only more intense electron beams have provoked dimerization due to thermal effects, which were mentioned earlier. Therefore, the phenomenon occurred under STM tip highlights a direct impact of the applied electric field. Ionization induced by scanning transmission microscopy[7] and direct polarization effect[29] may explain the facilitation of the dimer breaking, but does not explain the dependence of the reaction on the bias polarity. We suppose that this effect is connected with a governing role of the D–A interaction in the molecule dimerization.

Charge transfer states are obviously sensitive to the electric field and the effect, as mentioned in Pac et al.,[25] should be strong. Actually, only the latter can explain the electrochemical behavior of C_{60} films and C_{60} lipid films in ionic liquids,[30] as well as acceleration and deceleration of photoinduced electron transfer rates by an electric field in porphyrin-fullerene dyads.[31] It is important to note that a thorough study of volammetric behavior of C_{60}-lipid films in ionic liquids[30] highlights a close similarity in the C_{60} molecule ionization, thus proving both donor and acceptor activity of the molecule.

As for fullerene dimerization, intuition points to a significant role of the asymptotic behavior of the $E_{\text{int}}(A^+B^-)$ term determined by the gap $E_{gap} = I_A - \varepsilon_B$, whose change will undoubtedly influence the shape (including barriers) of the two-well IMI term in the ground state. Direct evidence of the field-influenced E_{gap} changing was found for boron-nitrogen-bridged ferrocene-donor organic-acceptor compounds.[32] As shown at negative sample bias (comparable by value to that applied in the above STM experiments), a considerable red shift is observed in the absorption spectrum attributed to the charge transfer states. In contrast, a positive sample bias causes a significant blue shift of the relative bands. The finding convincingly highlights the gap narrowing at the negative sample bias, while the positive bias caused enlarging of the gap. Evidently, the tendency should be similar in all D–A systems. This makes it possible to predict the expected transformation of IMI terms of a D–A binary system under an applied electric field.[9,33]

Schematically, changes in the IMI term shape caused by the static electric field are shown in Figure 8.5. Since field effect on states of neutral molecules is quite negligible, the main changes concern term $E_{\text{int}}(A^+B^-)$, which describes the interaction between ions. Due to the observed tendency of field stimulation either decreasing or increasing E_{gap}, the term will move either upward or downward with respect to the term $E_{\text{int}}(A^0B^0)$,

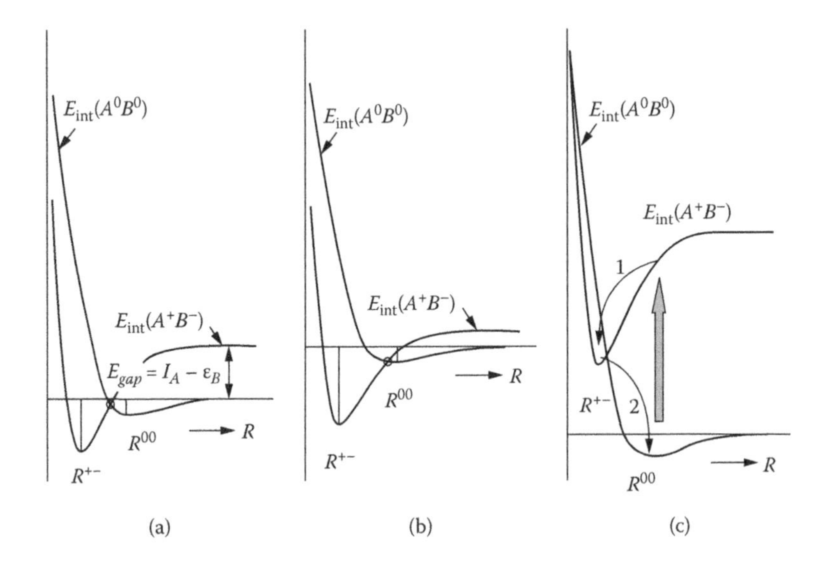

Figure 8.5 Static-electric-field effect on two-well IMI term of the ground state. (a) No field. (b) Negative sample bias. (c) Positive sample bias; photostimulation of the dimer formation (1) and dimer decomposition (2). (From Sheka, E. F., *Chem. Phys. Lett.*, 438, 119–26, 2007.)

depending on the sample bias polarity. This process is shown in the figure when assuming that the shape of the $E_{int}(A^+B^-)$ term does not change drastically under the field. Two main conclusions follow from the scheme. The first concerns the position of the R^{+-} minimum in the energy scale, which either becomes deeper (negative bias) or moves above (positive bias) the minimum at R^{00} point. In the latter case, the *AB* (dimer) formation becomes thermodynamically unfavorable. The second finding is related to the barrier. Evidently, the barrier considerably reduces at negative bias but becomes higher in the opposite case. Reducing the barrier obviously facilitates dimerization, and may make it even barrierless under a certain bias. This process appears to take place under the STM tip at –3 V sample bias.[7]

From this viewpoint we can explain the difference in the dimerization of pristine fullerene C_{60} and alkali fullerites AC_{60} (where A denotes K, Rb, or Cs).[3,34] If the C_{60} dimerization overcomes a significant barrier, as shown above, the AC_{60} dimerization occurs spontaneously under reversible solid-state transformation from a high-temperature phase. The difference becomes understood when comparing IP values for donors in a series C_{60}, KC_{60}, RbC_{60}, CsC_{60}. Since AC_{60} species are themselves adducts of D–A binary systems A + C_{60} with alkali metal atom donors, their IPs are similar to IPs of free alkali atoms according to a general tendency for D–A fullerene-based systems, pointed to earlier. Therefore, IPs for a set of molecules C_{60}, KC_{60}, RbC_{60}, CsC_{60} form a series 8.74 (see Table 8.1), 4.34, 4.18, 3.89 eV (handbook data). At the same time, EAs of the AC_{60} species coincide with that of C_{60}. Since IPs of alkali donors are less than that of pristine fullerene by more than 4 eV, a drastic reduction of the energy gap, E_{gap}, for AC_{60} + AC_{60} binary systems occurs (see Figure 8.5(b)), so that the R^{+-} minimum deepens and the barrier practically disappears, stimulating a barrierless dimerization.

When the sample bias is positive, the IMI term of the C_{60}-C_{60} dyad is transformed from that of type 1 to that of type 3, so that the C_{60} dimerization becomes energetically nonprofitable (similarly to the C_{60}-COANP dyad in Section 7.4). In this case, the dimer formation can be obtained only via a hidden photochemical reaction. These events were observed[35] when the C_{60} dimerization, not observed under positively biased STM tip, occurred when the visible light was switched on. On the other hand, when dimers are formed at a negative sample bias, according to the scheme in Figure 8.5(b), switching on the bias polarity causes a change in the energetic scheme, from scheme b in Figure 8.5 to scheme c, so that the dimer decomposition becomes energetically profitable due to lifting the energy of the minimum R^{+-} much above the R^{00} minimum. If the decomposition barrier, whose value depends on the applied field, is low, previously formed dimers will decompose. This very process has been observed experimentally as well.[7]

8.4 C_{60} oligomers

8.4.1 Polymerization grounds

Since the discovery of magnetism in all-carbon crystals consisting of polymeric layers of covalently bound C_{60} molecules,[36] there has been a splash of interest in the constituting polymer structures. Experimentally observed, there are three packing modes related to the polymer structure of C_{60}, namely, one linear mode characteristic for the orthorhombic (*O*) crystalline modification and two planar, or carpet packing, modes corresponding to the tetrahedral (*T*) and rhombohedral (*R*) crystal modifications.[28]

To construct the related model structures, let us first consider the polymerization as a stepwise reaction: $(C_{60})_n = (C_{60})_{n-1} + C_{60}$. Each successive step deals with a dyad of the $(C_{60})_{n-1} - C_{60}$ type. Similarly to dimers, each oligomer $(C_{60})_{n-1}$ has the related IP and EA, practically the same as those of the monomer C_{60}, which is why the IMI potential of type 1 governs the formation of the final adducts. As in the case of dimers, two products, $(C_{60})_{n-1} + C_{60}$ charge transfer complex and $(C_{60})_n$ oligomer, correspond to equilibrium positions at the R^{00} and R^{+-} minima of the IMI term. A coexistence of these two products was experimentally supported in the case of C_{60} trimers in both physical dry[7,14,18] and chemical wet[37] experiments.

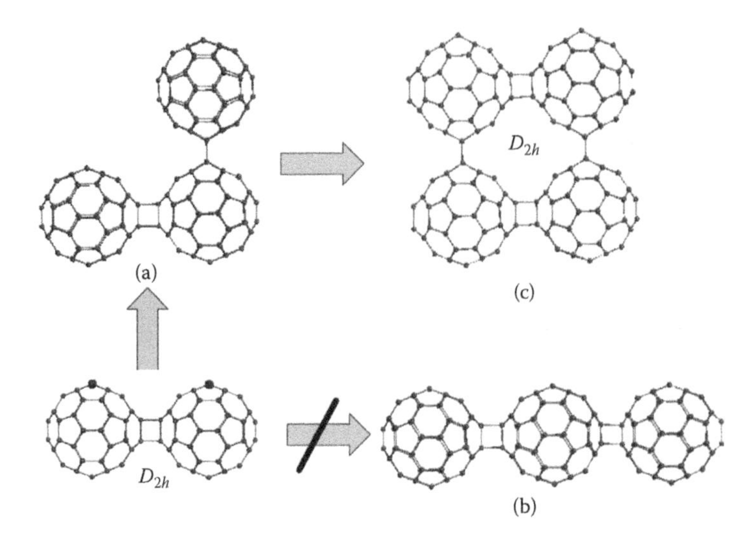

Figure 8.6 Scheme of the stepwise C_i-based C_{60} polymerization; favorable (a and c) and unfavorable (b) prolongations. Equilibrium structures of dimers, trimer, and tetramer. Black balls on dimer mark high-rank reactive atoms. UBS HF AM1 singlet state.

Following the concepts of computational chemistry of fullerenes presented in the previous chapters, let us consider the trimer formation in the same manner that was applied to fullerene derivatives. According to the ACS map of dimer $(C_{60})_2$, there are two groups of the highest-rank N_{DA} atoms. The first group combines the most reactive atoms, 3, 4, 5, and 6 (see Figure 8.2). Next, by reactivity, four atoms form two pairs located at the top of both monomers (see Figure 8.6). Enhanced ACS of the first four atoms is evidently connected with a considerable elongation of bonds 2 and 3 (see Table 8.2), which results in weakening the interaction of odd electrons connected with the related atoms. However, in spite of their high chemical reactivity, these atoms are nonaccessible in due course of the further oligomerization so that top atoms on the right monomer are the most important. Following these ACS indications, a right-angle-triangle trimer (90° trimer) must be produced. On the other hand, edge atoms located at the end of the horizontal axis are characterized by the least values on the ACS map, so that the formation of a linear trimer is highly improbable.

Wet chemical experiments[37] dealt with C_{180} species obtained in a chemical flask under convenient chemical conditions, and the selected product was afterwards investigated by using STM. Two fractions (A and B in a ratio of ~5:4) were obtained, the former predominantly (~60%) consisting of 90 trimers, while 100% of the latter fraction are presented by cyclic 60° trimers. Dry physical experiments deal with trimers produced under photoillumination of either C_{60} films preliminarily deposited on some substrates[7,14,18] or pristine C_{60} crystal.[17] Only linear three-ball chains were observed in these studies.

Therefore, the wet experiment, as might be expected, obviously supports the preference of 90° trimers predicted by ACS-guided covalent chemistry of C_{60}. Three other trimers, 108°, 120°, and 144°, observed experimentally within the A fraction,[37] are of higher energy, which follows from both our and previous[37,38] calculations. Kinetic conditions may allow for producing these adducts as well, but with less probability due to energetic unfavorableness. But what can be said about the availability of linear and 60° trimers? As for the former, since from the vacuum quantum chemistry viewpoint their formation is hardly possible, one must come up with other reasons that may facilitate the trimer production. As it turns out, the observation of similar events is not rare, and those are usually related to *topochemical reactions* whose occurrence is controlled by the reactants' packing. Thus, the polymerization of C_{70} fullerene in the solid state is explained as follows (p. 682)[39]: when "the alignment of molecules and their presumed orientational mobility facilitate polymerization via spatial adjustment of reactive double bonds of neighboring cages." All observed dry polymerization events should be obviously attributed to results of topochemical reactions. Consequently, chemical IMI is responsible only for a part of energetics in this case, while molecular packing plays a very important role. The latter will be followed by considerable strain of originated dimers, trimers, and other oligomers, which might, in its turn, considerably redistribute the ACS over molecule atoms.

The presence of the 60° trimers of fraction B in wet experiments has no relation to preliminary molecules' packing and should be explained on the basis of the covalent chemistry of C_{60}. However, it cannot be done within the framework of suggestion that C_{60} is presented in the action by only one isomer of C_i symmetry, which has been considered until now. The ACS map of this isomer well explains a preferable trimerization toward 90° trimers and a less profitable one in regards to 108°, 120°, and 144° in the framework of the scheme presented in Figure 8.6. Moreover, the D_{2h} symmetry of both the dimer itself and its monomer is in good correlation with the same symmetry of the tetramer and its monomer, so that the trimer in Figure 8.6 may be considered a precursor of tetragonal packed oligomers of higher order as well. As a whole, the scheme of polymerization suggested in Figure 8.6 can be considered a scheme of the C_i-based polymerization.

The fact that cyclic 60° trimers form a separate fraction indicates that their difference from the trimers of fraction A concerns both the origin and kinetics and must be deeply implanted into the monomer structure and properties. Actually, the trimers' formation can be understood if C_i symmetric isomer C_{60} is substituted by a D_{3d} symmetry one. We discussed

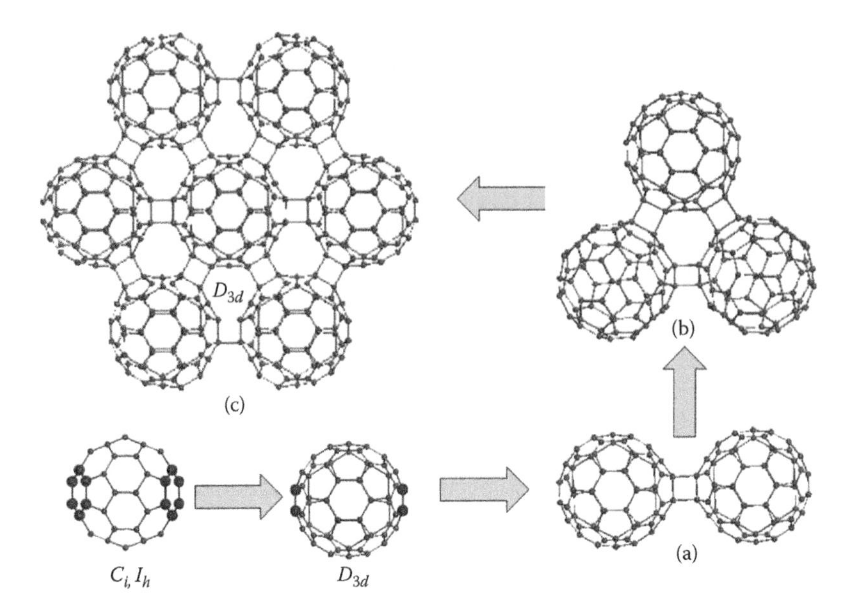

Figure 8.7 Scheme of the stepwise D_{3d}-based C_{60} polymerization. Equilibrium structures of dimer, trimer, and heptamer. Black balls mark high-rank reactive atoms of monomer.

a similar situation previously in the case of $C_{60}F_{48}$ fluoride (Section 4.4.5) or $C_{60}Cl_6$ chloride (Section 6.5). As shown in Section 3.5, there is such an isomer within the C_{60} isomer family that is isoenergetic with a C_i one and which may "go out of the shadow" to provide particular chemical reactions. Let us see how a D_{3d}-based C_{60} polymerization scheme might look. As seen in Figure 8.7, a change monomer symmetry results in another view on the high-rank N_{DA} ACS distribution, as well as in another manner of monomer coupling under oligomerization. UBS HF calculations show that the D_{3d}-based dimer is of the same energy as the C_i-based one. The same should be said concerning coupling energies per one [2 + 2] cycloaddition of C_i- and D_{3d}-based trimers, so that the cyclic trimer is highly stable. This is consistent with data obtained in the framework of restricted computational schemes.[37,38] Similarly to the precursor role of the C_i-based trimer in the formation of tetragonally packed (TP) higher oligomers, the D_{3d}-based trimer plays the same role for hexagonally packed (HP) ones.

The polymerization schemes in Figures 8.6 and 8.7 show a clear way how to computationally proceed with the formation of high-order C_{60} oligomers *prefactum*. Evidently, the way is extremely time-consuming, so that its undertaking should be stimulated by serious needs. At the same time, no evidence of experimental existence of free oligomers higher than trimers has been found so far. Well-studied TP and HP polymerization of C_{60} occurs on the basis of pristine cubic C_{60} crystal subjected to severe deformation at high temperatures and pressure (see, for example, Núñez-Regueiro et al.[28]), while linear polymerization is provided by a prolonged photoillumination.[17] In the former case, no size-restricted oligomers were found, and topochemical reactions were terminated by a complete polymerization of the solids followed by the formation of T and R crystalline modifications. In the second case, both high-order oligomers and fully polymerized orthorhombic crystal were obtained. A topochemical character of the products lowers considerably the significance of the *prefactum* consideration of free oligomers. At the same time, there is a strong necessity for an explanation on some delicate electronic properties of polymerized C_{60} crystals, such as their magnetism, which must be done at the quantum level. A quantum chemical cluster approach could be a good tool for solving the problem if cluster size and configuration allow for describing crystal properties adequately. A choice of such clusters, *postfactum* oligomers in other words, should naturally be done based on the real crystal structure.

A lot of efforts were undertaken to obtain structures of C_{60} polymerized crystals (see a full list of the relevant studies in Dzyabchenko et al.[40]). The available structural data allowed for suggesting a set of size-restricted *postfactum* oligomers (Figure 8.8). These oligomer studies lay the foundation of a quantum chemical description of magnetic properties of polymerized C_{60} crystal,[41] which will be discussed in detail in Chapter 13. Monomer molecules discriminated by color are surrounded by other molecules, the number of which is enough to considering the selected central units to be adequate in those in crystals. The structures of carpet and linear C_{60} oligomers are generally identical to the analogous structures studied by other researchers. The zigzag structure of C_{70} oligomer is similar to that followed from the covalent chemistry of the C_{70} fullerene subjected to the vacuum IMI. Actually, the concentration of the highest N_{DA} on the belt atoms located along short hexagon bonds (see Figure 3.11) causes a zigzag attaching C_{70} molecules to each other. A specific accommodation of the oligomer chains in crystal[39] permits the preservation of this structure.

8.4.2 *Structural data*

It is fairly clear that the consideration of magnetic properties needs a high confidence in the structural basis. Table 8.3 accumulates computational data for $(C_{60})n$ oligomers in terms of the structure of cyclobutane rings. The data are related to central monomers (discriminated by drawing in Figure 8.8), the properties of which are probably close to those of monomers in massive samples. Comparing data in Tables 8.2 and 8.3, one can see a full similarity in the behavior of the cyclobutane ring addition in all oligomers

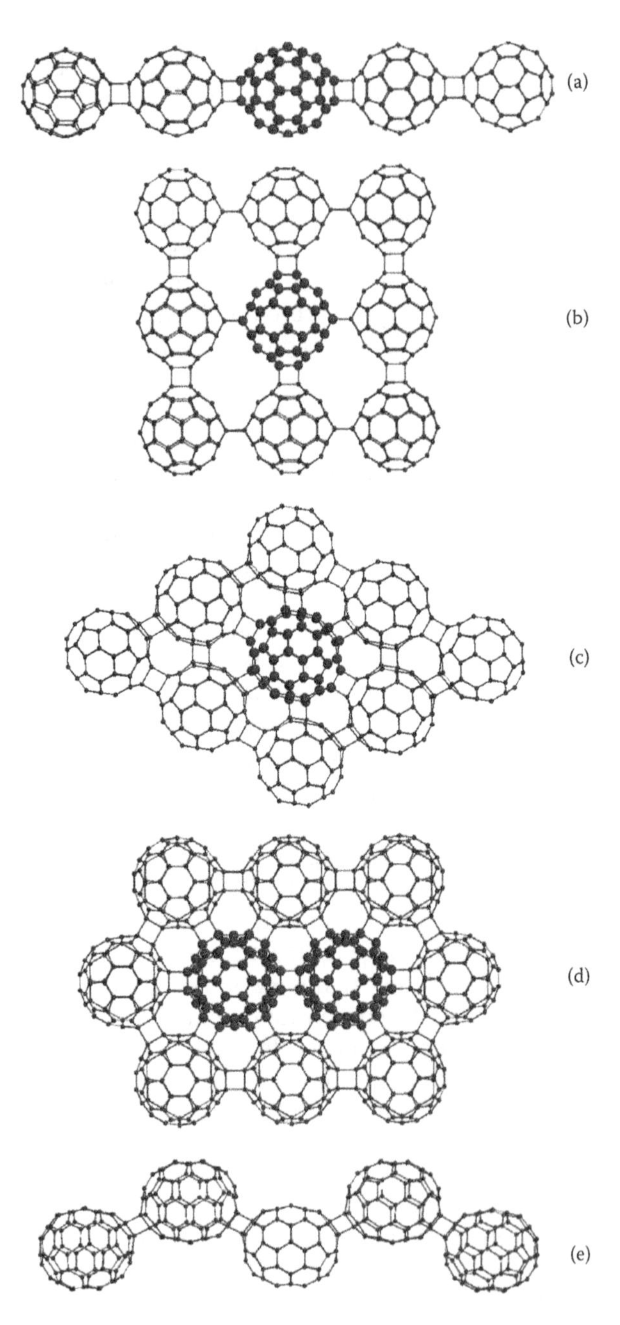

Figure 8.8 Equilibrium structure of *postfactum* C_{60} and C_{70} oligomers. (a and e) Linear packing. (b) Tetragonal packing. (c and d) Hexagonal packing. UBS HF AM1 singlet state, parallel calculations. (From Sheka, E. F., et al., *J. Exp. Theor. Phys.*, 103, 728–39, 2006.)

Table 8.3 Lengths of C-C bonds forming cyclobutane ring in C_{60} oligomers, Å

Oligomer	Data source [reference]	C-C distances (Å) in bond no. 1	2	3	4
3* C_{60} (*L*)	AM1_UBS HF [41]	1.596–1.597	1.515–1.516	1.515–1.516	1.548
	DCM AM1_ RHF [42][a]	1.606	1.509	1.509	1.555
	DCM PM3_ RHF [42][a]	1.597	1.510	1.510	1.560
4* C_{60} (*L*)	AM1_UBS HF [41]	1.603	1.515	1.515	1.546
	DCM AM1_ RHF [42][b]	1.603–1.604	1.507–1.515	1.507–1.515	1.554–1.556
	DCM PM3_ RHF [42][b]	1.596–1.597	1.511–1.516	1.511–1.516	1.558–1.559
5* C_{60} (*L*)	AM1_UBS HF [41]	1.596–1.597	1.514–1.516	1.514–1.516	1.546–1.552
5* C_{70} (*L*)	AM1_UBS HF [41]	1.593–1600	1.510–1.518	1.510–1.518	1.541–1.552
9*C_{60} (*Tg*)	AM1_UBS HF [41]	1.598–1.599			
		1.605–1.606	1.512–1.516	1.512–1.514	1.546–1.547
	Exp. [43]				1.56–1.58 1.60–1.62
10*C_{60} (*Hg*)	AM1_UBS HF [41]	1.598–1.599	1.504–1.505	1.512–1.514	1.548–1.549
9*C_{60} (*Hg*)	AM1_UBS HF [41]	1.597–1.599	1.503–1.505	1.512–1.514	1.548–1.551
	DCM AM1_ RHF [42][c]	1.596–1.598	1.509–1.514	1.509–1.514	1.560–1.562
	DCM PM3_ RHF [42][c]	1.596–1.598	1.509–1.514	1.509–1.514	1.560–1.562

[a] Trimer forms an equilateral triangle.
[b] Tetramer forms a rhombus.
[c] Heptamer has a hexagonal configuration.

without a remarkable dependence on the monomer packing. Table 8.4 lists data on the distances between the centers of monomer molecules in the oligomer structures. Let us again thoroughly consider the experimental data for dimers. The most accurate experimental data are those reported on the x-ray diffraction study of $(C_{60})_2$ dimers.[17] Calculations show that the

Table 8.4 Distance between centers of monomer molecules in oligomers

Oligomer	Data source [reference]	Distance, Å 1	2	3	4	5	6
2* C_{60}	AM1_UBS HF [41]	9.08					
	Exp. [19]	8.8 (4)					
	Exp. [18]	9 (0.4)					
	Exp. [17]	9.1					
3* C_{60} (*L*)	AM1_UBS HF [41]	9.08	9.08				
	Exp. [18]	9.2 (0.4)					
4* C_{60} (*L*)	AM1_UBS HF [41]	9.08	9.06	9.08			
5* C_{60} (*L*)	AM1_UBS HF [41]	9.08	9.07	9.09	9.08		
	Exp. [44]			9.1			
	Exp. [19]			9.2 (0.4)			
9*C_{60} (*Tg*)	AM1_UBS HF [41]	9.04	9.04	9.09	9.09		
	Exp. [28]			9.09			
10*C_{60} (*Hg*)	AM1_UBS HF [41]	9.11	9.11	9.12	9.11	9.11	9.12
9*C_{60} (*Hg*)	AM1_UBS HF [41]	9.11	9.11	9.11	9.11	9.11	9.11
	Exp. [28]			9.17			

subsequent steps of linear oligomer lengthening are accompanied by simply repeating the intermolecular spacings in the dimer, which agrees with the experimental data.[17,21,44] Analogous behavior is observed in the course of C_{70} oligomerization. In carpet oligomer structures, the intermolecular distances were determined relative to the central molecules of structures shown in Figure 8.8. The results of calculations well agree with the experimental data obtained by x-ray diffraction for *T* and *R* modifications of polymeric crystals.[28,40] In particular, the calculations correctly reproduce the tendency to increase the intermolecular distance on the passage from the *T* to *R* configuration, which is reflected by an increase in the linear parameter of the corresponding unit cell.

8.4.3 Binding energies in oligomers

Table 8.5 presents data on the binding energies in oligomers (scaled per one cyclobutane ring). As can be seen from these data, the binding energies fall within a rather broad interval, from –31 to –42 kcal/mol, although the cyclobutane rings in the oligomers studied are structurally very alike. Based on the data presented in the table, one can expect that crystals of the *O*, *T*, and *R* configurations must differ with respect to thermodynamic stability, which has to be maximum for the *O* phase and minimum for the *R* phase. This behavior was pointed out in Iwasa et al.[45] More recently, constructed using the experimental thermodynamic data, the equilibrium

Table 8.5 Coupling energies (per cyclobutane ring) in C_{60} and C_{70} oligomers (AM1(UBS HF) singlet)

Oligomer	E_{coupl},[a] kcal/mol
C_{60}	—
2* C_{60}	–40.86
3* C_{60}	–40.60
4* C_{60}	–39.51
5* C_{60}	–39.75
9*C_{60} (*Tg*)	–34.43
9*C_{60} (*Hg*)	–31.05
10*C_{60} (*Hg*)	–30.93
C_{70}	—
5* C_{70}	–41.93

Source: According to data from Sheka, E. F., et al., *J. Exp. Theor. Phys.*, 103, 728–39, 2006.

[a] $E_{cpl} = (\Delta H_{olig} - n\Delta H_{mol})/m$; n and m are the numbers of monomers and cyclobutane rings in oligomer, and ΔH_{olig} and ΔH_{mol} are the heats of formation for the oligomer and monomer, respectively. For the determination of the heat of formation, see in footnote b to Table 3.2.

phase diagram of polymerized C_{60} phases confirmed that the *R* phase is actually the least stable one, while the *O* and *T* modifications have close characteristics (with some preference for the *O* phase).[46] Based on the data of differential scanning calorimetry,[47] it was established that the average binding energy per cyclobutane ring varies in the following order (kcal/mol):

$$O(9.99) \geq T(9.96) > R\ (7.38).$$

Empirical values determined from experimental data are smaller than the calculated values. However, it should be borne in mind that these estimates have been obtained after a rather complicated mathematical processing of data, involving determination and evaluation of some unknown thermodynamic parameters.[46] And, if we cannot speak about a quantitative coincidence concerning the binding energies, we nevertheless have to point out an obvious similarity in trends and correlations for various types of structures. From this standpoint, the results of calculations fully agree with experimental observations concerning the minimum stability of the *R* modification.

8.5 Concluding remarks about the character of chemical reactions typical to fullerene

Fullerenes are exclusive molecular species that cannot be assigned to a particular class of compounds. Consequently, not a certain class of chemical reactions, as traditionally takes place, but a large set of reactions of different types are characteristic for them. Two main reasons lay the foundation for the behavior: high chemical susceptibility caused by effectively unpaired electrons and high donor and acceptor abilities of the species. Taking together theses factors considerably enlarges the variety of reactions than can occur with fullerene participation, from traditional addition reactions to complicated ones governed mainly by the D–A interaction.

Traditional addition reaction is not a proper term related to fullerene-based reactions since the kinetics of the latter is rigidly controlled by the regioselectivity caused by the effectively unpaired electron distribution over atoms. Seemingly, a multifold isomerism of the fullerene molecules needs to be taken into account as well. On the other hand, D–A reactions cannot be considered traditional since their particular features are connected with a multiwell structure of the ground-state energy term of the relevant binary system. A large variety of the D–A reactions of fullerenes can be explained by changing the mutual disposition of the well minima. Another feature of the fullerene-involving D–A reactions concerns the formation of not only heteroreactant but also homoreactant products. The latter implies dimers and higher fullerene oligomers. While in the heteroreactant case fullerenes are electron acceptors, the species play both acceptor and donor roles in the homoreactant case.

This bimode D–A ability is characteristic for not only fullerenes but also other sp^2 nanocarbons, such as carbon nanotubes (CNT) and graphene. Consequently, the reactions, or better, IMI between components of dyads such as fullerene-CNT, fullerene-graphene, CNT-CNT, and CNT-graphene, as well as more complicated triads, tetrads, and so forth, should be considered in terms of the D–A peculiarities discussed above. To conclude this part, let us consider a dyad constructed from fullerene C_{60} and a fragment of (4,4) single-walled CNT (SWCNT).[49] The intermolecular contact between two molecular constituents is formed by a pair of atoms of group 1 of a C_{60} molecule and an atom pair forming a short bond on the tubular partner wall. The location of this bond can be chosen quite arbitrarily due to practical homogeneous distribution of the ACS values over the tube wall atoms[48] (see Figure 11.3). Two final configurations of the dyad are shown in Figure 8.9. As follows from the calculated results, the IMI within the dyad is subordinated to the same rules that are typical for fullerene dyads so that a typical charge transfer complex is formed when the starting distance, R^{st}_{CC}, is 2.5 Å and the partners are located at R^{fin}_{CC} = 4.6 Å distance, while a [2 + 2] cycloaddition tightly bound composite is produced when R^{st}_{CC} is 1.8 Å. Similar configurations have been obtained for the dyad C_{60} + graphene[50] (Figure 8.10). A [2 + 2] cycloaddition provides the tight binding of the fullerene molecules with the graphene surface when R^{st}_{CC} = 1.7 Å and a charge transfer complex is formed when R^{st}_{CC} = 2.5 Å.

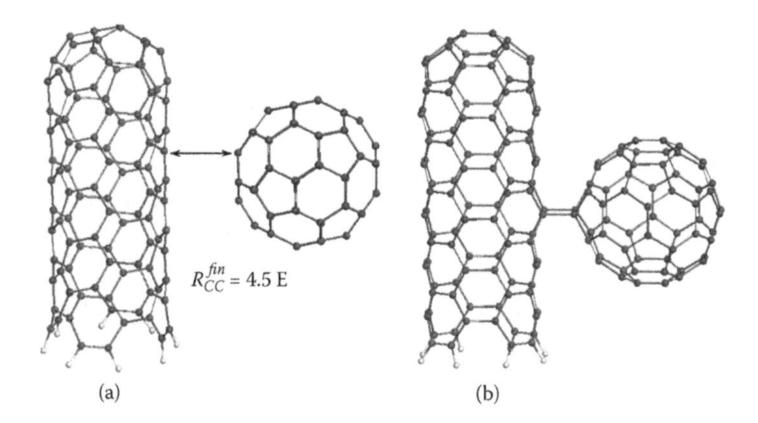

Figure 8.9 Equilibrium structures of the (a) C_{60}+ (4,4) SWCNT charge transfer complex, $R_{CC}^{st} = 2.5$ Å, $R_{CC}^{fin} = 4.62$ Å, and of the (b) C_{60}/(4,4) SWCNT composite, $R_{CC}^{st} = 1.8$ Å, $R_{CC}^{fin} = 1.57$ Å. UBS HF AM1 singlet state.

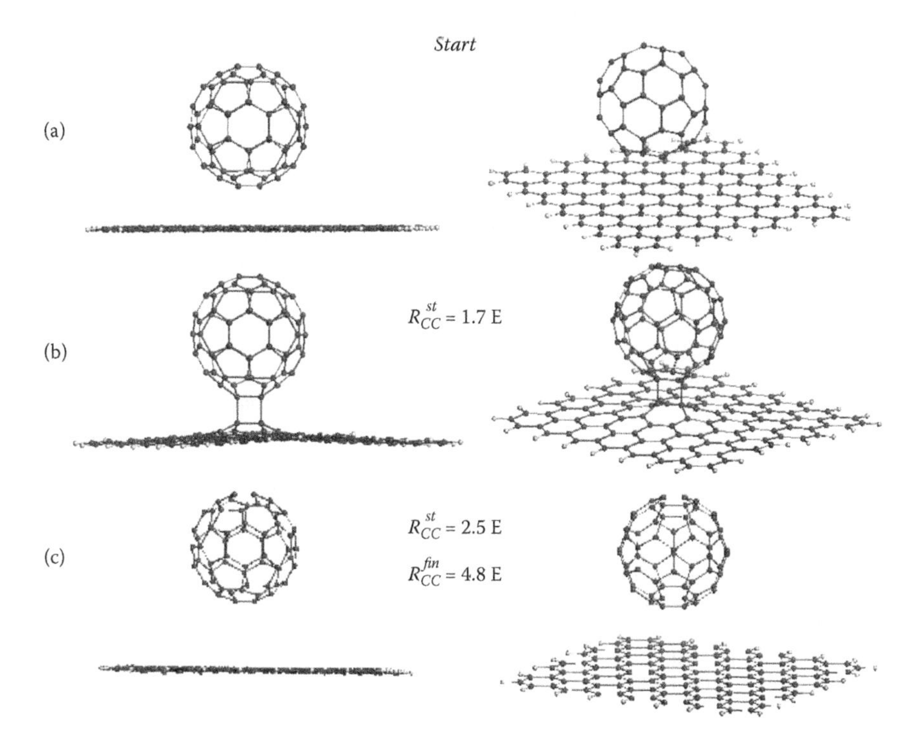

Figure 8.10 Two views of fullerene-graphene composites. (a) Initial composition of the C^{60}-graphene dyad. (b) Equilibrium structure of the [C_{60}] graphene composite. (c) Equilibrium structure of the C_{60} + graphene charge complex. UBS HF AM1 singlet state.

Composites formed by CNT and graphene and subordinated to the same rules will be considered in Chapter 12.

References

1. Rao, A. M., Zhou, P., Wang, K. A., Hager, G. T., Holden, J. M., Wang, Y., Lee, W. T., Bi, X.-X., Eklund, P. C., Cornett, D. S., Duncan, M. A., and Amster, I. J. 1993. Photoinduced polymerization of solid C_{60} films. *Science* 259:955–7.
2. Iwasa, Y., Arima, T., Fleming, R. M., Siegrist, T., Zhou, O., Haddon, R. C., Rothberg, L. I., Lyons, K. B., Carter, Jr., H. L., Hebard, A. F., Tycko, R., Dabbagh, G., Kraewski, J. J., Thomas, J. A., and Yagi, T. 1994. New phases of C_{60} synthesized at high pressure. *Science* 264:1570–72.
3. Pekker, S., Janossy, A., Mihali, L., Chauvet, O., and Forro, L. 1994. Single-crystalline $(KC_{60})_n$: A conducting linear alkali fulleride polymer. *Science* 265:1077–78.
4. Yamawaki, H., Yoshida, M., Kakudate, Y., Usuba, S., Yokoi, H., Fujiwara, S., Aoki, K., Ruoff, R., Malhotra, R., and Lorentz, D. 1993. Infrared study of vibrational property and polymerization of fullerene C_{60} and C_{70} under pressure. *J. Phys. Chem.* 97:11161–63.
5. Takahashi, N., Dock, H., Matsuzawa, N., and Ata, M. 1993. Plasma-polymerized C_{60}/C_{70} mixture films: Electric conductivity and structure. *J. Appl. Phys.* 74:5790–98.
6. Zhao, I. B., Poirier, D. M., Pechman, R. J., and Weaver J. H. 1994. Electron stimulated polymerization of solid C_{60}. *Appl. Phys. Lett.* 64:577–80.
7. Nakaya, M., Kuwahara, Y., Aono, M., and Nakayama, T. 2008. Reversibility-controlled single molecular level chemical reaction in a C_{60} monolayer via ionization induced by scanning transmission microscopy. *Small* 4:538–41.
8. Dzyabchenko, A. V., Agafonov, V. N., and Davydov, V. A. 1999. Theoretical molecular packings and the structural model of solid-phase polymerization of fullerene C_{60} under high pressures. *Crystallogr. Rep.* 44:18–24.
9. Sheka, E. F. 2007. Donor–acceptor interaction and fullerene C_{60} dimerization. *Chem. Phys. Lett.* 438:119–26.
10. Kitaygorodsky, A. I. 1973. *Molecular crystals and molecules.* New York: Academic Press.
11. Kovats, E., Oszlanyi, G., and Pekker, S. 2005. Structure of the crystalline C_{60} photopolymer and the isolation of its cycloadduct components. *J. Phys. Chem. B.* 109:11913–17.
12. Weaver, J. H., Martins, J. L., Komeda, T., Chen. Y., Ohno, T. R., Kroll, G. H., Troullier, N., Haufler, R., and Smalley, R. E. 1991. Electronic structure of solid C_{60}: Experiment and theory. *Phys. Rev. Lett.* 66:1741–44.
13. Wang, X.-B., Ding, C.-F., and Wang, L.-S. 1999. High resolution photoelectron spectroscopy of v_{60}^-. *J. Chem. Phys.* 110:8217–20.
14. Ecklund, P. C., Rao, A. M., Zhou, P., Wang, Y., and Holden, J. M. 1995. Photochemical transformation of C_{60} and C_{70} films. *Thin Solid Films* 257:185–203.
15. Adams, G. B., and Page, J. B. 2001. Theoretical studies of Raman spectra for planar polymerized C_{60}. *Phys. Stat Sol. B* 226:95–106.
16. Hedberg, K., Hedberg, L., Bethune, D. S., Brown, C. A., de Vries, M., Dorn, H. C., and Johnson, R. D. 1991. Bond lengths in free molecules of buckminsterfullerene C_{60} from gas-phase electron diffraction. *Science* 254:410–12.

17. Wang, G., Komatsu, K., Murata, Y., and Shiro, M. 1997. Synthesis and x-ray structure of dumb-bell-shaped C120. *Nature* 387:583–86.
18. Nakayama, T., Onoe, J., Nakatsuji, K., Nakamura, J., Takeuchi, T., and Aono, M. 1999. Photoinduced products in a C60 monolayer on Si(111) $(\sqrt{3}x\sqrt{3})$ -Ag: An STM study. *Surf. Rev. Lett.* 6:1073–78.
19. Hassanien, A., Gasperič, G., Demsar, J., Maševič, I., and Mihailovic, D. 1997. Atomic force microscope study of photo-polymerized epitaxial C_{60} films. *Appl. Phys. Lett.* 70:417–19.
20. Stafstrom, S., and Fagerstrom, J. 1997. Electronic structure and stability of fullerene polymers. *Appl. Phys. A* 64:307–14.
21. Wang, Y., Holden, J. M., Bi, X.-X., and Eklund, P. C. 1994. Thermal decomposition of polymeric C_{60}. *Chem. Phys. Lett.* 217:413–17.
22. Davydov, V. A., Kashevarova, L. S., Rakhmanina, A. V., Senyavin, V. M., Pronina, O. P., Oleynikov, N. N., Agafonov, V., Céolin, R., Allouchi, H., and Szwarc, H. 2001. Pressure-induced dimerization of fullerene C_{60}: A kinetic study. *Chem. Phys. Lett.* 333:224–29.
23. Sheka, E. F., Razbirin, B. S., Starukhin, A. N., and Nelson D. K. 2007. Electronic structure and spectra of *N*-methylfullerenepyrrolidine. *Optics Spectr.* 102:32–41.
24. Brant, J., Lecoanet, H., and Wiesner, M. R. 2005. Aggregation and deposition characteristics of fullerene nanoparticles in aqueous systems. *J. Nanopart. Res.* 7:545–53.
25. Pac, B., Petelenz, P., Slawik, M., and Munn, R. W. 1998. Theoretical interpretation of the electroabsorption spectrum of fullerene films. *J. Chem. Phys.* 109:7932–39.
26. Kazaoui, S., Minami, N., Tanabe, Y., Byrne, H. J., Eilmes, A., and Petelenz, P. 1998. Comprehensive analysis of intermolecular charge-transfer excited states in C_{60} and C_{70} films. *Phys. Rev. B* 58:7689–700.
27. Suzuki, M., Iida, T., and Nasu, K. 2000. Relaxation of exciton and photoinduced dimerization in crystalline C_{60}. *Phys. Rev. B* 61:2188–98.
28. Núñez-Regueiro, M., Markes, L., Hodeau, J.-L., Béthoux, O., and Perroux, M. 1995. Polymerized fullerite structures. *Phys. Rev. Lett.* 74:278–81.
29. Shen, H. 2006. Geometrical deformation and failure behavior of C_{60} fullerene dimer under applied external electric field. *Mol. Simulation* 32:59–64.
30. Li, Yi., Li, H., Jia, X., Hao, J., and Liu, W. 2006. Electrochemical behavior of C_{60} films and C_{60}/lipid films in ionic liquids. *Carbon* 44:894–99.
31. Ohta, N., Mikami, S., Iwaki, Y., Tsushima, M., Imahori, H., Tamaki, K., Sakata, Y., and Fukuzumi, S. 2003. Acceleration and deceleration of photoinduced electron transfer rates by an electric field in porphyrin-fullerene dyads. *Chem. Phys. Lett.* 368:230–35.
32. Thomson, M. D., Novosel, M., Roskos, H. G., Muller, T., Scheibitz, M., Wagner, M., Fabrizi de Biani, F., and Zanello, P. 2004. Electronic structure, photophysics, and relaxation dynamics of charge transfer excited states in boron–nitrogen-bridged ferrocene-donor organic-acceptor compounds. *J. Phys. Chem. A* 108:3281–91.
33. Sheka, E. F. 2007. Donor-acceptor origin of fullerene C_{60} dimerization. *Int. J. Quant. Chem.* 107:2361–71.
34. Stephens, P. W., Bortel, G., Faigel, G., Tegze, M., Janossy, A., Pekker, S., Oszlanyi, G., and Forro, L. 1994. Polymeric fullerene chains in RbC_{60} and KC_{60}. *Nature* 370:636–39.

35. Nakayama, T. 2005. Manipulation of intermolecular bonds between C_{60} molecules using a tip of STM. 87. In *37th ISTC Japan Workshop on Advanced Nanomaterials in Russia/CIS*, Tsukuba, Japan, December 12–13, 2005.
36. Makarova, T. L., Sundqvist, B., Höhne, R., Esquinazi, P., Kopelevich, Ya., Scharff, P., Davydov, V. A., Kashevarova, L. S, and Rakhmanina, A. V. 2001. Magnetic carbon. *Nature* 413:716–19.
37. Kunitake, M., Uemura, S., Ito, O., Fujiwara, K., Murata, Y., and Komatsu, K. 2002. Structural analysis of C_{60} trimers by direct observation with scanning tunneling microscopy. *Angew. Chem. Int. Ed.* 41:969–72.
38. Lee, K. H., Eun, H. M., Park, S. S., Suh, Y. S., Jung, K.-W., Lee, S. M., Lee, Y. H., and Osawa, E. 2000. Structures and energetics of regioisomers of C_{60} dimer and trimers. *J. Phys. Chem. B* 104:7038–42.
39. Soldatov, A. V., Roth, G., Dzyabchenko, A., Johnels, D., Lebedkin, S., Meingast, C., Sundqvist, B., Haluska, M., and Kuzmany, H. 2001. Topochemical polymerization of C_{70} controlled by monomer crystal packing. *Science* 293:680–83.
40. Dzyabchenko, A. V., Agafonov, V. N., and Davydov, V. A. 1999. Molecular interaction and crystal packing in products of polymerization of fullerene C_{60} under high pressures. *Crystallogr. Rep.* 44:11–17.
41. Sheka, E. F., Zaets, V. A., and Ginzburg, I. Ya. 2006. Nanostructural magnetism of polymeric fullerene crystals. *J. Exp. Theor. Phys.* 103:728–39.
42. Cabrera-Trujillo, J. M., and Robles, J. 2001. Theoretical study of the structural and electronic properties of two-dimensional polymerized fullerene clusters with 2, 3, 4, and 7 C_{60} molecules. *Phys. Rev. B.* 64:165408-1–6.
43. Davydov, V. A., Kashevarova, L. S., Rakhmanina, A. V., Narumbetov, B., Agafonov, V. N., Dzyabchenko, A. V., and Kulakov, V. I. 2004. Single crystal synthesis and refinement of the crystal structure of the polymerize tetragonal phase of C_{60}. *Fullerenes Nanotubes Carbon Nanostr.* 12:275–79.
44. Davydov, V. A., Kashevarova, L. S., Rakhmanina, A. V., Agafonov, V. N., Dzyabchenko, A. V., Dubua, P., Céolin, R., and Szwarc, H. 1997. Identification of polymerized orthorhombic phase of C_{60}. *JETP Lett.* 66:120–22.
45. Iwasa, Y., Tanoue, K., and Mitani, T. 1998. Energetics of polymerized fullerites. *Phys. Rev. B.* 58:16374–77.
46. Korobov, M. V., Bogachev, A. G., Senyavin,V. M., Popov, A. A., Davydov, V. A., Rakhmanina, A. V., and Markin, A. V. 2006. Equilibrium phase diagram of polymerized C_{60} and kinetics of decomposition of the polymerized phases. *Fullerenes Nanotubes Carbon Nanostr.* 14:401–7.
47. Korobov, M. V., Bogachev, A. G., Popov, A. A., Senyavin, V. M., Stukalin, E. B., Dzyabchenko, A. V., Davydov, V. A., Kashevarova, L. S., Rakhmanina, A. V., and Agafonov, V. 2005. Relative stability of polymerized phases of C_{60}: Depolymerization of a tetragonal phase. *Carbon* 43:954–61.
48. Sheka, E. F., and Chernozatonskii, L. A. 2010. Broken symmetry approach and chemical susceptibility of carbon nanotubes. *Int. J. Quant. Chem.*, 110:1466–80.
49. Sheka, E. F., Nikitina, E. A., and Shaymardanova, L. A. 2010. Private communication 1.
50. Sheka, E. F., Nikitina, E. A., and Shaymardanova, L. A. 2010. Private communication 2.

chapter nine

Nanomedicine of fullerene C_{60}

9.1 Introduction

Nanomedicine of fullerene seems to be a beautiful platform to illustrate the synergetics of chemistry, biology, and physics in fullerene science. Modern medicinal fullerenics is a largely explored field, actively developing and enlarging. We are not going to go in depth on the topic, but refer readers to some recent exhaustive reviews.[1–3] Our purpose is to show that the basic grounds that lay the foundation of biological and medicinal applications of fullerenes are tightly connected with the concepts discussed in this book. This concerns first of all the mechanism of the therapeutic action of fullerenes.

For today, a large number of efficient medicobiological actions of fullerenes have been found, among which there are antiviral, anticancer, neuroprotective, enzymatic, antiapoptopic, and many others.[4–6] It is worthwhile to supplement this list with the latest sensational news on fullerene-based gene delivery in mice.[7] Expert judgments suggest that this work opens a large way to test the efficacy of fullerenes for *in vivo* applications, such as insulin gene delivery to reduce blood glucose levels for diabetes treatment, and so forth.

Empirically estimated, fullerenes fulfill therapeutic functions acting as either antioxidant or oxidative agent, thus revealing seemingly two contradictory behaviors. However, this two-mode behavior is just the manifestation of two appearances of fullerenes that are, on the one hand, radicals due to the availability of a considerable number of effectively unpaired electrons, N_D, and, on the other hand, an efficient donor–acceptor (D–A) agent. Actually, the consideration of chemical behavior of fullerenes discussed in the previous chapters clearly shows that they must willingly interact with other radicals, forming tightly bound compositions, and thus providing efficient radical scavenging. In full agreement with this statement, the first exhibited therapeutic function of fullerene C_{60} was its action as a radical scavenger.[8] Later on, this laid the foundation of the antioxidant administrating of fullerenes in medical practice.[9–11] Establishing the preservation of antioxidant properties in C_{60} derivatives in general, as well as its dependence on the chemical structure and, mainly, on the number of attached chemical groups, with a clear preference toward monoderivatives, is in a complete accordance with the expected behavior of molecular chemical susceptibility and can be quantitatively described in terms of N_D. It is enough to remain a clearly justified working out this pull of effectively unpaired electrons under successive fluorination and hydrogenation. Therefore, the antioxidant therapeutic function of fullerenes is intimately connected with the electronic structure of the molecule itself.

In contrast to the individual-molecule character of the antioxidant action, the oxidative action of fullerenes occurs under photoexcitation of their solutions in both molecular and polar solvents in the presence of molecular oxygen. The difference in the behavior of singlet and triplet oxygen is obviously connected with the difference in the pairing of the molecule electrons caused by different spin multiplicity. A quantitative characteristic of the pairing can be expressed in terms of the total number of unpaired electrons N_D. Calculations performed within the framework of the UBS HF approach expose $N_D = 2$ for both spin states. But, if for the triplet state this finding just naturally reflects two electrons that are responsible for maintaining the molecule spin multiplicity, in the singlet state the availability of two effectively unpaired electrons evidences a biradical character of the molecule, which explains $^1O_2$ high oxidative activity. Therefore, the photostimulated $^3O_2 \rightarrow {}^1O_2$ transformation in the presence of fullerene molecule just means exempting the molecule two electrons from the spin multiplicity service thus transforming chemically inactive molecule into a biradical. The action consists in the oxidation of targets by either singlet oxygen, $^1O_2$, or reactive oxygen-containing species, such as hydroxyls (^-OH), superoxyanions ($O_2^{\bullet}$-), and hydrogen peroxide (H_2O_2), produced in due course of photoexcitation of fullerene solutions.

The presence of fullerene for the $^3O_2 \rightarrow {}^1O_2$ transformation is absolutely mandatory, so that the treatment is called photodynamic fullerene therapy.[12,13] For this reason alone, that the action is provided by a complex involving fullerene and solvent molecules, as well as molecular oxygen, it becomes clear that it results from a particular intermolecular interaction. However, until now, the mechanism of photodynamic therapy has been hidden behind a slogan 'triplet state photochemical mechanism' that implies the excitation transfer over a chain of molecules according to a widely accepted scheme.[12–14]

$$^1C_{60} \xrightarrow{\hbar\nu} {}^1C_{60}^* \rightarrow {}^3C_{60}^* \xrightarrow{^3O_2 \rightarrow {}^1O_2} {}^1C_{60}$$

Scheme 9.1

The scheme implies the energy transfer from the singlet photoexcited fullerene to the triplet one that further transfers the energy to convenient triplet oxygen thus transforming the latter into singlet oxygen. The first two stages of this 'single-fullerene-molecule' mechanism are quite evident while the third one, the most important for the final output, is quite obscure in spite of a lot of speculations available [14]. Obviously, the stage efficacy depends on the strength of the intermolecular interaction between fullerene and oxygen molecules. Numerous quantum chemical calculations show that pairwise interaction in the C_{60}-O_2 dyad in both singlet and triplet state is practically absent. The AM1 UBS HF computations fully support the previous data disclosing the coupling energy of the dyad E_{cpl}^{f-o} equal to zero in both cases. This puts a serious problem for the explanation of the third stage of the above scheme forcing to suggest the origination of a peculiar intermolecular interaction between C_{60} and O_2 molecules in the excited state once absent in the ground state. However, as we know from the previous chapters, the exclusive D–A ability of fullerenes strongly influences intermolecular interaction (IMI) and cannot be omitted when considering intermolecular events, particularly under photoexcitation. Let us look at oxidative fullerene-based solutions from this viewpoint.

9.2 *Spin-flip in the oxygen molecule in fullerene solutions*

The system under consideration consists of fullerene C_{60}, solvent, both polar (water, etc.) and nonpolar (benzene, etc.), and oxygen. Molecules of fullerene and solvent are in a singlet ground state while the ground state of the oxygen molecule is triplet. Let us call this system photodynamic (PD) solutions. There are a few types of IMI in the solutions, among which we will be interested in the IMI between fullerene molecules (*f-f*), between fullerene and solvent molecules (*f-s*), and between fullerene and oxygen (*f-o*). So far, we have pointed out the *f-s* IMI, which might be important in some cases (see the influence of this interaction on nanophotonics of fullerene solutions in Chapter 10). In the case of such solvents as benzene and water, it is very weak and can be neglected.

Consequently, instead of an ideal solution, the PD presents a conglomerate of clusterized C_{60} molecules, as shown schematically in Figure 9.1, which is experimentally proven in many cases (see, for example, Nath et al.[16] and Samal and Geckeler,[17] as well as Chapter 10).

As shown in Chapter 8, any cluster of C_{60} molecules has properties of a charge transfer complex, so that under excitation by the UV-visible light a pair of molecular ions is produced, which quickly relax into the ground state of the neutral molecule after the light is switched off. In contrast to the neutral C_{60}, both molecular ions C_{60}^- and C_{60}^+ interact with the oxygen molecule, giving a coupling energy, E_{cpl}, of –10.03 and –10.05 kcal/mol, respectively, referring to the $^3O_2$ molecule, and –0.097 and –0.115 kcal/mol in regards to $^1O_2$. Therefore, the oxygen molecule is quite strongly held in the vicinity of both molecular ions, forming (C_{60}^- + O_2) and (C_{60}^+ + O_2) complexes as

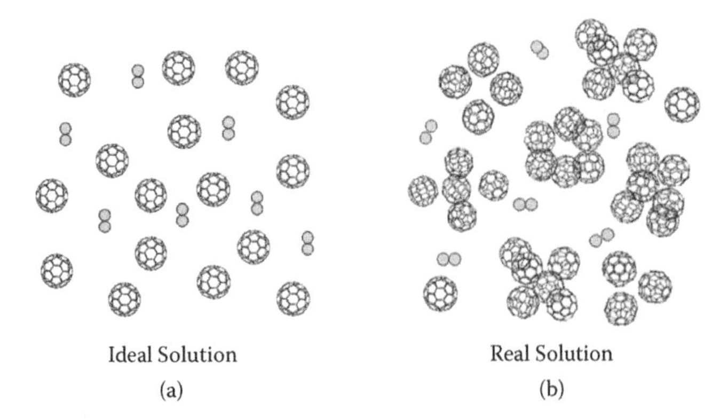

Figure 9.1 Schematic presentation of an ideal (a) and real (b) fullerene solution in the presence of molecular oxygen.

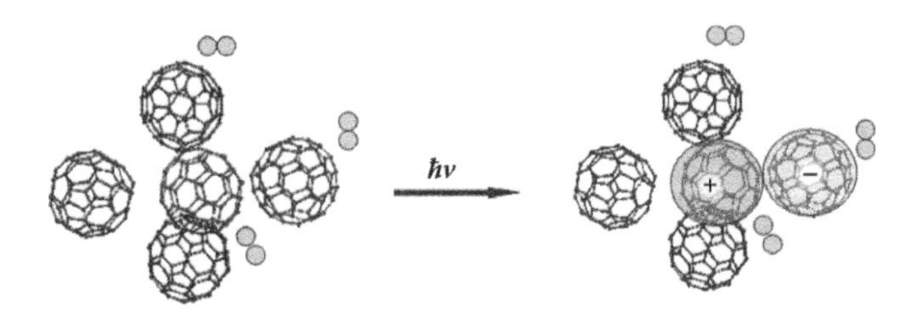

Figure 9.2 The formation of $^2[C_{60}^- + O_2]$ and $^2[C_{60}^+ + O_2]$ complexes under photoexcitation of $(C_{60})_5$ cluster.

schematically shown in Figure 9.2. UBS HF AM1 calculations for the corresponding pairs show that the complexes are of $^2[C_{60}^- + O_2]$ and $^2[C_{60}^+ + O_2]$ compositions of the doublet spin multiplicity. Since both fullerene ions take responsibility over the complex spin multiplicity, so that two electrons of the oxygen molecule that were on the service of triplet spin multiplicity of $^3[(C_{60})_n + O_2]$ dyads in the ground state are not more needed for the job and become effectively unpaired thus adding two electrons to the N_D pool of unpaired electrons of complexes $^2[C_{60}^- + O_2]$ and $^2[C_{60}^+ + O_2]$. The distribution of effectively unpaired electrons of both complexes over their atoms, which displays the distribution of the atomic chemical susceptibility of the complexes, is shown in Figure 9.3. A dominant contribution of electrons located on oxygen atoms 61 and 62 is clearly seen thus revealing the most active sites of the complexes. It should be noted that these distributions are intimate characteristics of both complexes so that not oxygen itself but both complexes as a whole provide the oxidative effect. The effect is lasted until the complexes exist and is practically immediately terminated when the latter disappear when the light is switched off.

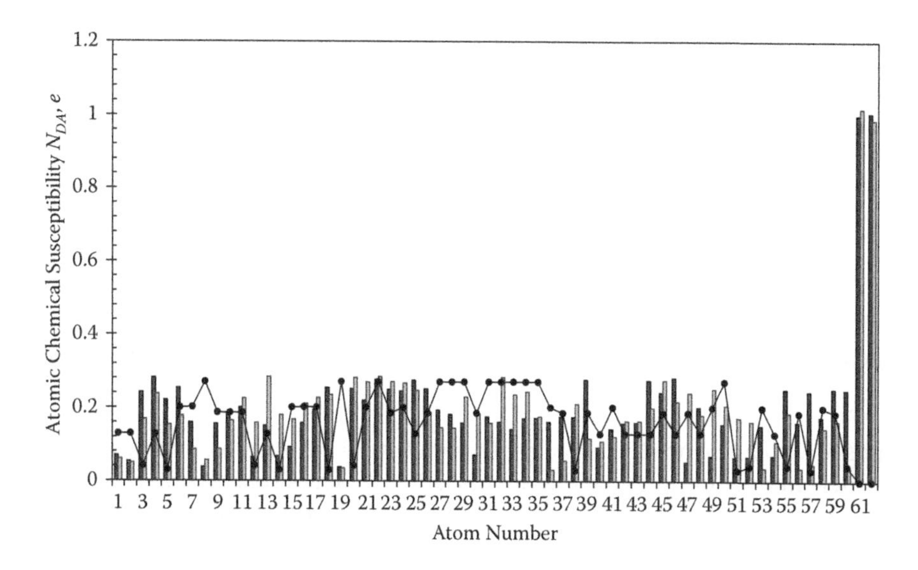

Figure 9.3 Distribution of atomic chemical susceptibility, N_{DA}, over atoms of $^2[C_{60}^- + O_2]$ (black bars) and $^2[C_{60}^+ + O_2]$ (light gray bars) complexes. Curve with black dots plots distribution over atoms of C_{60}. UBS HF AM1 doublet and singlet states. (From Sheka, E. F., arXiv:1005.2383 (cond-mat.mes-hall).)

The obtained results make it possible to suggest the following mechanism that lays the foundation of the photodynamic effect of fullerene solutions

$$^3\left[(C_{60})_n + O_2\right] \xrightarrow{h\nu} \begin{cases} ^2\left[(C_{60})_{n-2}C_{60}^{+} + O_2\right] \\ ^2\left[(C_{60})_{n-2}C_{60}^{-} + O_2\right] \end{cases}$$

Scheme 9.2

Here $(C_{60})_{n-2}\,C_{60}^{+}$ and $(C_{60})_{n-2}\,C_{60}^{-}$ present fullerene clusters incorporating molecular ions. The transformation of the triplet ground state complex into two doublet ones under photoexcitation is accompanied with a spin flip of the oxygen molecule electrons in the presence of fullerene molecule which is shown scematically in Figure 9.4. This approach allows attributing phodynamical effect of fullerene solutions to a new type of chemical reactions in the modern spin chemistry.

Since fullerene derivatives preserve D–A properties of pristine fullerene, Figure 9.2 is fully applied in this case as well. Thus, not only parent C_{60} or C_{70} themselves, but their derivatives can be used in PD solutions. However, parameters of the photodynamic therapy are different, depending on the fullerene derivative structure.[19] Changing solute molecules, it is

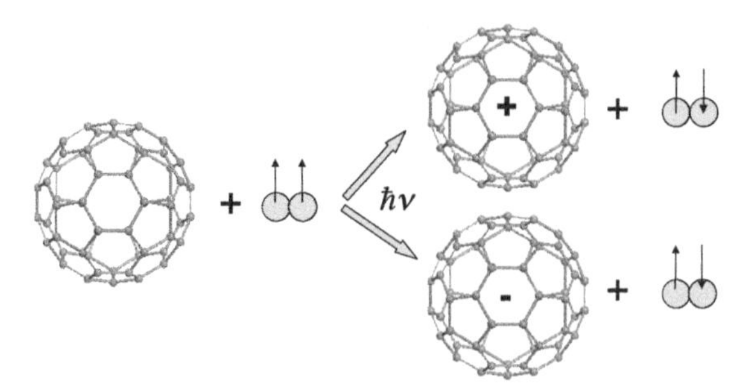

Figure 9.4 Schematic presentation of the spin-flip in oxygen molecule under photoexcitation.

possible to influence the efficacy of their clusterization, which in turn may either enhance or press the therapeutic effect.[2,14] The situation appears to be similar to that which occurs in nanophotonics of the fullerene solution.

9.3 Fullerene-silica complexes for medicinal chemistry

If photodynamical therapy is mainly administered by using an aqueous solution, the delivery of fullerene-based antioxidant in a living body presents a serious problem. A few types of techniques have been suggested for medical practice, and among them the following are mostly used:[2]

1. Films or fullerene-coated surfaces containing immobilized fullerene[20]
2. Aqueous suspensions of micronized crystalline fullerene[21,22]
3. Stable colloidal fullerene solutions in water[23]
4. Water-soluble fullerene-based complexes[24]
5. Water-soluble fullerene derivatives[25]

Each of these techniques presents a large field of investigations and has its own advantages and disadvantages. The author's previous experience in amorphous silica study[26] complemented by knowledge of the high medicinal activity of nanosize silica (NSS)[27] forced to think about a possibility of conjugation of silica and fullerene to provide an easy delivery of the medicament in the body just using NSS as a carrier of immobilized fullerene molecules, as well as about enhancement of therapeutic effects of each component in a synergetic manner.

As known, there are a few technological polymorphs of NSS,[26,28] among which the most popular are pyrogenic nanosized silica (PNSS, or Aerosil), silica gel (SCG), and aerogel. Either component of a possible NSS-C_{60} complex exhibits an appreciable medicobiological effect; for example, SCG-based enterosorbents are widely used in medicine. More versatile, the medicinal

Table 9.1 Main factors underlying the medicobiological activity of pyrogenic nanosized silica (PNSS) and C_{60} fullerene

SILICS[29]	C_{60} fullerene
High hydrophobicity of the PNSS surface	Antioxidant activity
High efficiency in the sorption of proteins	Neuroprotective activity
Agglutination of a large number of microorganisms and microbial toxins	Antivirus and antimicrobial effect
Adsorption of low-molecular-weight compounds	Inhibition of enzymatic activity
Enhancement of the action of immunoactive drugs	Gene delivery
Inhibition of the aggregation of thrombocyte, etc.	

chemistry of PNSS has made even more impressive progress.[27] One result of these studies is SILICS,[29] a wide-spectrum-effect drug that proved to be not only a highly efficient enterosorbent, superior to all known sorbents, but also an effective medicinal agent for monotherapy of various diseases.[27] The medicobiological activity of fullerenes has been discussed earlier. The main factors that make them biologically active are summarized in Table 9.1. In this connection, it seems natural to find out what effect these two components would produce when combined.

The idea of creating complex drugs on the basis of NSS is not new. For example, experiments with PNSS covered with various medicinal agents, such as amphotericin and highly dispersed medicinal plants, demonstrated[30] that the use of such composite systems with a prolonged action of the drug may decrease its dose and enhance its bioaccessibility, features indicative of a synergistic action of the ingredients. Moreover, composite systems on the basis of fullerene and highly dispersed silica were also used.[31–34] The carrier was a highly porous SCG. It was demonstrated that appreciable amounts of fullerene are adsorbed or retained in the SCG pores. Fullerene-SCG composites (fullerenized SCG, in the terminology of Podosenova et al.[31]) selectively adsorb low-density lipoproteins, a property that makes them effective immunosorbents for treating atherosclerosis. But it remains unknown how C_{60} fullerene is bound to the carrier and how its properties change because of this binding.

Empirically it is known that C_{60} is a poor adsorbate in regards to NSS substrates. Thus, the specific amount of C_{60} fullerene (a hydrophobic substance) adsorbed on unmodified PNSS, a hydrophilic carrier, proved extremely low. When the surface of PNSS was modified, for example, with amines, the amount of fullerene adsorbed increased substantially.[35] The specific amount of fullerene on aerogel was also very low.[36] And only SCG, according to investigations,[31–34] seemed to be promising. Let us look at what is happening on the PNSS surface or inside the SCG pore in the presence of fullerene C_{60}.

9.3.1 C_{60} fullerene–highly dispersed silica composite

NSSs are formed during the condensation or polymerization that accompanies the hydrolysis of silicon tetrahalides, their organic orthoesters, and silicic acid salts.[28] The commercial products manufactured in these ways are known as Aerosil, aerogel, and SCG. A special series of experiments aimed at examining the vibrational spectra of these products (summarized in Sheka et al.[26]) showed that the spectra of the frameworks and surface zones of these three types of silicas differ so radically that they should be considered different structural formations. The finding has led to exhibiting new structural phenomenon called technological polymorphism of NSS. A comprehensive understanding of the motivation leading to the formation of different polymorphs gave rise to a new algorithmic approach to modeling NSSs.[26] Let us briefly consider the main point of the approach. A set of cluster models that simulate different NSS polymorphs is presented in Figure 9.5.

9.3.1.1 Aerosil (PNSS)

Aerosil is prepared by hydrolysis of silicon tetrachloride in an oxygen-hydrogen flame with the subsequent polycondensation of orthosilic acid formed at the first stage. Agglomerated solid-phase nuclei look like virtually ideal particles. The particles are composed of closely packed silicon-oxygen tetrahedra (SOT) with Si-O lengths lying in a narrow interval (Figure 9.5(a)). That the frequencies of bending (Si-O-Si and O-Si-O) and torsional vibrations are small allows the corresponding angles to vary within wide limits, leading to the amorphization of the substance. The most abundant functional group on the surface of the particle is the isolated silanol group. Silanediol groups are located at structural defects of the surface, but their concentration is below 10 to 15%.[37] Later, based on the principles underlying the modeling of structures, Sheka and coworkers[38–40] developed models of the interfaces between PNSS particles and polysiloxane polymers, models that made it possible, in particular, to understand why the polymer becomes stiffer upon being filled with PNSS.

9.3.1.2 Silica gel (SCG)

Silica gel is normally prepared in aqueous solutions of silicates of alkali metals in the presence of an acid.[28] The hydrolysis of a metal silicate in an aqueous medium produces SOT chains of varying length, such that each silicon atom is bonded to two hydroxyl groups (Figure 9.5(b)). At a high concentration, chains close to form cycles composed of different numbers

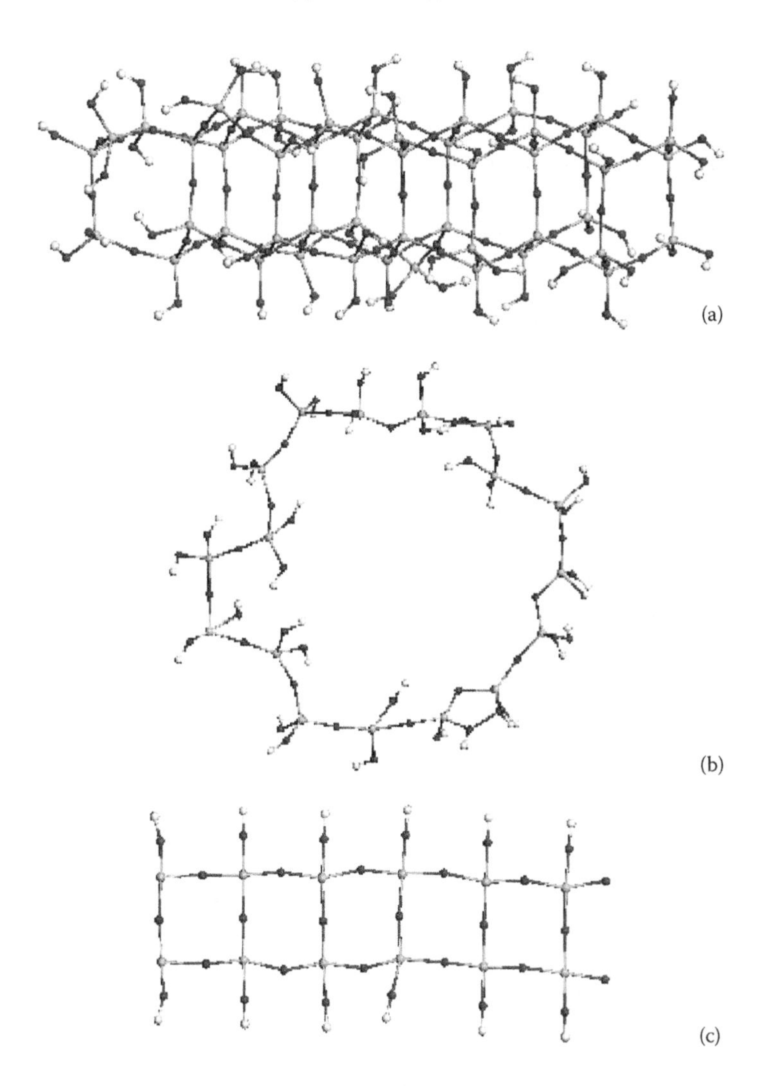

Figure 9.5 Cluster models of nanosized silica: (a) fragment of an Aerosil particle comprised of 48 SOT (Si48), (b) siloxane cycle of silica gel composed of 17 SOT (Si17sg), and (c) polymer chain of aerogel composed of 12SOT (Si12ag). Light gray, dark and white balls mark silicon, oxygen, and hydrogen atoms, respectively.

of silicon atoms. Brought in contact, such cycles form a silica gel pore in the form of a deflated football, with faces of different sizes.

9.3.1.3 Aerogel

The industrial technology for manufacturing this product is the hydrolysis of tetraethylorthosilicate catalyzed by an acid or alkali.[41] The hydrolysis is accompanied by the formation of a silicate polymer in which each atom is bonded to a hydroxyl group (Figure 9.5(c)). Intertwining and bonding to each other, such chains form a gel.

As can be seen in Figure 9.5, distinctions in silicon-oxygen structures give rise to the diversity in the structure of the hydroxyl covering of these products, a feature that manifests itself through vibrational spectra recorded by means of inelastic neutron scattering.[26] NSS models presented in Figure 9.5 allow for examining the interaction of a C_{60} molecule with PNSS, modeled by a Si48 cluster (Figure 9.5(a)), and with SCG, modeled by one or two Si17sg linear cycles. The main focus is on the possibility of formation of *fullerosil* and *fullerosilica gel.*

9.3.2 *Fullerosil*

Given that the hydroxyl covering of the cluster is heterogeneous, let us examine how a fullerene molecule is adsorbed at two areas of its surface with different compositions of hydroxyl groups. In the first case, the molecule position is characterized by the shortest distance, C_f-O_{siln}, between one of its atoms and the oxygen atom of a silanol group. In the second case, the initial distance from C_f to the oxygen atom of a silanediol group, O_{sild}, was determined. The attacking carbon atom was selected among those related to group 1 according to the N_{DA} map. It turned out that the parameters of the equilibrium structures of the complex obtained by full optimization of the initial structures depend on the initial distances C_f-O_{siln} and C_f-O_{sild}. At C_f-O_{siln} ≤ 1.2Å and C_f-O_{sild} ≤ 1.6 Å, the fullerene molecule is bonded to the particle surface in the configurations displayed in Figure 9.6. At larger initial distances, it is not bonded to the particle surface.

As seen in Figure 9.6, in both cases the binding of the C_{60} molecule to the surface occurs via the formation of a Si-C_f bond. The carbon atom substitutes previously bound hydroxyl, which after release is coupled to another carbon atom that is a partner by a short bond to the first one. The final configuration shown in Figure 9.6(b) differs from that shown in Figure 9.6(a) in that the silicon atom has a second hydroxyl attached, a factor that produces a substantial effect on the coupling energy of the C_{60} molecule to the surface, +10.38 and −6.42 kcal/mol for the structures shown in Figures 9.6(a) and 9.6(b), respectively. Since silanediol groups reside predominantly in defective areas of the particle hydroxyl covering, their characteristics vary, which manifests itself as a ±1.32 kcal/mol variation in the energy of coupling the fullerene molecule to the model cluster.

Thus, for the configuration shown in Figure 9.6(a), the attachment of a fullerene molecule to a PNSS particle is an endothermic process, and therefore cannot occur under normal conditions. Exothermic as it is, the formation of the second configuration (Figure 9.6(b)) is feasible. We believe that

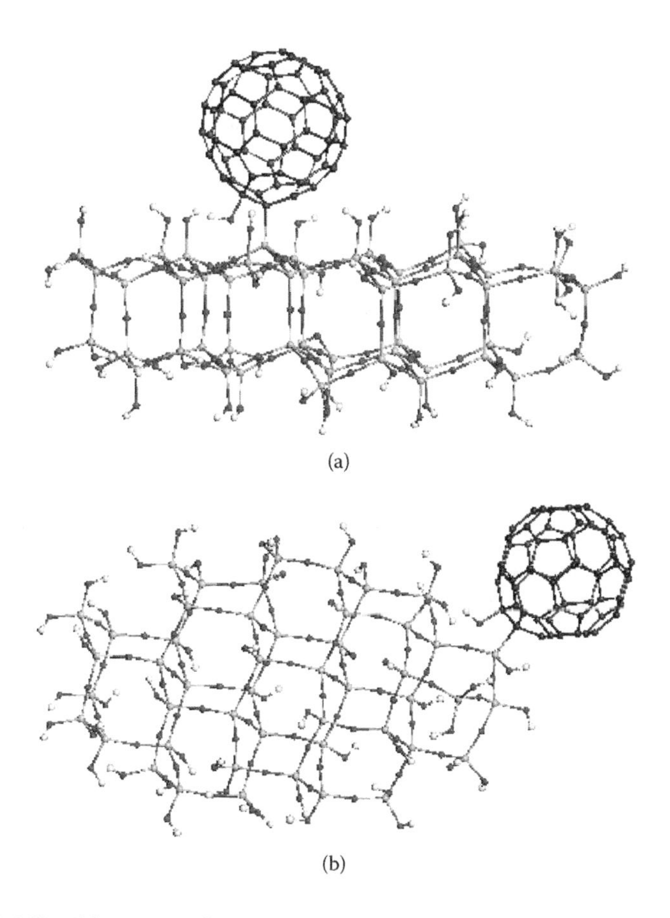

Figure 9.6 Equilibrium configurations of the Si48-C_{60} complex, with the fullerene molecule located near (a) silanol and (b) silanediol groups. UBS HF AM1 singlet state. (From Sheka, E. F., *Russ. J. Phys. Chem.*, 81, 959–66, 2007.)

this process is responsible for the adsorption of small amounts of C_{60} on PNSS observed in Davydov et al.[35] According to these experimental data, the coupling energy of C_{60} with this substrate is on the order of a few kcal/mol. The product formed can be termed *fullerosil*. The sharp dependence of whether the addition occurs or not on the initial distance between the molecule and the particle surface is suggestive of the existence of a substantial barrier. The reaction is accompanied by a 0.42 au charge transfer from the particle to the surface. Owing to the presence of C_{60} molecules, fullerosil exhibits high donor–acceptor properties, with the ionization potential and electron affinity being 9.58 and 2.35 eV, respectively.

The obtained characteristics of fullerosil suggest that it may prove to be a highly efficient medicinal agent. A relatively low energy of binding of the C_{60} molecule to the surface signifies that it can be readily detached in a biological medium from the particle on which it was transported. Thus, the pharmacological activity of either component can reveal itself in the best way. At the same time, limited by the concentration of silanediol groups, the concentration of C_{60} molecules is low, a characteristic that may prove favorable for the possible pharmacological uses of the complex in light of the specificity of the action of medicinal agents administered in extremely small concentrations.[43]

9.3.3 Fullerosilica gel

As discussed above, the surface covering of nanosized pores in silica gels is largely formed by silanediols. Since the presence of silanediols was demonstrated to be the necessary prerequisite for the formation of fullerosil, we assumed that C_{60} fullerene would readily attach to siloxane rings; nevertheless, quantum chemical calculations did not support this assumption. Figure 9.7 shows the initial and equilibrium configurations of the NSS-C_{60} complex imitated by a C_{60} molecule bonded to the Si17sg ring. At all reasonable values of the initial C_f-O_{sild} distances, the C_{60} molecule was ejected out of the ring, irrespective of whether the atoms comprising the Si-O-Si chain were allowed to optimize their positions. This means that if silica gel were composed of individual rings, it could have not retained the C_{60} fullerene.

However, SCG has a porous structure, and Figure 9.8 displays calculation results for an element of an SCG pore modeled by two Si17sg siloxane rings. During the optimization of the geometry of the complex, the structures of the siloxane chains were fixed to reproduce the properties of the actual SCG silica gel framework, while the positions of the hydroxyl groups were optimized. As can be seen in Figure 9.8(a), the fullerene molecule remains inside the pore. No chemical contacts with the pore body via Si-C bonds, as in the PNSS case, are formed. To within 0.0004 au, no charge transfer occurs between the C_{60} molecule and the surrounding siloxane rings. Nevertheless, the coupling energy of this complex is quite noticeable and constitutes –4.13 kcal/mol.

To make the fullerene molecule chemically interact with one of the siloxane rings comprising the pore, the rings were made to approach each other, as shown in Figure 9.8(b). Under these conditions, the fullerene molecule remains inside one of the rings, borrowing a hydroxyl group and a hydrogen atom from its silanediol covering. However, the coupling energy of such a complex has a large positive quantity (41.64 kcal/mol),

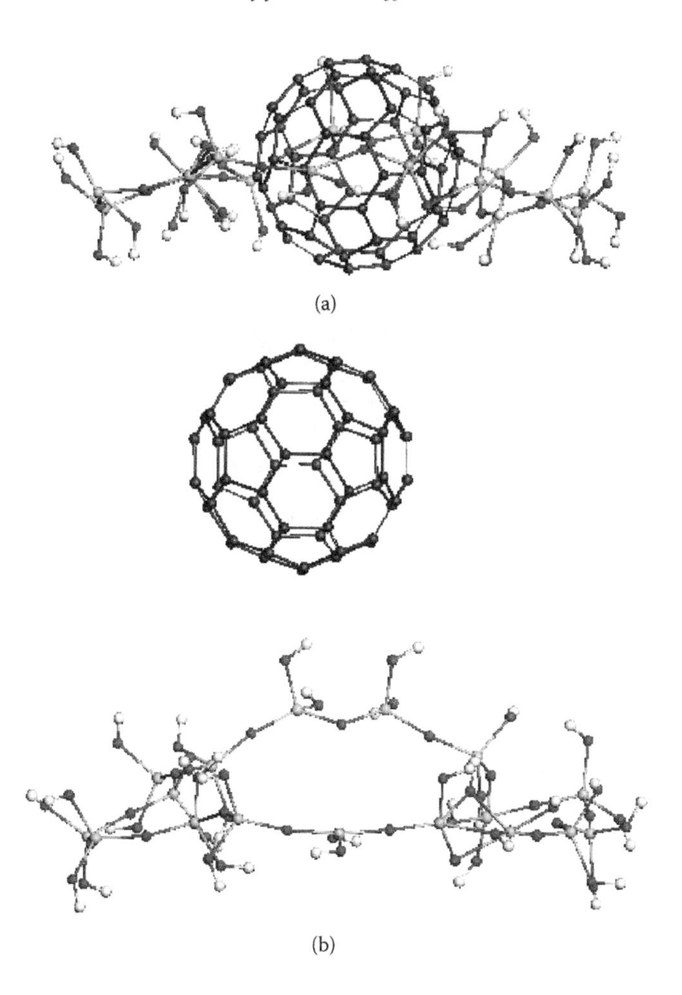

Figure 9.7 (a) Initial and (b) equilibrium configurations of the Si17sg-C_{60} complex. (From Sheka, E. F., *Russ. J. Phys. Chem.*, 81, 959–66, 2007.)

rendering this configuration energetically unfavorable. Therefore, when examining the retention of fullerene molecules in an SCG pore, such ring configurations should be excluded from consideration. Thus, the retention of a C_{60} molecule in an SCG is a result of the balanced forcing out of the molecule from each of the rings comprising the pore without the formation of new chemical bonds. There is good reason to believe that such a situation will be realized in an SCG pore of arbitrary shape. Clearly, in some pores, the resultant force acts so as to eject the molecule, while in others, the molecule is retained, with the coupling energy dependent on the characteristics of the pore.

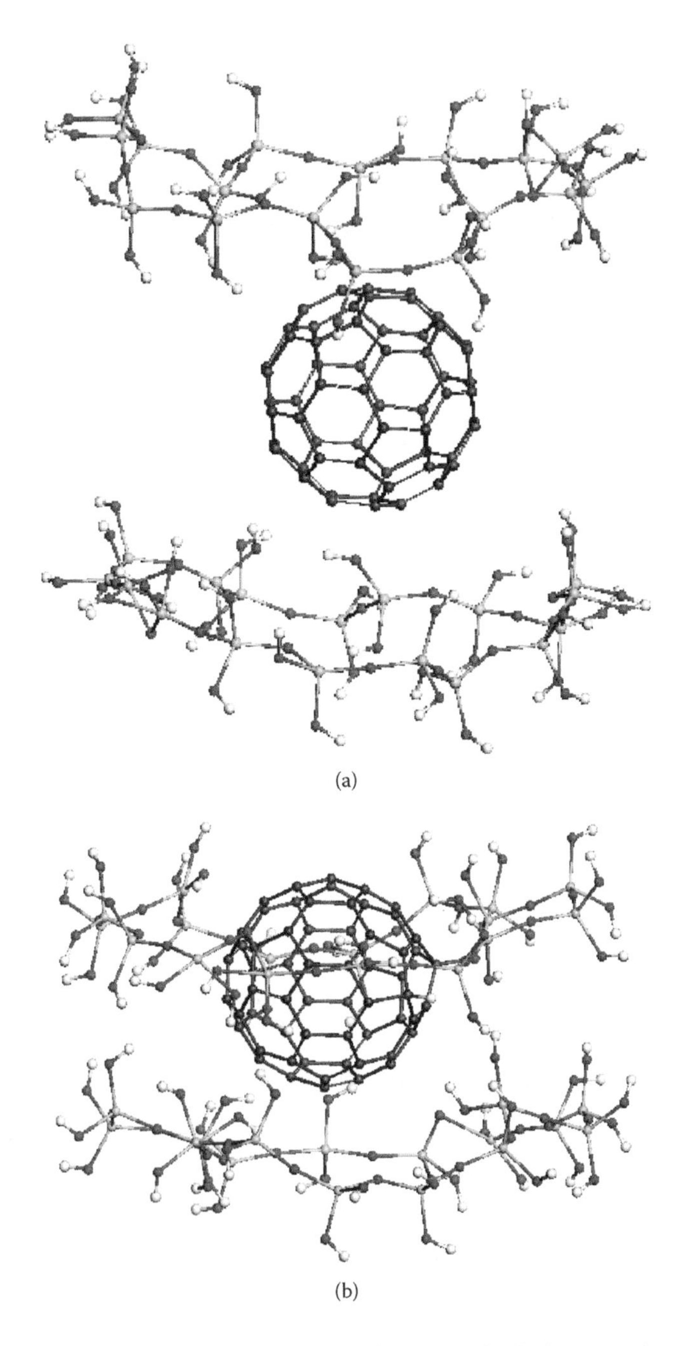

Figure 9.8 Equilibrium configurations of the Si17sg-C_{60}-Si17sg complex at a normal (a) and compressed (b) configuration of siloxane rings. (From Sheka, E. F., *Russ. J. Phys. Chem.*, 81, 959–66, 2007.)

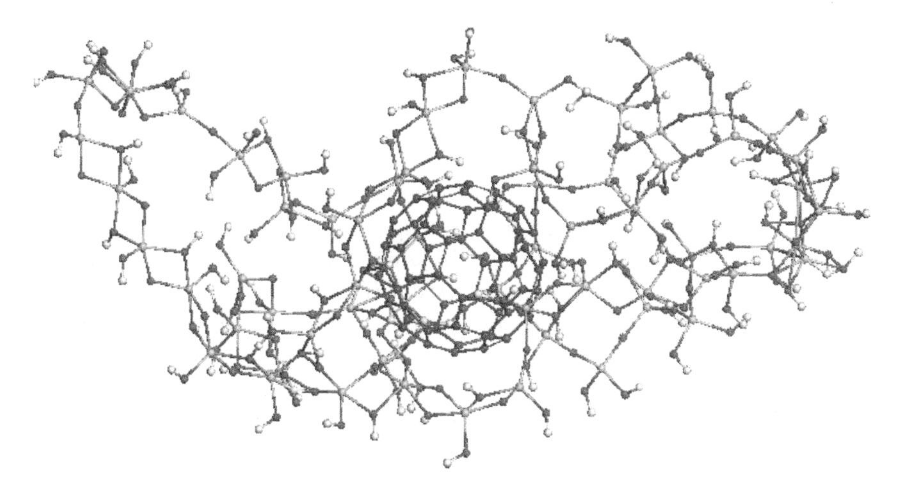

Figure 9.9 A model of fullerosilica gel. The pore in silica gel was built of three cycles: Si34sg, Si28sg, and Si17sg.

Based on these results, a generalized model of fullerosilica gel is presented in Figure 9.9. Let us look at how this construction makes it possible to explain the main experimental observations for fullerized SCG:[31,32]

1. The two-step adsorption isotherms for low-density lipoprotein (LDLP) on fullerosilica gel, in contrast to single-step ones for SCG modified by aromatic molecules, were explained in Podosenova et al.[31,32] by the existence of a 3D adsorption element. This explanation is in full agreement with our concept that a 3D SCG pore with a fullerene molecule retained in it can be considered a single whole.
2. That the adsorption of LDLP increases with the fullerene concentration points to the fullerene molecule being incorporated into the composition of a complex adsorption element in the monomeric form. This finding is fully coherent with the suggested general view on fullerene molecules incorporating inside the SCG pores. Consequently, the number of elements increases with the concentration of molecules introduced, being limited only by the number of pores suitable for accommodating fullerene.
3. That LDLP is adsorbed by fullerosilica gel more effectively than lipoproteins with other structures can also be explained by the spatial structure of the adsorption element. As demonstrated in Iler,[28] the size of linear siloxane cycles only rarely exceeds twenty to thirty units. As can be seen in Figure 9.9, the internal size of a pore composed of cycles comprised of 17, 28, and 34 SOT is commensurate

with the diameter of the fullerene molecule, so that the latter occupies a significant fraction of the pore, a configuration that prevents high-molecular-weight lipoproteins from penetrating into the pore.

4. Another feature favorable for the selective adsorption of LDLP on fullerosilica gel is the donor–acceptor interaction between LDLP and C_{60}. The observed electron exchange adsorption of LDLP[32,33] is a direct result of this interaction, in which LDLP and C_{60} act as a donor and an acceptor, respectively. According to calculations, the high donor–acceptor characteristics of the C_{60} molecule experience virtually no change upon its inclusion into the composition of fullerosilica gel. For example, the ionization potential and electron affinity for the adsorption element shown in Figure 9.8(a) were found to be 9.60 and 2.41 eV, respectively, compared to 9.86 and 2.66 eV for the free molecule (see Table 3.2). It is its high electron affinity that makes the fullerene molecule so effective in donor–acceptor interactions with both LDLP and simple amines.

9.4 *Concluding remarks on the nature of the biological activity of fullerene*

To imagine the character of the medicinal efficiency of drugs based on NSS-fullerene compositions, let us come back to the basics of the chemical activity of fullerenes. It is known that reactive oxygen-containing species, such as singlet oxygen ($^1O_2$) and superoxide ($O_2^{\bullet-}$), hydroxy (HO•), and hydroperoxy (HOO•) radicals, play an important role in regulating a predominant majority of biological processes in a living body. Normal or pathological conditions of the vital activity of biosystems are characterized by the corresponding levels of these species.[44] As other vitally important human being characteristics, such as temperature, blood pressure, glucose level in blood, and so forth, a normal level of the reactive oxygen-containing species must be kept within a rather narrow interval. Thus, many pathological conditions are associated with an anomalously high level of overoxidation of biomolecules,[44] especially lipids in cellular membranes. At the same time, it is known that a decrease in the overoxidation level is accompanied by the attenuation of inflammatory processes. That is why the antioxidant therapy is the most effective if it can support a normal overoxidation level and thus treat a wide spectrum of pathological conditions.

To analyze the therapeutic activity of compositions based on NSS-fullerene, it is necessary to compare their characteristics at the atomic level with those related to fullerene-based drugs experienced in practice. Molecular-colloid solutions containing hydrated C_{60} molecules (C_{60}HyFn = C_{60} {H_2O}n)[23,45] seem to be a proper analog. A wide spectrum of positive therapeutic effects of C_{60}HyFn administered in small doses, sometimes comparable with homeopathic ones, led the authors of Andrievsky et al.[46] to assume that "the C_{60}HyFn show 'wise' and long-term anti-oxidative activity, maybe due to the universal mechanism of the level regulation of free radicals (FR) in aqueous medium that is determined by properties of ordered water structures."

The above peculiarities of the antioxidant action of C_{60}HyFn are undoubtedly associated with the C_{60} molecule being virtually free within the hydrate complex. Fullerosil and fullerosilica gel are expected to exhibit a similar antioxidant activity due to the weak binding between the fullerene molecules and solid carrier. As to the specific and "wise" action of C_{60} fullerene, it is clearly associated with the radical type properties of the C_{60} molecule, so that the mechanism of the scavenging action of the C_{60} molecule and its hydrate complex is obvious. Given that each C_{60} molecule is capable of trapping tens of radicals, it becomes self-evident why its antioxidant ability increases with the concentration of reactive oxygen-containing species.

What remains unclear is the regulatory function of the fullerene. In the case of C_{60}HyFn hydrates, the feature is connected with a particular role of the ordered water structure that influences a recombination of free radicals. As for NSS compositions, some clues as to how this function is realized can be obtained by examining the interaction of a fullerene molecule with a PNSS particle. As can be seen in Figure 9.6, the fullerene molecule tears the hydroxyl group away from the surface silicon atom while interacting with either a silanol or a silanediol group. Note, however, that the energies of these reactions differ significantly, even in sign. Thus, while in the former case it is energetically more favorable for the fullerene molecule to return the hydroxy group back to the surface, in the latter one it is more profitable to retain it. Since the electronic characteristics of the radical on the surface (the charges on the atoms, bond lengths, and valence indices) are similar in both cases, the distinctions in the character of the intermolecular interaction are probably associated with a cooperative effect, the characteristics of which are determined by the configuration of the atoms surrounding the attacked hydroxyl radical. In our opinion, it does not seem far-fetched to assume that similar cooperative effects in a biological medium will accompany the absorptive and regulatory functions of a fullerene molecule as an antioxidant.

References

1. Piotrovski, L. B. 2006. Biological activity of pristine fullerene C60. In *Carbon nanotechnology*, ed. L. Dai, 235–53. Amsterdam: Elsevier.
2. Piotrovski, L. B., and Kiselev, O. I. 2006. *Fullerenes in biology* [in Russian]. St. Petersburg: Rostok.

3. Da Ros, T. 2008. Twenty years of promises: Fullerene in medicinal chemistry. In *Medicinal chemistry and pharmacological potential of fullerenes and carbon nanotubes*, ed. F. Cataldo and T. Da Ros, 1–21. Berlin: Springer.
4. Jensen, A. W., Wilson, S. R., and Schuster, D. I. 1996. Biological application of fullerenes. *Biorg. Med. Chem.* 4:767–79.
5. Bianco, A., Da Ros, T., Prato, M., and Toniolo, C. 2001. Fullerene-based amino acids and peptides. *J. Pept. Sci.* 7:208–19.
6. Piotrovski, L. B., and Kiselev, O. I. 2004. Fullerenes and viruses. *Fullerenes Nanotubes Carbon Nanostr.* 12:397–403.
7. Maeda-Mamiya, R., et al. 2010. *Proc. Natl. Acad. Sci.* DOI: 107:5339–44.
8. Krusic, P. J., Wasserman, P. N., Keiser, P. N., Morton, J. R., and Preston, K. F. 1991. Radical reaction of C_{60}. *Science* 254:1183–85.
9. Morton, J. R., Negri, F., and Preston, K. F. 1998. Addition of free radicals to C_{60}. *Acc. Chem. Res.* 31:61–69.
10. Wang, I. C., Tai, L. A., Lee, D. D., Kanakamma, P. P., Shen, C. K.-F., Luh, T.-Y., Cheng, C. H., and Hwang, K. C. 1999. C_{60} and water-soluble fullerene derivatives as antioxidant against radical initiated lipid peroxidation. *J. Med. Chem.* 42:4614–20.
11. Chabri, N., Pressac, M., Hadchouel, M., Szwarc, H., Wilson, S. R., and Moussa, F. 2005. [60] Fullerene is an *in vivo* powerful antioxidant with no acute or subacute toxicity. *Nano Lett.* 5:2578–85.
12. Kasermann, F., and Kempf, C. 1997. Photodynamic inactivation of enveloped viruses by buckminsterfullerene. *Antiviral Res.* 34:65–70.
13. Kasermann, F., and Kempf, C. 1998. Buckminsterfullerene and photodynamic inactivation of viruses. *Rev. Med. Virol.* 8:143–51.
14. Mroz, P., Tegos, G. P., Gali, H., Wharton, T., Sarna, T., and Hamblin, M. R. 2008. Fullerenes as photosensitizers in photodynamic therapy. In *Medicinal chemistry and pharmacological potential of fullerenes and carbon nanotubes*, ed. F. Cataldo and T. Da Ros, 79–106. Berlin: Springer.
15. Beck, M. T., and Mandi, G. 1997. Solubility of C_{60}. *Full. Sci. Technol.* 5:291–310.
16. Nath, S., Pal, H., Palit, D. K., Sapre, A. V., and Mitta, J. P. 1998. Aggregation of fullerene, C_{60}, in benzonitrile. *J. Phys. Chem. B* 102:10158–64.
17. Samal, S., and Geckeler, K. E. 2001. Unexpected solute aggregation in water on dilution. *J. Chem. Soc. Chem. Commun.* 2224–25.
18. Sheka, E. F. 2010. Spin flip in oxygen molecule under photoexcitation of photodynamic fullerene solutions. arXiv.1005.2383v1. [cond-mat.mes-hall].
19. Da Ros, T., Spalluto, G., and Prato, M. 2001. Biological application of fullerene derivatives: A brief overview. *Croatica Chem. Acta* 74:743–55.
20. Richmond, R. C., and Gibson, U. J. 1994. Fullerene coated surfaces and uses thereof. U.S. Patent 5,310,669, May 10.
21. Moussa, F., Chretien, P., Dubois, P., Chuniaud, L., Dessante, M., Trivin, F., Sizaret, P. Y., Agafonov, V., Ceolin, R., Szwarc, H., Greugny, V., Fabre, C., and Rassat, A. 1995. The influence of C_{60} powders on cultured human leukocytes. *Full. Sci. Technol.* 3:333–42.
22. Lyon, D. Y., Adams, L. K., Folkner, J. C., and Alvarez, P. J. J. 2006. Antibacterial activity of fullerene water suspensions: Effects of preparation method and particle size. *Environ. Sci. Technol.* 40:4360–66.

23. Andrievsky, G. V., Kosevich, M. V., Vovk, O. M., Shelkovsky, V. S., and Vashchenko, L. A. 1995. On the production of an aqueous colloidal solution of fullerene. *J. Chem. Soc. Chem. Commun.* 1281–82.
24. Andersson, T., Nilsson, K., Sundahl, M., Westman, G., and Wenerstrom, O. 1992. C_{60} embedded in g-cyclodextrin: A water soluble fullerene. *J. Chem. Soc. Chem. Commun.* 604–5.
25. Yevlampieva, N. P., Biryulin, Yu. F., Melenevskaya, E. Yu., Zgonnik, V. N., and Rjumtsev, E. I. 2002. Aggregation of fullerene C_{60} in *N*-methylpyrrolidone. *Colloid Surf. A* 209:167–71.
26. Sheka, E. F., Khavryuchenko, V. D., and Markichev, I. V. 1995. Technological polymorphism of disperse silicas: Inelastic neutron scattering and computer modelling. *Russ. Chem. Rev.* 64:389–414.
27. Chuiko, A. A., ed. 2003. *Medicinal chemistry and clinical application of silicon dioxide* [in Russian]. Kiev: Naukova dumka.
28. Iler, R. K. 1979. *The chemistry of silica: Solubility, polymerization, colloid and surface properties, and biochemistry of silica.* New York: Interscience.
29. Chuiko, A. A., ed. 2003. *SILICS: A bioregulating enterosorbent. Properties and clinical applications* [in Ukrainian]. 2003. Kiev: Biofarma.
30. Chuiko, A. A., and Pentyuk, A. A. 1998. Composites of silica with drugs [in Ukrainian]. In *Proceedings of Scientific Session of the Chemical Department of the National Academy of Science of Ukraine*, 38–39. Kharkov: Osnova.
31. Podosenova, N. G., Sedov, V. M., Andozhskaya, Yu. S., and Kuznetsov, A. S. 1997. Adsorption of low density lipoproteins. *Russ. J. Phys. Chem.* 71:1315–18.
32. Podosenova, N. G., Sedov, V. M., Kuznetsov, A. S., and Knyazev, A. S. 1999. New sorbents for electron-exchange adsorption of low-density lipoproteins. *Russ. J. Phys. Chem.* 73:97–100.
33. Podosenova, N. G., Sedov, V. M., Sharonova, L. V., and Drichko, N. V. 2001. Fullerene effects on the adsorption properties of silica gel with respect to low-density lipoproteins. *Russ. J. Phys. Chem.* 75:1871–75.
34. Sedov, V. M., Podosenova, N. G., and Kuznetsov, A. S. 2002. Oxidation of low-density lipoproteins in the presence of a fullerene-containing silica gel. *Kinet. Catal.* 43:56–60.
35. Davydov, V. Y., Sheppard, N., and Osawa, E. 2002. An infrared spectroscopic study of the hydrogenation and dehydrogenation of the complexes of aromatic compounds and of fullerene C_{60} with silica-supported platinum. *J. Catal.* 211:42–52.
36. Piwonski, I., Zajac, J., Jones, D. J., Rozière, J., Partyka, S., and Plaza, S. 2000. Adsorption of [60]fullerene from toluene solutions on MCM-41 silica: A flow microcalorimetric study. *Langmuir* 16:9488–92.
37. Brei, V. V. 1993. Hydroxils accommodation on silica surface [in Russian]. In *Khimia, fizika i tekhnologia poverkhnosti* [Chemistry, Physics, and Technology of Surface], 75–83. Kiev: Naukova dumka.
38. Sheka, E., Khavryutchenko, V., and Nikitina, E. 1999. From molecule to particle. Quantum-chemical view applied to fumed silica. *J. Nanopart. Res.* 1:71–81.
39. Sheka, E. F. 2003. Intermolecular interaction and vibrational spectra at fumed silica particles/silicone polymer interface. *J. Nanopart. Res.* 5:419–37.

40. Nikitina, E. A., Khavryutchenko, V. D., Sheka, E. F., Barthel, H., and Weis, J. 1999. Deformation of poly(dimethylsiloxane) oligomers under uniaxial tension. Quantum-chemical view. *J. Phys. Chem. A* 103:11355–65.
41. Brinker, C. J., Keefer, K. D., Schaefer, D. W., and Ashley, C. S. 1982. Sol-gel transition in simple silicates. *J. Non-Cryst. Solids* 48:47–64.
42. Sheka, E. F. 2007. Fullerene–silica complexes for medical chemistry. *Russ. J. Phys. Chem.* 81:959–66.
43. Burlakova, E. B. 2000. Specific effects of superlow doses of biologically active substances and low-level physical factors. *Misaha Newsletter*, July–December, pp. 2–10, 30–31.
44. Wang, I. C., Tai, L. A., Lee, D. D., Kanakamma P. P., Shen C. K.-F., Luh T.-Y., Cheng, C. H., and Hwang, K. C. 1999. C_{60} and water-soluble fullerene derivatives as antioxidants against radical-initiated lipid peroxidation. *J. Med. Chem.* 42:4614–20.
45. Andrievsky, G., Klochkov, V., Bordyuh, A., and Dovbeshko, G. 2002. Comparative analysis of two aqueous-colloidal solutions of C_{60} fullerene with help of Ft-IR reflectance and UV-Vis spectroscopy. *Chem. Phys. Lett.* 364:8–17.
46. Andrievsky, G., Klochkov, V., and Derevyanchenko, L. 2005. Is C_{60} fullerene molecule toxic? *Fullerenes Nanotubes Carbon Nanostr.* 13:363–76.
47. Andrievsky, G. V., Bruskov, V. I., Tykhomyrov, A. A., and Gudkov, S. V. 2009. Peculiarities of the antioxidant and radioprotective effects of hydrated C_{60} fullerene nanostuctures *in vitro* and *in vivo*. *Free Rad. Biol. Med.* 47:786–93.

chapter ten

Nanophotonics of fullerenes

10.1 Introduction

Nanophotonics of fullerenes nowadays is associated with changes in properties of nonlinear optical materials low doped with fullerenes and the ensuing changes in characteristics of nonlinear optical devices.[1] Since enhanced nonlinear optical effects are obviously of an electromagnetic nature and can be directly controlled by electric field, these effects must involve a contribution of charge-separated states. The existence of these states was attributed to the formation of charge transfer complexes resulting from the donor–acceptor (D–A) interaction between fullerenes and host matrices.[1] Further optical studies provided direct evidence of charge transfer complex formation in materials where fullerenes act as both electron donors and acceptors (e.g., see Kamanina et al.[2,3] and Hosoda et al.[4]). However, even though there is no doubt that the D–A interaction is responsible for the enhanced nonlinear optical properties of fullerene-doped materials, the microscopic mechanism of this effect has remained unclear for a long time. And only recently have sophisticated spectral studies of the fullerene solutions allowed for proving this basic point and disclosing its role in the enhancement of nonlinear effects.[5–8] As shown, the key role is played by an inhomogeneous structure specific to fullerene-doped nonlinear optical materials, where D–A pairing leads to nanoclustering of fullerenes and their clustering with host compounds. The clustering manifests itself by visible emission (hereinafter called blue emission) observed in addition to the red photoluminescence of fullerenes. The nature of the emission differs between different matrices, while its intensity provides a direct measure of the effect of fullerene doping on nonlinear optical properties and can serve as a qualitative criterion for selecting efficient nonlinear optical materials. Let us consider the grounds of this effect.

10.2 Schematic characterization of the excited states and optical spectra of fullerenes in solution

A fullerene-doped nonlinear optical material is a dilute solution of a fullerene in a liquid, a liquid crystal, or a solid matrix. We rule out any direct chemical interaction between the fullerene and the matrix and restrict analysis to solutions in inert molecular matrices. In what follows, *fullerene* means either C_{60} or its derivatives. The energy spectrum of impurity states in a dilute molecular solution is generally represented as the single-molecule spectrum of the impurity embedded in the band gap of the host material, with a weak shift and a broadening due to interactions both between impurity molecules and between impurity and matrix molecules. This representation is suggested by the empirical observation that most impurities have molecular type spectra. However, the spectra of fullerenes in solution (for example, their absorption spectra in toluene, xylene, hexane, and other conventional solvents) are much more complex than a mere combination of the single-molecule spectrum of a fullerene and the spectrum of the solvent. To sort out these spectra, we first recall the general structure of the excitation spectrum of a doped molecular crystal.

As follows from the solid-state theory, the excitation spectrum of a doped molecular crystal is represented as a combination of dissociated (two-particle) and bound (single-particle) states of an electron-hole pair.[9] The bound states are split off from free electron-hole states when the electron-hole binding energy reaches a critical value (see Broude et al.[9] for the principles underlying the structure of such bound states). An electron and a hole localized on molecules of the host material make up a bound excitonic state. Analogous localization on impurity molecules leads to the formation of bound impurity states. Bound states of both kinds may differ in terms of electron-hole pair localization. The bound state of an electron and a hole localized on the same molecule corresponds to Frenkel excitons in the host crystal or an impurity state in the oriented-gas model of a molecular crystal. When an electron and a hole are localized on different sites, the energy spectrum of the crystal combines charge transfer excitonic states (CT excitons) and impurity local CT states. Photoexcitation of states of the latter kind always involves the transfer of an electron from one molecule to another. Thus, two branches corresponding to nonpolar and polar excitations, respectively, appear in the spectrum.

Two excitation branches are characteristic of any molecular crystal. One essential distinction of the excitonic states in crystalline fullerenes or fullerenes in solution from excited states in other molecular materials is a significant overlap between the nonpolar and polar branches.[10] The excitation spectra of a fullerene crystal and a fullerene embedded in a crystalline matrix are schematized in Figure 10.1. The crystal spectrum is represented by Frenkel excitons (nonpolar branch) and CT excitons (polar branch). The lowest-lying singlet and triplet bands are shown for completeness. It should be recalled here that dissociated electron-hole states of different multiplicities generally have different energies. The energies of Frenkel excitons are determined by the excitation spectrum of the fullerene molecule. The Frenkel exciton bandwidths are determined by the intermolecular

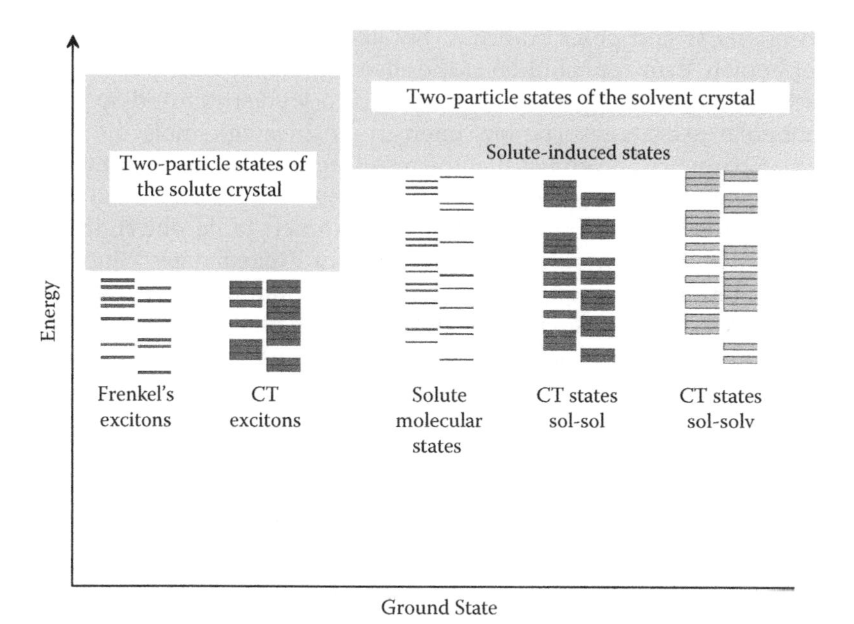

Figure 10.1 Scheme of singlet and triplet excited states of crystalline fullerene (left) and fullerene in a crystalline matrix (right). (From Sheka, E. F., et al., *J. Exp. Theor. Phys.*, 108, 738–50, 2009.)

resonance interaction in the crystal, which is known to be weak.[11] The energies of the CT excitons depend on the electron-hole pair localization in the lattice, and the corresponding bandwidths are determined by both the resonance interaction and the polarization energy, which depends on the relative electron-hole position. Because of the latter contribution, the CT exciton bands are wider than the Frenkel exciton bands.[12]

The excitation spectrum of a fullerene in a crystalline matrix is schematized, as an impurity spectrum, in the right-hand part of Figure 10.1. The Frenkel excitons are transformed into molecular impurity states split off from the two-particle electron-hole continuum of the matrix. The CT excitons are replaced by two branches of impurity local CT states, with electron-hole pair localization on solute molecules (*sol–sol* branch) and on solute and solvent molecules (*sol–solv* branch), respectively. Due to a wide distribution over electron-hole distances, the corresponding energy levels are broadened.

Since the nonpolar and polar branches of fullerene excitation spectra are close in energy, fullerenes in solution have very complex optical spectra. Moreover, their absorption and emission spectra are complex in different ways. As follows from the energy spectrum in Figure 10.1, the absorption spectrum of a solution is the sum of transitions to states of both nonpolar and polar branches. Because of broadening states in the latter branch, their contribution to the absorption spectrum is manifested as a broad-band background (pedestal) for a well-structured spectrum of nonpolar excitations. The low intensity of the single-molecule excitation spectrum of C_{60}[13] implies that the resulting spectrum is dominated by polar excitations, as observed experimentally.[14]

The emission spectra of solutions are known to be determined by the radiative and nonradiative decay times of excited states. Since nonpolar excitations decay quickly, the emission is dominated by transitions from the lowest-lying singlet or triplet states. The corresponding well-structured molecular fluorescence and phosphorescence spectra lie in the red and near-infrared regions (red spectrum below). Emission from polar excited states has been discussed in a recent study,[14] and it was suggested that the single-molecule fluorescence and phosphorescence spectra of fullerene sit on top of a bell-shaped background attributed to radiative decay of the lowest-lying polar excited states.

10.3 Electromagnetic theory of enhanced optical effects

Our first goal is to find characteristic features in absorption and emission spectra of fullerenes in solutions that might be indicative of the efficacy of nonlinear optical properties. The fact that the spectral and nonlinear optical properties are interconnected follows from the electromagnetic theory of enhanced optical effects (local field enhancement model).[15] The theory predicts enhancement of both linear and nonlinear optical effects and describes the enhancement of incident and outgoing light in terms of the local field acting on the polarized object, which may be much stronger than the incident field. These effects are generally described in the dipole approximation by representing the microscopic induced dipole moment, p, as a series in the local field strength $E_{loc}(\omega)$:

$$p = \alpha E_{loc} + \beta E_{loc} E_{loc} + \gamma E_{loc} E_{loc} E_{loc} + \ldots. \tag{10.1}$$

The first-, second-, and third-order polarizabilities, α, β, and γ, describe different optical effects. In particular, the linear polarizability, α, is responsible for refraction, one-photon absorption and luminescence, and spontaneous Raman scattering. (The last phenomenon is caused by α modulations of the form

$$\alpha = \alpha_0 + \left(\frac{\partial \alpha}{\partial Q}\right) Q$$

where Q is displacement along a normal vibrational coordinate of the object.) The hyperpolarizability β controls the second harmonic generation, sum frequency emission, and optical rectification processes. The hyperpolarizability γ determines third harmonic generation, four-wave mixing, optical Kerr effect, two-photon absorption and luminescence, and stimulated Raman scattering. The macroscopic optical polarization, P, of the object is calculated by taking into account all induced dipoles and collective interactions as follows:

$$\begin{aligned} P(\omega) &= L'(\omega)\chi^{(1)}L(\omega)E(\omega) \\ &+L'(\omega)\chi^{(2)}L(\omega_1)L(\omega_2)E(\omega_1)E(\omega_2) \\ &+L'(\omega)\chi^{(3)}L(\omega_1)L(\omega_2)L(\omega_3)E(\omega_1)E(\omega_2)E(\omega_3)+ \end{aligned} \tag{10.2}$$

The local factors L' and L are introduced to describe the enhancement of the incident and outgoing fields E, respectively, and $\chi^{(i)}$ is the ith-order optical susceptibility (see Heritage and Glass[15] and Meliu[16] for details). Note that all three susceptibilities contribute to the total polarization. The susceptibilities $\chi^{(i)}$ result from summing microscopic polarizations in the interaction region. In the electromagnetic model, the local factors L' and L quantify the relationship between linear and nonlinear optical properties of the polarized object. Thus, an analysis of enhanced linear optical effects can throw light on the microscopic mechanisms of nonlinear effects, and vice versa.

One should expect enhanced spectral effects of three types to be observed: (1) one-photon absorption/luminescence and spontaneous Raman scattering determined by $\chi^{(1)}$, (2) second harmonic generation due to $\chi^{(2)}$, and (3) stimulated Raman scattering controlled by $\chi^{(3)}$. Let us focus on first-order effects. Among these, the most widely known are the giant Raman scattering by molecules adsorbed on metal particles or micro- and nanostructured rough metal surfaces (surface-enhanced Raman scattering (SERS))[17,18] and on the tip of a tunneling microscope (tip-enhanced Raman scattering (TERS)).[19] These phenomena occur because incident and outgoing light waves having similar frequencies are enhanced simultaneously by the polarization effects, resulting from the excitation of an electron-hole continuum on a metal particle, a tip, or a rough surface. Absorption or luminescence intensity can be enhanced only when the corresponding spectral region matches that of an electron-hole continuum.[15,20] An additional condition for luminescence enhancement is the existence of sufficiently strong molecular transitions leading to luminescence. Observation of enhanced absorption is a technically difficult task.

The local factors L and L' depend on the size and shape of the polarized object, the orientations and positions of molecules in the object, and a variety of other structural characteristics. These factors can hardly be determined exactly. However, estimates show that these factors depend resonantly on the incident frequency in the case of an ellipsoidal nanosize object.[21] The resonance condition is

$$\mathrm{Re}\left[\varepsilon(\omega_{res})Q_1(\xi_0)-\xi_0 Q_1'(\xi_0)\right]=0 \tag{10.3}$$

where ξ_0 is a shape-dependent parameter and Q_1 and Q'_1 are the Legendre function of the second kind and its derivative, respectively. For a spherical particle, this condition reduces to

$$\mathrm{Re}\left[\varepsilon(\omega_{res})\right]+2=0 \tag{10.4}$$

and can be used to obtain an order of magnitude estimate for the real part of the dielectric function, which determines resonance frequencies. At resonance, the enhancement of linear effects depends on the imaginary part of the dielectric function of the object as $[\mathrm{Im}\varepsilon(\omega_{res})]^{-4}$ [15]; i.e., a small imaginary part makes it easier to observe enhanced linear effects, as is the case with a silver substrate that is used in the majority of SERS experiments.

To apply the local field enhancement model to optical characteristics of fullerenes in solution, one must (1) identify the polarized object in the system, (2) find the solution component responsible for the excitation of the electron-hole plasma, (3) determine the spectral region of plasma resonance, and (4) verify the resonance conditions for local factors.

As concerns the first prerequisite, the first observations of enhanced Raman spectra of solute molecules and enhanced luminescence of solvents in solutions containing C_{60} and its derivatives in crystalline toluene and carbon tetrachloride matrices[5] made it possible to relate these phenomena to fullerene clustering in solutions. Thus, we are facing the clustering problem in fullerene solutions for the second time. The first discussion concerning photosensitive clusters of fullerene molecules was related to photodynamic therapy of fullerenes in Section 9.2. As shown in Section 8.2, the clustering between C_{60} molecules is caused by the presence of the R^{00} minimum at the two-well intermolecular interaction (IMI) potential shown in Figure 8.1. As for C_{60} derivatives, their clustering will occur if their IMI potentials are of type 1. Since it is not a general case (see Chapter 7), the type of the IMI potential related to a particular derivative should be checked. As for the pristine C_{60}, besides observations mentioned in the previous chapter, its clustering has been revealed in toluene, benzene, carbon disulfide, and carbon tetrachloride solutions.[22–26] According to small-angle neutron scattering,[24] the cluster size in toluene constitutes ~30 Å.

The second and third prerequisites are met by the energy spectra schematized in Figure 10.1. If *sol-sol* and *sol-solv* cluster formation is possible, then the cluster excitation spectrum must involve local CT excitons, whose key role in the generation of electron-hole plasma is analogous to that played by the surface plasmons excited in metal particles. The increase in resonance enhancement observed experimentally as the excitation frequency is tuned through the CT exciton band spectrum can be interpreted as direct evidence of the role played by the CT excitons localized on fullerene clusters in the enhancement.

The arguments in support of the electromagnetic nature of field enhancement summarized in items 1 to 3 above concern necessary conditions for the observation of enhanced spectral effects. However, they are not sufficient unless the resonance conditions for local factors that follow from Equation 10.3 are fulfilled. According to a simplified version of this relation in Equation 10.4, the near-resonance conditions hold for the local factors L and L' when the dielectric function satisfies the condition $\mathrm{Re}[\varepsilon(\omega_{res})] \approx -2$. For most molecular substances, $\mathrm{Re}[\varepsilon(\omega)]$ is always positive and its value is high. In contrast, $\mathrm{Re}[\varepsilon(\omega)]$ in the pristine C_{60} crystal can vary from +3 to –5 in the λ_{ex} interval of interest, 2.4 to 2.6 eV, as seen from the data in Figure 10.2.[27] It is hoped that this exceptional fact for molecular compounds will promote the fulfillment of resonance or at least close-to-resonance conditions for not only the C_{60} crystal, but also solutions of C_{60} and its derivatives.

If assume that nanosized fullerene clusters in solutions can act as light wave enhancers, like gold or silver colloid particles, it is possible to evaluate a possible attainable magnitude of enhancement K. It is known[17–20] that K for silver colloid particles is 10^4 to 10^6. In terms of the local field theory, enhancement of a light wave is due not only to the fulfillment of the resonance conditions for $\mathrm{Re}[\varepsilon(\omega)]$ discussed above. As mentioned earlier, the spectrum intensity is controlled by the dielectric loss factor of the object, causing its dependence on $\mathrm{Im}[\varepsilon(\omega_{res})]^{-4}$.[15] The difference determined by these two circumstances for fullerene clusters and silver particles[28] can be revealed by comparative analysis of the dielectric losses of the C_{60} fullerene crystal and silver shown in Figure 10.2. As seen in the figure, $\mathrm{Re}[\varepsilon(\omega)]$ values coincide for both enhancers in the visible light region, while the values of $\mathrm{Im}[\varepsilon(\omega)]$ differ by a factor of about 15 in favor of silver particles; that is, there is a ~10^4-fold difference in amplification. According to these data, the maximal enhancement attainable with the use of fullerene clusters cannot be higher than 10^2.

Even though the first experiment[5] has already shown that fullerene clusters in crystalline toluene and carbon tetrachloride matrices enhance emission spectra, playing the role of a colloidal metal sol, some major issues remained unresolved. First, it was not clear whether the observed

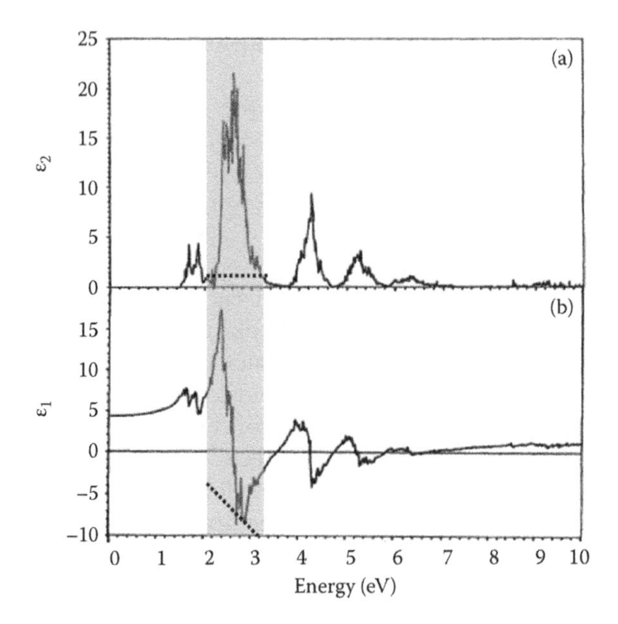

Figure 10.2 Dielectric loss spectrum of the C_{60} crystal. The shaded area corresponds to the visible region of the spectrum. The dashed segments show fragments of the dielectric loss spectrum of silver. (Modified from Ching, W. Y., et al., *Phys. Rev. Lett.*, 67, 2045–48, 1991; Johnson, P. B., and Christy, R. W., *Phys. Rev. B*, 6, 4370–79, 1972.)

manifestation of the spectrum enhancement was unique and universal for linear optical properties of solutions. Second, it was not obvious how much computations could be useful to find characteristics of fullerene clusters that would quantify the enhanced nonlinear optical response of solutions.

Experiments and computations performed for solutions of pristine C_{60} and three its derivatives in crystalline matrices of toluene and tetrachloride[6–8] gave answers to these questions, which we are going to consider below.

10.4 Absorption and emission spectra of fullerenes in solution

Experimental studies were performed on solutions involving pristine C_{60}, N-methyl-2(4-pyridine)-3,4-[C_{60}]fulleropyrrolidine (I), fullerene azyridine [C_{60}] (II), and ethyl ester of [C_{60}]fullerene acetic acid (III) (see Sections 3.2.4 and 3.2.5) as solutes. The equilibrium structures of the molecules are presented in Figure 3.3. Hereinafter, these roman numbers and C_{60} notation denote the corresponding solutions. Figure 10.3 presents raw absorption spectra of the toluene solutions. The spectra consist of an ill-structured (polar) background and a well-structured (nonpolar) component.

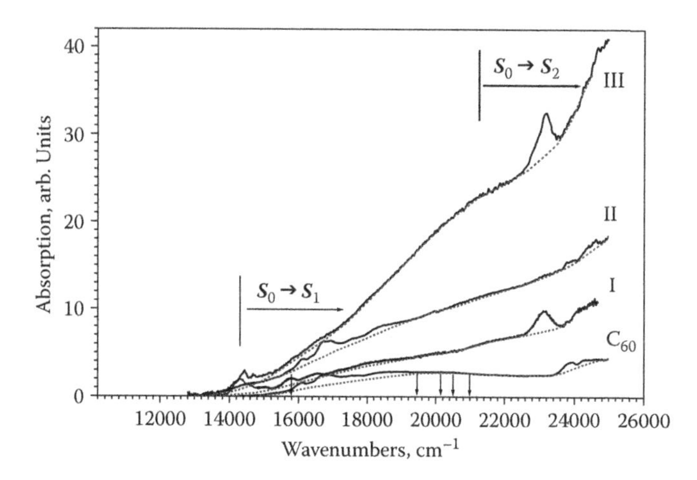

Figure 10.3 Absorption spectra of C_{60} and fullerenes I, II, and III (see structures in Figure 3.3) in toluene at 80 K. Dotted curves represent polar background components. Arrows indicate excitation laser frequencies used in experiments. (From Sheka, E. F., et al., *J. Exp. Theor. Phys.*, 108, 738–50, 2009.)

Extraction of this background from the raw spectra made it possible to get background-free molecular spectra, shown in Figures 3.4 and Fig.3.5. As discussed in Section 3.2.5, the lowest-lying molecular absorption spectra of C_{60} and II consist of the Herzberg-Teller vibronic series, while those spectra of I and III are formed by the Franck-Condon vibronic series. Despite this substantial difference, the disparity in the nonpolar components between spectra in Figure 10.3 is relatively small compared to their polar component, whose intensity is higher by an average factor of 2, 3, and 5 for fullerenes I, II, and III, respectively, than for C_{60}. The nonpolar component of the spectrum of each solution is well reproduced in different experiments, whereas the polar component varies significantly, suggesting that this part of the spectrum depends on the crystallization conditions of the sample under investigation.

Figure 10.4 shows the solution luminescence spectra excited at λ_{ex} = 632.8 nm. Similarly to the absorption spectra, these red spectra also have a well-structured molecular component sitting on top of a bell-shaped background (dotted curves). This is most clearly seen in the emission spectra presented in Figures 3.4 and 3.5 after the background extraction. These well-structured luminescence spectra are distinct series of vibronic lines originating from purely electronic transition bands resonantly matching those in the absorption spectra. A detailed discussion on them is given in Section 3.2.5. The bell-shaped background in the fluorescence and phosphorescence regions (11,500 to 1,5000 and 10,000 to 11,500 cm^{-1}, respectively) corresponds to transitions from the lowest-lying polar states. The

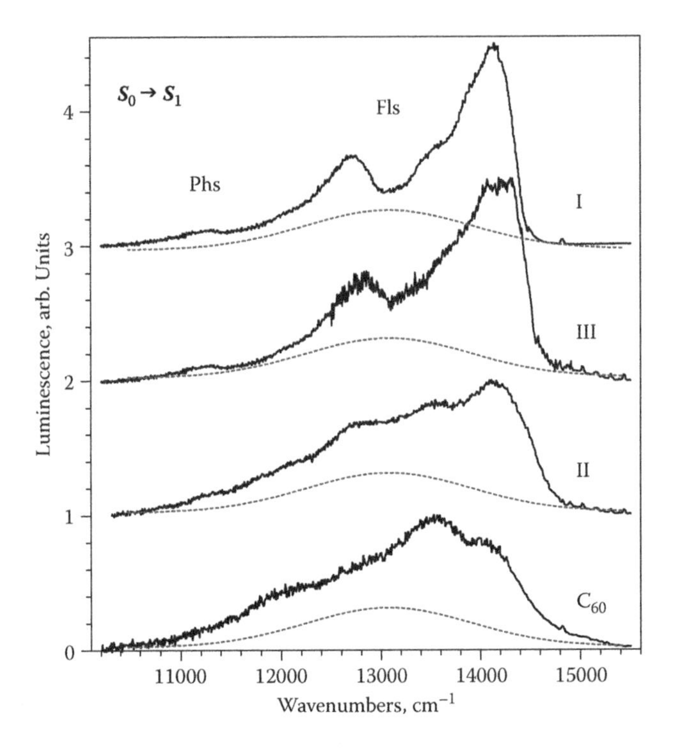

Figure 10.4 $S_1 \rightarrow S_0$ luminescence spectra of C_{60} and fullerenes I, II, and III in toluene at 80 K; $\lambda_{ex} = 632.8$ nm. Dotted curves represent polar background components. (From Sheka, E. F., et al., *J. Exp. Theor. Phys.*, 108, 738–50, 2009.)

two-component red spectra were used as reference ones in the analysis of emission spectra of solutions presented below.

In addition to the red spectra described above, the spectra of all solutions excited at λ_{ex} from 514.5 to 476.5 nm include visible portions (blue emission) varying in intensity and structure. Figure 10.5 shows spectra excited at $\lambda_{ex} = 476.5$ nm. As seen in the figure, the blue spectra are quite different. Thus, the blue emission spectrum of solution I is a spontaneous Raman scattering of toluene.[5] Its intensity amounts to only 5% of the red spectrum intensity and is invariant with respect to λ_{ex}. The blue emission from solution C_{60} is also the Raman spectrum of toluene, but its intensity is higher by a factor of more than 3 and is sensitive to excitation wavelength, decreasing with increasing λ_{ex}. The spectrum of strong blue emission from solution III consists of two parts. One of these is definitely the Raman spectrum of toluene, but its intensity is higher than that from C_{60}. The position of the other part that was attributed to photoluminescence[6,7] does not change as λ_{ex} varies from 476.5 to 514 nm, while its intensity decreases therewith (see Figure 10.6(a)). The spectrum is intimately connected with

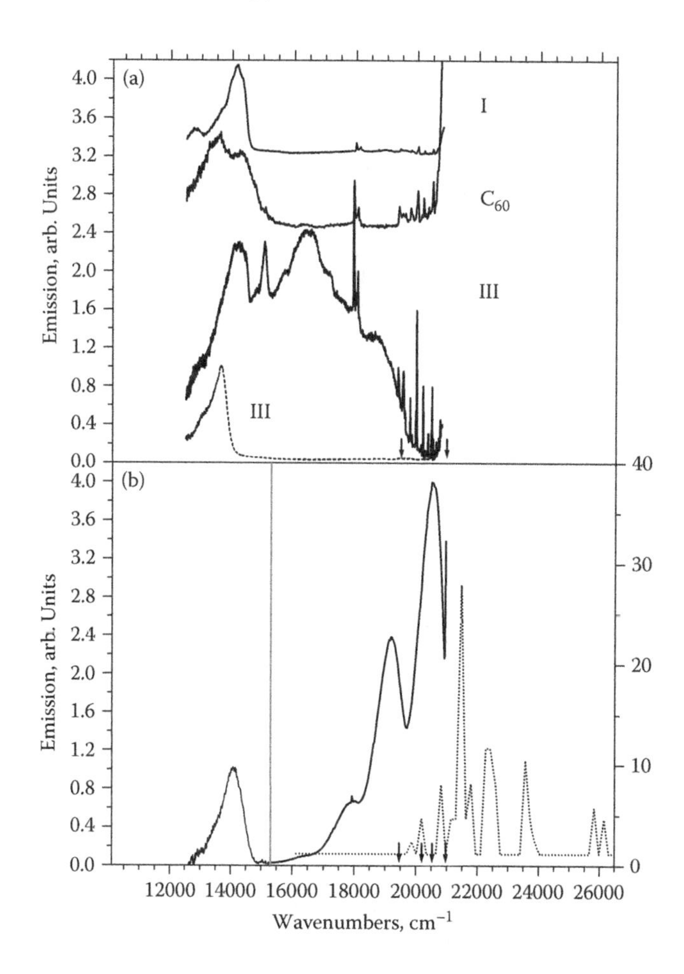

Figure 10.5 Emission spectra of C_{60} and fullerenes I, III (a), and II (b) in toluene at 80 K [7]; λ_{ex} = 476.5 nm. Dotted curves correspond to emission from fullerene III in carbon tetrachloride (a) and CT exciton density of states in crystalline C_{60} (b). (Courtesy by Dr. A. Eilmes, Jagellonian University, Krakow.)

the toluene solvent since it disappears when toluene is substituted by carbon tetrachloride (dotted curve in Figure 10.5(a)). The blue emission is the strongest in the spectrum of solution II (see Figure 10.5(b)). Its profile is characteristic of enhanced Raman scattering, and its intensity decreases with increasing λ_{ex}. A detailed analysis of this spectrum presented in Razbirin et al.[5] attributed it to Raman scattering from fullerene clusters (fullerene-enhanced Raman scattering (FERS)).

The blue emission vanishes at λ_{ex} = 337.1 nm. However, not only the red spectrum of fullerenes is emitted by the toluene solutions, but also photoluminescence from the solvent is observed whose spectrum is

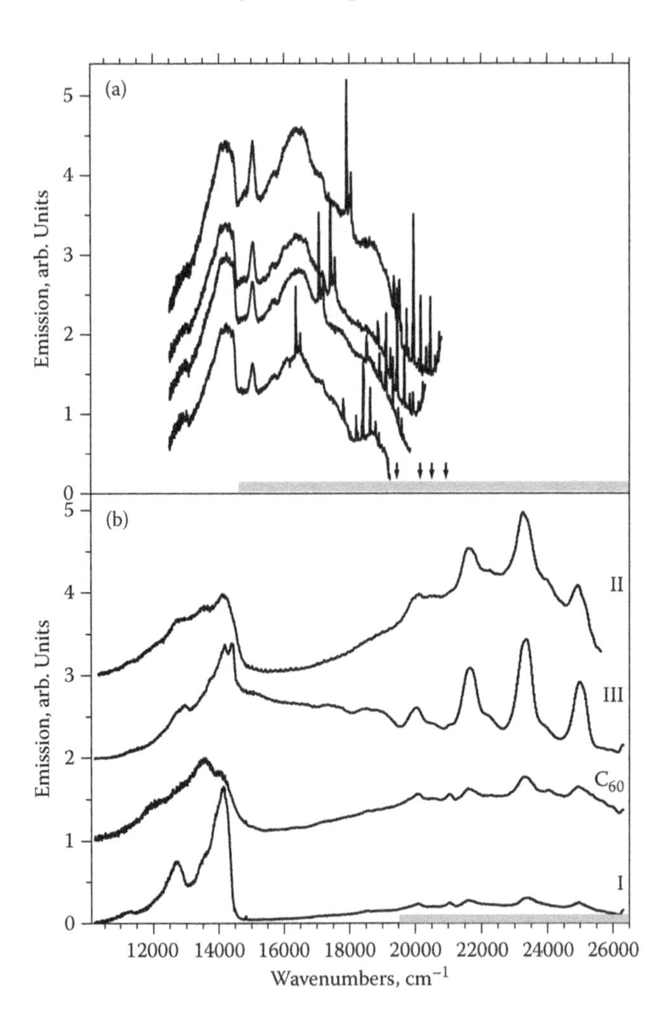

Figure 10.6 (a) Emission spectra of fullerene III in toluene at 80 K; bottom to top: λ_{ex} = 514.5, 496.5, 488.0, and 476.5 nm (indicated by arrows). (b) Emission spectra of fullerenes I, C_{60}, II, and III (bottom to top) in toluene at 80 K; λ_{ex} = 337.1 nm. (From Sheka, E. F., et al., *J. Exp. Theor. Phys.*, 108, 738–50, 2009.)

characteristic of benzaldehyde emission[30] (see Figure 10.6(b)). The shape of the photoluminescence spectrum is similar for all solutions, whereas its intensity relative to the red spectrum of fullerene emission increases in the sequence of I → C_{60} → III → II. Because this behavior of the spectrum is characteristic of enhanced photoluminescence, it was attributed to solvent-enhanced luminescence (SEL).[7] Solvent-enhanced Raman scattering (SOERS; to distinguish from the widely known SERS), observed for solutions C_{60} and III, as well as fullerene-enhanced luminescence (FEL) in solution III have completed so far a set of the observed enhanced spectral phenomena.

A summarized picture of the emission spectra of fullerenes in solution looks like the following:

- At $\lambda_{ex} = 623.8$ nm, which is located at the red end of absorption spectra, all solutions emit red spectra combining the two-component fluorescence and phosphorescence spectra of the dissolved fullerenes and emission from the low-lying CT states.
- At λ_{ex} ranging from 514.5 to 476.5 nm, which are located within both polar and nonpolar parts of absorption spectra, the red spectrum emitted by the solute is observed in combination with a visible blue emission whose spectral profile differs between different solutions. The blue emission is attributed to normal Raman scattering from toluene for solution I, to enhanced Raman scattering from solvent for solutions C_{60} and III (in the latter case, combined with an enhanced luminescence spectrum of the solution), and to solute-enhanced Raman scattering for solution II. The relative intensity of the blue emission increases as λ_{ex} varies from 514.5 to 476.5 nm for all solutions with the exception of I. The observed dependence of the blue emission intensity on the excitation wavelength is attributed to the excitation of impurity CT states of solutions satisfying resonance conditions.
- At $\lambda_{ex} = 337.1$ nm, which is located in the gap of the energy spectrum of charge transfer states, the red spectrum of fullerenes is combined with the solvent luminescence spectrum, whose relative intensity increases in the sequence of I $\rightarrow$ C_{60} $\rightarrow$ III $\rightarrow$ II. Its position coincides with the spectrum of CT states, which provides a significant enhancement of the spectrum.

Since the excitation spectrum of a solution contains both *sol-sol* and *sol-solv* impurity CT states, whose spectra may strongly overlap (see Figure 10.1), it is natural to hypothesize that varying degrees of overlap and excitation of these states in solutions of different composition can explain the observed variations in blue emission spectrum with fullerene structure. This hypothesis was validated by examining the nature and structure of the entities where these CT states are localized.

10.5 Quantum chemical analysis of intermolecular interactions in solutions of fullerenes

There is no doubt that a major contribution of D–A pairing of the *sol-sol* type to intermolecular interaction is responsible for the observed aggregation or clustering of solute molecules. Even though no direct evidence is available of any effect of D–A pairing of the *sol-solv* type on solute clustering or solution structuring, only this type of intermolecular interaction can explain the fact that crystallization of various derivatives of C_{60} in toluene solution involves irreversible incorporation of solvent molecules into the resulting crystals.[31] Assuming that the impurity CT states in the solutions are localized on various *sol-sol* and *sol-solv* aggregates, let us look at a numerical analysis of this scenario.

UBS HF AM1 calculations showed that all three derivatives under discussion are characterized by the IMI potential of type 1 so that the clustering occurs in the region of the R^{00} minimum (see Figure 1.2(b)). A detailed analysis of the contribution of their CT states to emission spectra requires knowledge of the energy range of these states and the intermolecular interaction energies in systems of *sol-sol* and *sol-solv* clusters. Only a qualitative analysis of the expected changes in the solutions spectrum, compared to that of crystalline C_{60}, is possible, since direct calculations of the energy spectrum of CT excitation states are practically not feasible. The dotted curve in Figure 10.5(b) represents the energy distribution of the singlet density of CT exciton states of the crystal calculated by A. Eilmes and B. Pack[29] following Pac et al.[11,12] Generally, the CT exciton energies are determined by solving a set of equations subsumed under the general formula

$$\hbar\omega \approx I - \varepsilon + P \tag{10.5}$$

where $E_{gap} = I - \varepsilon$ is the energy gap between the electron and hole band edges, and P is the polarization energy, depending on relative electron-hole position. Since exact calculation of the polarization energy presents the main difficulty, we analyze below only the dependence of the shift in the $\hbar\omega$ spectrum on the energy gap, E_{gap}, assuming that the polarization energy is approximately constant.

Table 10.1 lists ionization potentials (IPs), electron affinities (EAs), and E_{gap} for a number of clusters based on the considered fullerenes. Each cluster listed in the table is a charge transfer complex. Photoexcitation of clusters always involves the transfer of an electron from one molecule to another. Model *sol-sol* clusters of n fullerene molecules are denoted by $(X)_n$, where X = I, C_{60}, II, or III. Model *sol-solv* clusters of n fullerene molecules and m toluene molecules are denoted by $(X)_n(T)_m$. Semiempirical calculations generally overestimate ionization potentials, whereas the calculated electron affinities are in good agreement with experimental data. Since the present analysis focuses on the difference in E_{gap} of different solutions or clusters as evaluated by using the same approach, errors in absolute ionization potentials are practically irrelevant. The coupling energies $E_{cpl}\{(X)_n\}$ and $E_{cpl}\{(X)_n(T)_m\}$ are associated with $(X)_n$ and $(X)_n(T)_m$ clusters, respectively.

Table 10.1 Ionization potentials, electron affinities, and coupling energies of fullerene *Sol-Sol* and *Sol-Solv* clusters: UBS HF AM1 singlet state

Clusters	Coupling energy, kcal/mol	*I*, eV	ε, eV	E_{gap}, eV
	$E_{cpl}\{(X)_n$	*Sol-Sol* clusters		
I		9.68	2.48	7.20
$(I)_2$	>0	9.68	2.50	7.18
C_{60}		9.87	2.66	7.21
$(C_{60})_2$	−0.52	9.87	2.66	7.21
$(C_{60})_6$	−2.74	9.87	2.66	7.21
II		9.79	2.57	7.22
$(II)_2$	−1.26	9.80	2.60	7.20
$(II)_3$	−2.73	9.80	2.60	7.20
$(II)_4$	−4.32	9.77	2.65	7.12
$(II)_5$	−7.94	9.75	2.72	7.03
$(II)_6$	−8.42			7.12
III		9.84	2.59	7.25
$(III)_2$	−3.66	9.66	2.69	6.97
$(III)_3$	−8.01	9.66	2.79	6.87
$(III)_4$	−3.21	9.49	2.81	6.67
	$E_{cpl}\{(X)_n(T)_n\}$	*Sol-Solv* clusters		
Toluene		9.34	−0.56	9.90
$(I)_1(T)_1$	>0	9.44	2.45	6.99
$(C_{60})_1(T)_1$	>0	9.35	2.66	6.69
$(II)_1(T)_1$	−2.75	9.64	2.53	7.11
$(II)_2(T)_1$	−4.16	9.58	2.58	7.00
$(II)_3(T)_2$	>0	9.33	2.66	6.67
$(III)_1(T)_1$	−0.64	9.08	2.61	6.47
$(III)_2(T)_1$	−4.23	9.32	2.68	6.64
$(III)_2(T)_2$	−6.02	9.29	2.69	6.60
$(III)_3(T)_2$	−9.25	9.47	2.75	6.72
$(III)_4(T)_4$	−5.81	9.03	2.80	6.23

Source: Modified from Sheka, E. F., et al., *J. Exp. Theor. Phys.*, 108, 738–50, 2009.

The data on energy gaps, E_{gap}, presented in the table suggest that the energy spectra of the *sol-sol* CT states of fullerenes I and C_{60} are almost identical, while those of fullerenes II and III may be shifted down in energy by approximately 0.1 and 0.2 eV, respectively. Taking the CT excitons in crystalline C_{60} as a reference spectrum, we can expect that the energy spectrum of singlet CT states of *sol-sol* clusters in the solutions under investigation extends from approximately 19,500 cm^{-1} to higher frequencies, as represented by a gray strip in Figure 10.6(b). As seen in the figure, the low-frequency edge of this energy spectrum coincides with that of the λ_{ex} interval between 514.5 and 476.5 nm. Thus, varying λ from 514.5 to 476.5 nm means tuning through the spectrum of CT states, which explains the resonant behavior of the blue emission intensity observed as λ_{ex} decreases.

As for the energy spectrum of CT states of *sol-solv* clusters, such clusters can be formed only in a toluene solution. The intermolecular interactions between fullerenes and carbon tetrachloride are repulsive, with positive binding energies. In the clusters incorporating toluene molecules, whose ionization potential is lower than those of fullerenes, the energy spectra of CT states are shifted down in energy by 0.2 and 0.5 eV on average for fullerenes II and III, respectively. The resulting spectrum of solution III is represented by a gray strip in Figure 10.6(a). The shift in energy spectrum is significant and very important. If *sol-sol* and *sol-solv* clusters coexist in a solution, then the luminescence of *sol-sol* clusters is enhanced by its matching the energy spectrum of *sol-solv* CT states in a manner similar to the enhancement of the observed toluene luminescence spectrum by its matching the spectrum of *sol-sol* CT states (see Figure 10.6(b)).

Analyzing the coupling energy data given in Table 10.1, one can conclude that the IMI total strength between fullerenes I does not ensure stable *sol-sol* cluster formation. Since the corresponding coupling energy is positive, the interaction between molecules is predominantly repulsive. By contrast, the pairwise coupling energy of C_{60} is negative, which favors clustering. For a growing cluster, the binding energy is roughly equal to the pairwise coupling energy multiplied by the number of pairs in the cluster. As mentioned above, C_{60} cluster formation in a toluene matrix has been observed experimentally.[24,25] The interaction within the assemblage of fullerenes II and III is much stronger than between C_{60}. Obviously, the pairwise interaction depends on the relative position of the molecular groups attached to the C_{60} core, being strongest for the configurations with equilibrium structures shown in Figure 10.7. Cluster

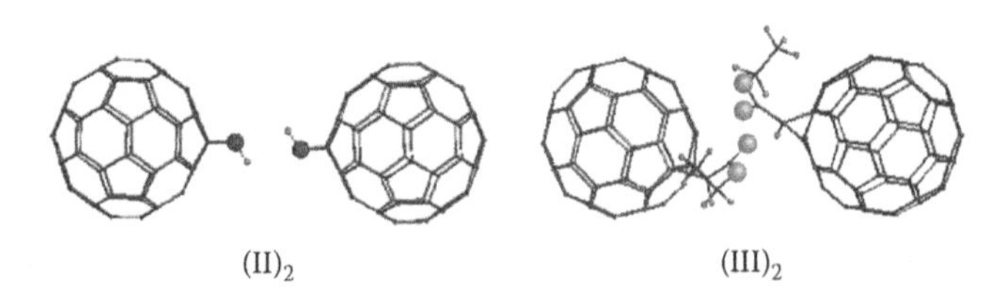

Figure 10.7 Equilibrium structures of clusters $(II)_2$ and $(III)_2$. UBS HF AM1 singlet state. (From Sheka, E. F., et al., *J. Exp. Theor. Phys.*, 108, 738–50, 2009.)

growth may not entail an increase in the total coupling energy, because the latter depends on the relative positions of molecules. The sterical barriers due to the complex structure of the addend protuberances prevent close packing of fullerene II and III molecules, which leads to a variety of energetically stable configurations. As a consequence, there exist groups of cluster isomers with similar compositions, but different coupling energies. In particular, this explains the variability of the polar components of the absorption spectra of solutions observed in repeated experiments. This explains as well the broad-band shape of the blue spectra of II and III caused by inhomogeneous broadening. Even though the presented configurations may not be optimal in terms of composition, they demonstrate significant coupling energies, which evidence the formation of stable configurations. The equilibrium structures of some *sol-sol* clusters are shown in Figure 10.8.

Sol-solv cluster configurations are stable only for fullerenes II and III in toluene solution. Table 10.1 lists coupling energies related to corresponding clusters, part of which is shown in Figure 10.9. The coupling energy is greatly influenced by the way of incorporation of toluene molecules into a fullerene cluster. Thus, $(II)_n(T)_m$ stable clusters could be formed only when toluene molecules are outside the interaction range of the protuberances of fullerene II. When a toluene molecule is within this range, the coupling energy is reduced and may even change its sign. Conversely, clusters $(III)_n(T)_m$ are stable for any relative position of fullerene and toluene molecules.

Therefore, the numerical results show that the molecules of fullerene I in toluene solution are uniformly distributed over the matrix and widely separated from one another.

In solution C_{60}, the formation of relatively stable *sol-sol* clusters is possible, which leads to toluene solution structuring.[25] The energies of impurity CT states should be close to the CT exciton spectrum of crystalline C_{60}.

In solutions II and III, *sol-sol* and *sol-solv* clusters coexist. This explains the high intensity of the polar components of the absorption spectra of these solutions (see Figure 10.2). As concerns the impurity CT states, the corresponding energy spectra of *sol-sol* and *sol-solv* clusters in solution II are similar up to a minor red shift of the *sol-sol* CT states. However, the corresponding states in solution III are separated by a significant gap (approximately 0.5 eV).

10.6 Blue emission, pairwise interaction, and efficacy of nonlinear optical behavior

A characteristic common feature of the emission spectra of fullerenes in solutions is a blue emission whose compositions depend on the structure

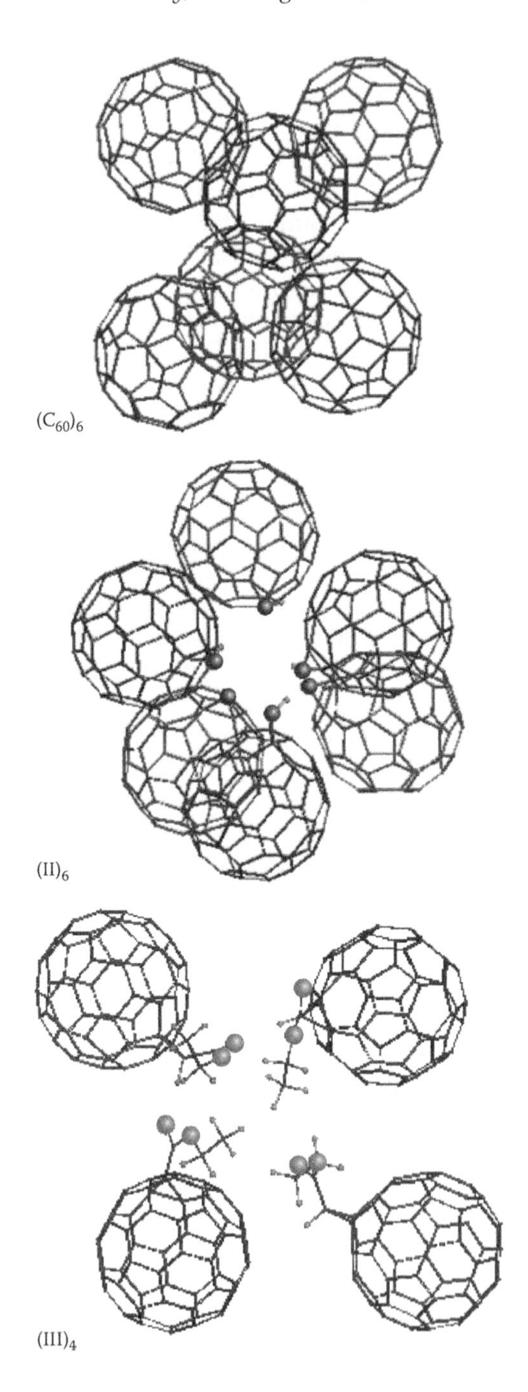

Figure 10.8 Equilibrium structures of clusters $(C_{60})_6$, $(II)_6$, and $(III)_4$. (From Sheka, E. F., et al., *J. Exp. Theor. Phys.*, 108, 738–50, 2009.)

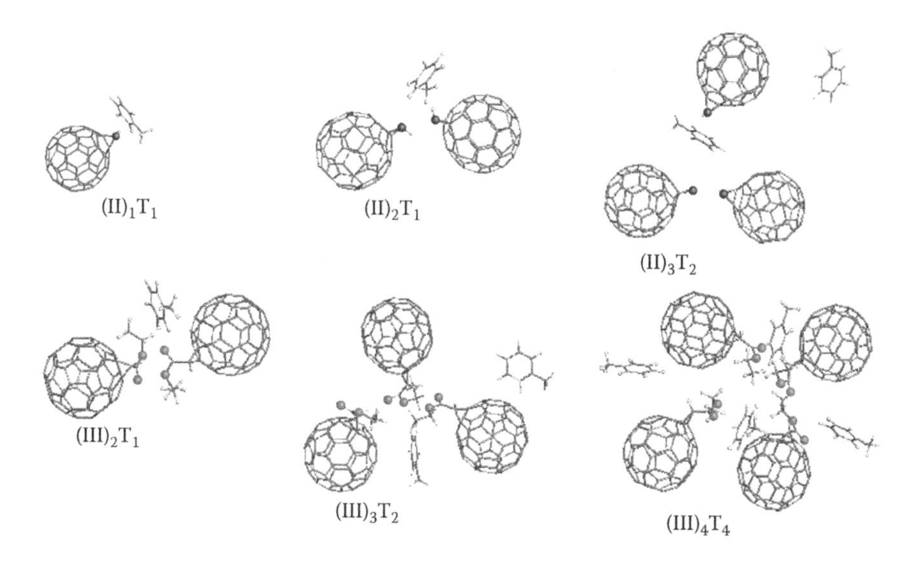

Figure 10.9 Equilibrium structures of clusters $(II)_1(T)_1$, $(II)_2(T)_1$, $(II)_3(T)_1$, $(III)_2(T)_1$, $(III)_3(T)_2$, and $(III)_4(T)_4$. (From Sheka, E. F., et al., *J. Exp. Theor. Phys.*, 108, 738–50, 2009.)

of the solute molecule. This emission is typical only for solutions, because the single-molecule emission spectrum of the solute lies in the red and near-infrared region and that of the solvent lies in the ultraviolet region.

Another common feature is the dependence of the blue emission intensity on the efficiency of fullerene clustering. Actually, if we characterize the clustering efficacy by the relevant pairwise coupling energy, either $E_{cpl}\{(X)_2\}$ or $E_{cpl}\{(X)_1(T)_1\}$, one can see a direct energy-clustering relationship. Thus, in solution I, both energies are positive and the absence of clustering just follows from the fact that the blue emission, as might be expected, corresponds to the normal spontaneous Raman spectrum of toluene and is very weak. When the energy $E_{cpl}\{(X)_2\}$ is negative, but not large (–0.52 kcal/mol) and $E_{cpl}\{(X)_1(T)_1\}$ is positive (solution C_{60}), *sol-sol* cluster formation occurs, which is supported by the blue emission spectrum that is a SOERS of toluene with obvious signs of enhancement. A further increase in the absolute value of $E_{cpl}\{(X)_2\}$ as well as a change of sign and an increase in the absolute value of $E_{cpl}\{(X)_1(T)_1\}$ are obviously manifested by efficient *sol-sol* and *sol-solv* cluster formation (solutions II and III). The resulting blue emission has a high intensity and a two-component spectrum. Its major components are FERS from clusters in solution II[5] and enhanced luminescence from *sol-sol* clusters FEL in solution III.[6,7] In both cases, the other component is SOERS from toluene.

A third feature lies in the excitation wavelength dependence of the blue emission intensity and its increase with λ_{ex} varying from 514.5 to 476.5 nm.

It is shown that this interval of λ_{ex} adjoins the red edge of the CT exciton band spectrum of *sol-sol* clusters, whose excitation increases the resonance enhancement of optical effects.

A fourth feature of emission from solutions is SEL,[5] with an intensity whose dependence on clustering efficiency follows that of the blue emission intensity. Since SEL is due to matching between the emission spectrum of toluene and the spectrum of impurity CT states of *sol-sol* clusters, this similarity implies a high degree of similarity between the energy spectra of CT states.

In view of the local field enhancement model, the grounds of explanation of the spectral features of fullerenes in solution look like the following:

1. The polarized objects in fullerene solutions are *sol-sol* and *sol-solv* nanosize clusters of the solute and solute-solvent molecules, respectively. Clustering efficiency depends on the pairwise coupling energies $E_{cpl}\{(X)_2\}$ and $E_{cpl}\{(X)_1(T)_1\}$. These quantities control cluster formation, and cluster structure is therefore determined by the structure of the fullerene molecule. When the absolute value of this energy is less than 0.5 kcal/mol, the efficacy of the cluster formation is practically negligible.
2. The excitation spectrum of clusters is the sum of those of localized Frenkel excitons and localized CT excitons (impurity CT states). The former, after relaxation, provide a molecular-like well-structured red emission spectrum. The relaxation of the latter results in a broadband red emission, but their main role consists in the generation of the electron-hole plasma, whose polarization enhances both incident and outgoing electromagnetic waves, thus resulting in blue emission.
3. The spectral region of where plasma resonance conditions for enhancement of incident and outgoing light are satisfied coincides with the energy range of localized CT excitons. Just the very localization of local CT excitons in the visible region is responsible for the blue emission appearing there. Each particular instance of the blue emission spectrum is a specific combination of enhanced Raman scattering, either SOERS or FERS, and enhanced luminescence, SEL and FEL, respectively. Note that the contribution of the solvent is of key importance for the observation of not only SOERS and SEL (which is natural), but also of FERS and FEL. The dependence of the latter effects on the solvent is due to the existence of two spectral regions of localized CT excitons corresponding to *sol-sol* and *sol-solv* clusters, respectively. A narrow gap between the energies of these states favors the observation of FERS, whereas a wide gap is favorable for FEL.
4. The results discussed above have shown that nanosized fullerene clusters in solutions act as light wave enhancers, like gold or silver colloid particles. However, as shown earlier, the attainable magnitude of enhancement K cannot be more than 10^2. In the experiments described

above, the coefficient *K* varied from 2.6 in solution C_{60} to 733 in solution II. It seems that the latter value is close to the limit. Since enhancement of linear optical properties is highly indicative for the efficacy of the medium nonlinear optical (NLO) behavior, the *K* value may be considered a quantitative characteristic of such efficacy. From this viewpoint, solutions II and III should be considered good candidates for revealing peculiarities of the nonlinear optical properties.

The fact that both the energies of CT excitons localized on fullerene clusters and the energy interval where plasma resonance conditions for local factors are satisfied lie in the visible spectral region makes fullerene-doped matrices perfect enhancers of nonlinear optical effects of practical importance. The manifestation of the blue emission provides strong evidence for their applicability in nonlinear optical devices. The high intensity of enhanced spectra points to their efficiency as nonlinear optical materials. Therefore, blue emission provides an empirical criterion for selecting nonlinear optical materials with desired properties. As an auxiliary test, one may consider poor solubility of a fullerene in a matrix, which generally favors fullerene clustering.[26] Additional to these empirical tests, predetermined pairwise coupling energies $E_{cpl}\{(X)_2\}$ and $E_{cpl}\{(X)_1(T)_1\}$ for *sol-sol* and *sol-solv* clusters exceeding 0.5 kcal/mol by absolute value are a firm indication of an efficient clusterization and, consequently, a high ability of the mixture to act as an efficient nonlinear optical material. These criteria were validated by application to a C_{70}-doped cyanobiphenyl-based matrix,[32] whose nonlinear optical properties are widely employed in optical limiters, displays, and diffractive media for reversible holographic recording.[33]

10.7 And again about blue emission, photodynamic therapy, and nanophotonics of fullerene solutions

Blue emission stimulated by fullerene clusters in crystalline toluene matrix[5–8] and discussed in the previous sections was not the first observation of events of this kind, since a similar emission was observed earlier in aqueous solutions of fullerene C_{60} hydrates known as C_{60}HyFn[34] and discussed in Section 9.4. An additional broad emission band that has a maximum at about 530 nm was observed under excitation by a number of laser lines with λ_{ex} from 441.6 to 514.5 nm. Its relative intensity with respect to convenient red band depends on the excitation wavelength, and the maximum position is red shifted when λ_{ex} increases. The authors[34] did not suggest any explanation of the emission behavior, only definitely attributing the additional band to C_{60} clusters inside C_{60}HyFn hydrates.

This assumption fully correlates with the basic concept discussed above, which connects spectral behavior of diluted fullerene solutions with solute clustering. According to this, the observed band should be attributed to FEL. Based on the regularities in the blue emission behavior discussed in the previous section, let us try to explain the peculiar behavior of the FEL band in this case.

Two peculiarities, namely, the band shift and its intensity enhancement, depending on λ_{ex}, should be explained. As for the former, this can be connected with the inhomogeneous broadening of the band caused by a large variation in the cluster composition of the hydrates. The same inhomogeneous broadening should be attributed to the spectrum of CT states. Consequently, a change in λ_{ex} just means the selective excitation of a particular fraction of the clusters, which in turn causes a redistribution of the emission intensity in favor of the excited fraction within the broad band, thus simulating its shift.

As for the intensity enhancement of the band, a mandatory requirement concerns the presence of electron-hole excitation spectra in the spectral region of the emission. Since in the discussed case the matter can be about *sol-sol* clusters only, such a red-shifted spectrum can be attributed to triplet CT states. A reason to pay attention to this part of the energy spectrum gives a peculiar λ_{ex}-dependent luminescence of singlet oxygen in C_{60} films and crystal.[35] The explanation seems quite reliable, and coming back to the peculiarities of the broad-band blue emission in solution II, one should think about a possible combination of FERS and FEL in this case.

One more observation of the blue emission is related to the bicapped [60]fullerene-γ-cyclodextrin complex in aqueous solutions.[36] The authors did not connect the event with the cluster structure of the solutions, but the connection seems to be straightforward. It should be noted that the complex is largely explored in antioxidant therapy as well as in C_{60}HyFn hydrates.

This close similarity between physical and medical fullerene-based objects forces us to think about the possibility of extending the understanding obtained in physical laboratories to more complicated events studied by biologists and physicians. In particular, this might concern fullerene photodynamic therapy. As discussed in Section 9.2, a reliable explanation of the action can be proposed on the basis of fullerene clustering in aqueous solutions, thus forming charge transfer complexes responding to the applied photoexcitation. It seems natural to suppose that the drug action output must depend on the clustering efficacy, which in turn depends on the fullerene structure. If so, looking for blue emission of the solutions, analyzing its parameters, and computationally determining coupling energy of pairwise interaction might be suggested as reliable criteria for the choice of proper fullerene candidates to optimize the drug action.

To conclude the chapter, it is worthwhile to mention that the nature is sage and saves on efforts directed, for example, to the creation of different materials, being drastically different at first glance, but being based on the same basic grounds, which causes a deep similarity in their behavior under particular conditions. Thus, there is seemingly no connection between blue emission of fullerene solution and graphene. However, hydrothermally cut graphene sheets can be transformed in blue-luminescent graphene quantum dots of 5 to 13 nm in size.[37] It is important to note that the blue luminescence of the dots is λ_{ex} dependent, similarly to that of C_{60}HyFn hydrates, even in details. There has been no understanding of the emission nature so far. However, if the dots have a complicated structure and consist of sets of graphene pieces of smaller size, then each dot can be similar to a C_{60}HyFn droplet, thus being a cluster of small pieces with IP and EA characteristics typical for fullerenes and providing photoluminescence caused by peculiarities of D–A interaction between the pieces that is characteristic for fullerene clusters.

10.8 *Nanophotonics of fullerenes in chemistry, medicine, and optics*

A deep similarity of photostimulated effects occurring in physical and biological objects involving fullerene forces us to raise the question: What is meant by nanophotonics of fullerene, and should we not imply under this conventional term, usually restricted to optical events, something more general? Actually, fullerenes are extremely active photosensitive objects. Is not this extreme activity just the very keystone that should be laid on the foundation of the generalized nanophotonics of fullerenes? Let us try to answer these questions.

As we know by now, the photoactivity of fullerenes is manifested via three groups of photosensitive phenomena:

1. Photostimulated chemical reactions, including a particular reaction of dimerization or oligomerization of fullerenes
2. Photodynamic therapy of diluted fullerene aqueous solutions
3. Photoinduced enhancement of spectral properties of fullerene solutions

Although these phenomena have different appearances and are related to different scientific topics, there is a strong feeling of a common origin of the fullerene behavior concerning all of them. As discussed in this book, this makes it possible to suggest that the formation of a positive-negative fullerene ion pair at each photon absorption act is common for all events, and evidently provides their common origin.[38] In view of this, the three phenomena differ only by the manner of this action implementation.

Thus, in the case of photochemical reactions, positive-negative ion pairs are provided by the photoexcitation of either hetero- or homocomponent binary systems involving fullerene molecules. The reaction is either promoted or made possible due to ionization of the system components providing a drastic facilitation of the interaction between them via Coulomb attraction, which leads to the formation of a chemically bound *AB* adduct.

Each elementary act during photodynamic effects related to the fullerene aqueous solution therapy evidently concerns the formation of a positive-negative ion pair under photoexcitation within a clusterized fullerene component of the solution. The act is followed by capturing an oxygen molecule by each of the fullerene molecular ions, thus changing the multiplicity of the ground state of the complex $C_{60} + O_2$ from a triplet to a doublet, which causes the spin-flip of the electron of the oxygen molecule, thus transforming the triplet oxygen into the singlet one.

In the case of enhanced spectroscopy of fullerene solution, ion pairs that occurred under photoexcitation of fullerene clusters, as in the case of the photodynamic effect, provide effective polarization of the object. The clusters fulfill an additional role of confining electromagnetic fields, thus acting as amplifiers of optical effects.

The exclusive D–A ability of fullerenes is undoubtedly the main factor that makes the fullerene nanophotonics so peculiar. Due to deep similarity between fullerenes, carbon nanotubes, and graphene in regards to this property, similar nanophotonic behavior should be expected in the case of various compositions of these nanocarbons as well.

References

1. Kafafi, Z. H., ed. 1997. *Fullerenes and photonics IV. Proceedings of SPIE*. Vol. 3142. San Diego: SPIE.
2. Kamanina, N. V., and Sheka, E. F. 2004. Optical limiters and diffraction elements based on a COANP-fullerene system: Nonlinear optical properties and quantum-chemical simulation. *Opt. Spectr.* 96:599–612.
3. Kamanina, N. V. 2003. Non-linear optical study of fullerene-doped conjugated systems: New materials for nanophotonics applications. In *Organic nanophotonics*, ed. F. Charra, V. M. Agranovich, and F. Kajzar, 177–92. NATO Science Series 100. Dordrecht: Kluwer Academic.
4. Hosoda, K., Tada, R., Ishikawa, M., and Yoshino, K. 1997. Effects of C_{60} doping on electrical and optical properties of poly[(disilanylene)oligophenylenes]. *Jpn. J. Appl. Phys.* 36:L372–75.
5. Razbirin, B. S., Sheka, E. F., Starukhin, A. N., Nelson, D. K., Troshin, P. A., and Lyubovskaya, R. N. 2008. Enhanced Raman scattering provided by fullerene nanoclusters. *JETP Lett.* 87:133–39.
6. Sheka, E. F., Razbirin, B. S., Starukhin, A. N., Nelson, D. K., Degunov, M. Yu., Lyubovskaya, R. N., and Troshin, P. A. 2009. Fullerene-cluster amplifiers and nanophotonics of fullerene solutions. *J. Nanophot.* 3:033501.

7. Sheka, E. F., Razbirin, B. S., Starukhin, A. N., Nelson, D. K., Degunov, M. Yu., Lyubovskaya, R. N., and Troshin, P. A. 2009. The nature of enhanced linear and nonlinear optical effects of fullerenes in solution. *J. Exp. Theor. Phys.* 108:738–50.
8. Sheka, E. F., Razbirin, B. S., and Nelson, D. K. 2009. Fullerene nanoclusters as enhancers in linear spectroscopy and nonlinear optics. *High Energy Chem.* 43:628–33.
9. Broude, V. L., Rashba, E. I., and Sheka, E. F. 1985. *Spectroscopy of molecular excitons*. Springer: Berlin.
10. Kazaoui, S., Minami, N., Tanabe, Y., Byrne, H. J., Eilmes, A., and Petelenz, P. 1998. Comprehensive analysis of intermolecular charge-transfer excited states in C_{60} and C_{70} films. *Phys. Rev. B* 58:7689–700.
11. Pac, B., Petelenz, P., Eilmes, A., and Munn, R. W. 1998. Charge-transfer exciton band structure in the fullerene crystal-model calculations. *J. Chem. Phys.* 109:7923–31.
12. Pac, B., Petelenz, P., Slawik, M., and Munn, R. W. 1998. Theoretical interpretation of the electroabsorption spectrum of fullerene films. *J. Chem. Phys.* 109:7932–39.
13. Orlandi, G., and Negri, F. 2002. Electronic states and transitions in C_{60} and C_{70} fullerenes. *Photochem. Photobiol. Sci.* 1:289–308.
14. Razbirin, B. S., Starukhin, A. N., Nelson, D. K., Sheka, E. F., and Prato, M. 2007. Optical spectra and covalent chemistry of fulleropyrrolidines. *Int. J. Quant. Chem.* 107:2787–802.
15. Heritage, J. P., and Glass, A. M. 1982. Nonlinear optical effects. In *Surface enhanced Raman scattering*, ed. R. K. Chang and F. E. Furtak, 391–412. New York: Plenum Press.
16. Metiu, H. 1982. A survey of recent theoretical works. In *Surface enhanced Raman scattering*, ed. R. K. Chang and F. E. Furtak, 1–34. New York: Plenum Press.
17. Chang, R. K., and Furtak, F. E. 1982. *Surface enhanced Raman scattering*. New York: Plenum Press.
18. Lal, S., Grady, N. K., Kundu, J., Levin, C. S., Lassiterde, J. B., and Halas, N. J. 2008. Tailoring plasmonic substrates for surface enhanced spectroscopies. *Chem. Soc. Rev.* 37:898–911.
19. Pettinger, B., Picardi, G., Schuster, R., and Ertl, G. 2002. Surface-enhanced and STM-tip-enhanced Raman spectroscopy at metal surfaces. *Single Mol.* 3:285–94.
20. Ritchie, G., and Chen, C. Y. 1982. In *Surface enhanced Raman scattering*, ed. R. K. Chang and T. E. Furtak, 361–78. New York: Plenum Press.
21. Gersten, J. I., and Nitzan, A. 1980. Electromagnetic theory of enhanced Raman scattering by molecules adsorbed on rough surfaces. *J. Chem. Phys.* 73:3023–37.
22. Ahn, J. S., Suzuki, K., Iwasa, Y., Otsuka, N., and Mitani, T. 1997. Nanoscale C_{60} aggregates in solution. *Proc. SPIE* 3142:196–204.
23. Suzuki, K., Ahn, J. S., Ozaki, T., Morii, K., Iwasa, Y., Otsuka, N., and Mitani, T. 1998. Photoinduced transformation of C_{60} aggregates to carbon nanoballs. *Mol. Cryst. Liq. Crysts.* 314:221–32.
24. Török, Gy., Lebedev, V. T., and Cser, L. 2002. Small-angle neutron-scattering study of anomalous C_{60} clusterization in toluene. *Phys. Sol. State* 44:572–73.

25. Ginzburg, B. M., and Tuichiev, S. 2005. On the supermolecular structure of fullerene C_{60} solutions. *J. Macromol. Sci. B* 44:1–14.
26. Avdeev, M. V., Aksenov, V. L., and Tropin, T. V. 2010. Models of cluster formation in solutions of fullerenes. *Russ. J. Phys. Chem.* 84:1405–16.
27. Ching, W. Y., Huang, M.-Zh., Xu, Y.-N., Harter, W. G., and Chan, F. T. 1991. First-principles calculation of optical properties of C_{60} in the fcc lattice. *Phys. Rev. Lett.* 67:2045–48.
28. Johnson, P. B., and Christy, R. W. 1972. Optical constants of the noble metals. *Phys. Rev. B* 6:4370–79.
29. Eilmes, A., and Pack, B. 2007. Private communication.
30. Kanda, Y., and Sponer, H. 1958. Triplet-singlet emission spectra of solid toluene at 4°K and 77°K and in EPA solution at 77°K. *J. Chem. Phys.* 28:798–806.
31. Konarev, D. V., and Lyubovskaya, R. N. 1999. Donor-acceptor complexes and ion-radical salts based on fullerenes. *Russ. Chem. Rev.* 68:19–38.
32. Sheka, E. F., Razbirin, B. S., Starukhin, A. N., Nelson, D. K., Lyubovskaya, R. N., Troshin, P. A., and Kamanina, N. V. 2009. Nonlinear photonics of fullerene solid solutions. arXiv:0901.3728v1 [cond-mat.mtrl-sci].
33. Kamanina, N. V. 2005. Fullerene-dispersed nematic liquid crystal structures: Dynamic characteristics and self-organization processes. *Physics-Uspekhi* 48: 419–28.
34. Andrievsky, G. V., Avdeenko, A. A., Derevyanchenko, L. I., Fomin, V. I., Klochkov, V. K., Kurnosov, V. S., and Peschanskii, A. V. 2005. Peculiarities for luminescence in systems with fullerene C_{60}-water interface. In *Spectroscopy of emerging materials*, ed. E.C. Faulques, D. L. Perry, and A. V. Yeremenko, 151–60. NATO Science Series 165. Dordrecht: Kluwer Academic Publishers.
35. Nissen, M. K., Wilson, S. M., and Thewalt, M. L. W. 1992. Highly structured singlet oxygen photoluminescence from crystalline C_{60}. *Phys. Rev. Lett.* 69: 2423–26.
36. Ala-Kleme, T., Mäki, R., Laaksonen, P., and Haapakka, K. 2002. Blue and red photoluminescence of bicapped [60]fullerene-γ-cyclodextrin complex in aqueous solutions. *Anal. Chim. Acta* 472:83–87.
37. Pan, B. D., Zhang, J., Li, Z., and Wu, M. 2009. Hydrothermal route for cutting graphene sheets into blue-luminescent graphene quantum dots. *Adv. Mater.* 21:1–5.
38. Sheka, E. F. 2010. Synergistic nanophotonics of fullerenes. arXiv.1008.1690v1 [physics.chem-ph].

chapter eleven

Odd electron–enhanced chemical reactivity of carbon nanotubes

11.1 Introduction

Peculiarities of chemical reactivity, which we considered in the previous chapters, were related mainly to C_{60}. Obviously, the disclosed regularities can be attributed to other fullerenes as well since they are intimately connected with the sp^2 origin of the molecule's electronic structure. Within the framework of the unrestricted broken symmetry Hartree-Fock (UBS HF) approach, the features of the C_{60} electron system are connected with effectively unpaired electrons in the case of weakening interaction between odd electrons when the C-C bond length exceeds 1.395 Å. Since the C-C bond lengths in all fullerenes fill the same region, one might expect that the atomic chemical susceptibility approach, which turned out to be successful for describing chemical reactivity of C_{60}, can be expanded over other fullerenic molecules or, in other words, over a fullerenic class of sp^2 nanocarbons. However, fullerenes are not the only representatives of the latter, among which carbon nanotubes (CNTs) and graphene are the most popular today. And a natural temptation arises to tackle the two new classes of sp^2 nanocarbons on the platform suitable for fullerenes. Standard C-C bonds of the tubes and graphene exceed the above critical value, which might lay the foundation of the description of their chemical reactivity in terms of effective unpairing of their odd electrons. It is interesting as well to compare the three classes of nanocarbons differing by shape to elucidate the role of the latter, in particular.

11.2 Chemical reactivity of carbon nanotubes

The computational consideration of the chemical reactivity of carbon nanotubes has a long history. This is due to the increasing number of physical, chemical, and biomedical applications of the species, which require chemical modification of the latter to make them more amenable to rational and predictable manipulation.[1] Two main strategies concerning noncovalent[2] and covalent[3] functionalizations have been elaborated and have gained large practical success and achievements (see comprehensive reviews[4–11]). Empirically exploited, the two approaches have revealed the two-well structure of the intermolecular interaction (IMI) potential in dyads with a CNT as one partner. And here we meet the first indication of intrinsic similarity in the behavior of fullerenes and CNTs in regards to intermolecular interaction. This was expected since CNTs, similarly to fullerenes, are both good donors and acceptors of electrons, with values of ionization potential and electron affinity comparable to those of fullerenes. Controllable functionalizing techniques, which can provide CNTs tailoring in a determinable manner, are far from being completed; more investigations are required to elucidate the nature and locality of covalently attached moieties. Due to extreme complications of native species as well as difficulties in their separation and individual studying, a considerable load connected with looking for answers was put on the shoulders of computational approaches. Starting in Niyogi et al.,[5] a number of calculations have been performed.[12–22]

Not very long ago the main concept of the connection between the electronic structure and chemical reactivity of CNTs was based on the Haddon approach,[23] which attributed the reactivity of fullerenes to the strain engendered by their spherical geometry as reflected in pyramidalization angles of carbon atoms. It was suggested that the curvature-induced pyramidalization and misalignment of the π orbitals of the carbon atoms[24,25] induces a local strain in a defect-free CNT. This concept made allowance for explaining the difference between the reactivity of a CNT end cap and that of a sidewall in favor of the former, and gave a simple explanation of the reactivity increase while the CNT diameter decreases. Further development of the approach turned out to be useful for considering the reactivity of the convex and concave sidewalls of CNTs toward addition reactions.[12,13]

The Haddon approach was quite productive but basically empirical, while more sophisticated theoretical approaches were needed. Various density functional theory (DFT) techniques actively explored in recent years with respect to the chemical functionalization of carbon nanotubes, to name a few,[15–22] were a natural response to the requirement. However, a lot of severe limitations were usually implied when using the technique. This concerns:

1. The aromaticity concept attributed to the electronic structure of the tubes, which results in the closed-shell approximation for wave functions (that concerns the restricted DFT approach)
2. The periodic boundary conditions along the tubes, which restrict the tube area consideration to the sidewall only, leaving the tube ends, the most active spots on the tube, as was shown,[26] outside the consideration

3. The absence of atomically matched characteristics that could exhibit the chemical reactivity of the tube atoms
4. A *postfactum* character of simulations that were aimed at obtaining energy and structural data consistent with experimental findings by adapting the functionals used

As was the case with fullerenes, the most crucial is the first assumption in view of weakly interacting odd electrons in carbon nanotubes. When the problem is considered in the framework of the aromaticity concept, this means strong coupling between the electrons so that odd electrons are covalently coupled in pairs, similarly to π electrons of the benzene molecule. The only UBS DFT calculation that is known so far[27] concerns a thorough analysis of the singlet state of cyclacenes and short zigzag CNTs. It was shown that the energy of the UBS DFT singlet state was lower than that of the restricted DFT singlet that pointed to the open-shell character of the singlet state. This finding is of great importance, exhibiting the falsity of aromaticity-based concept for quantitative description of the electronic states of carbon nanotubes on the DFT platform.

The UBS HF approach was first applied to two fragments of (4,4) defect-free and (4,4)[5–7] defect single-walled CNTs (SWCNTs).[26] It was convincingly shown that N_{DA} values appeared and followed in synchronism, with the excess of C-C distance that separates two odd electrons over the limit value $R_{cov} \cong 1.395$ Å. Therefore, the N_{DA} map that describes the tube atoms' atomic chemical susceptibility is tightly connected with the tube structure, thus highlighting the distribution of the C-C bond length excess over the tube. To prove the statement, an extended set of fragments of (n,n) and (m,0) CNTs was considered later on.[28]

11.2.1 (4,4) Single-walled carbon nanotubes

To manifest the main consequences that follow from the consideration of the electronic structure of CNTs in the framework of the UBS HF approach, let us consider a collection of results related to a set of fragments of (4,4) SWCNT. Equilibrium structures of the fragments are presented in Figures 11.1 and 11.2. Numbered from NT1 to NT9, the fragments form three groups. The first group (Figure 11.1) involves three fragments related to short defect-free (4,4) SWCNTs different by the end structure. One end of all fragments is capped, while the other is either open but hydrogen terminated (NT1) and empty (NT2), or capped (NT3). The attached hydrogens are not only service terminators, but also present real hydrogenation of the tube open end. The collection is added to by a defect (4,4) SWCNT (NT4) with a Stone-Wales pentagon-heptagon pair. The second group (Figure 11.2)

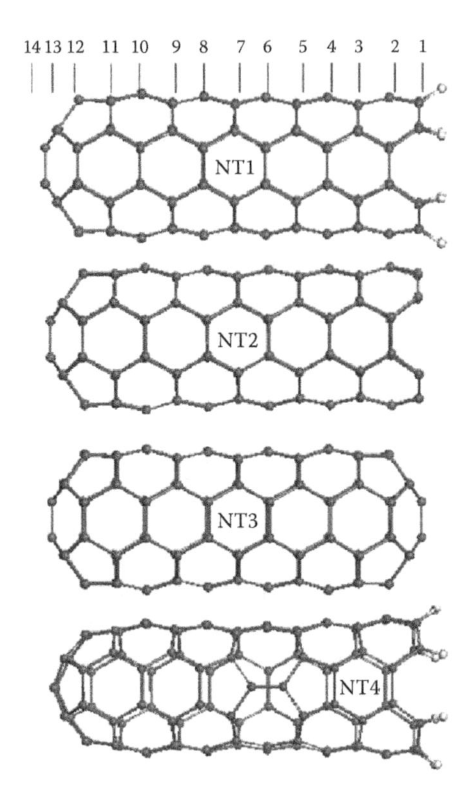

Figure 11.1 Equilibrium structures of (4,4) single-walled carbon nanotube fragments. Group 1 (NT1, NT2, NT3, and NT4). UBS HF AM1 singlet state. (From Sheka, E. F., and Chernozatonskii, L. A., *Int. J. Quant. Chem.*, 110:1466–80, 2010.)

involves fragments NT5 and NT6, which represent capped-end and open-end (4,4) SWCNTs, but different by the termination of the open end. They both differ from fragments NT1 and NT3 by four atom rows' elongation. The third group (Figure 11.2) combines NT7, NT8, and NT9 fragments of (4,4) SWCNTs with both open ends, but different by hydrogen termination. The fragment sidewall is longer by one atom row than that of the second group. To sum up, the selected set of fragments has allowed exhibiting the following structure effects on the N_{DA} distribution caused by (1) end-capping, (2) terminating or emptying open ends, (3) introducing a pair of pentagon-heptagon defects, and (4) the tube sidewall elongation.

11.2.1.1 Fragments of group 1

As shown in Sheka and Chernozatonskii,[26,28] the dominating majority of C-C bonds of the tube are longer than the limit R_{cov} = 1.395 Å, so that the large total number of effectively unpaired electrons for the tube N_D = 32.38

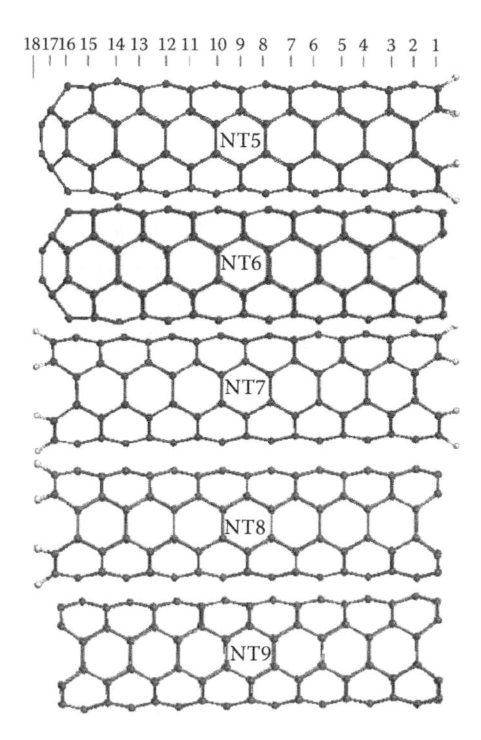

Figure 11.2 Equilibrium structure of (4,4) single-walled carbon nanotube fragments of group 2 (NT5 and NT6) and group 3 (NT7, NT8, and NT9). UBS HF AM1 singlet state. (From Sheka, E. F., and Chernozatonskii, L. A., *Int. J. Quant. Chem.*, 110:1466–80, 2010.)

does not look strange. The distribution of these electrons over the atoms forms the ACS N_{DA} map that is shown in Figure 11.3(a). As seen in the figure, the tube can be divided into three regions. The first is related to the cap with adjacent atoms and covers rows 9 to 14. The second concerns mainly the tube sidewall and covers rows 4 to 8. The third refers to the open end terminated by hydrogen atoms and covers rows 1 to 3. The biggest nonuniformity of the N_{DA} distribution is characteristic of the cap region. One should pay attention to the fact that the largest N_{DA} values belong to atoms that form the longest bonds. As for the sidewall region, the N_{DA} distribution is practically uniform, with the N_{DA} value scatter not bigger than 0.5%, which is consistent with a quite uniform distribution of C-C bond lengths as well. The N_{DA} distribution in the end region is significantly affected by the hydrogenation. Therefore, the C-C bond length is actually a controlling factor in the distribution of the density of the effectively unpaired electrons over atoms. The curve with dots in Figure 11.3(a) presents the free valency of the tube atoms calculated in accordance with Equation 3.6 that well coincides with the atomic chemical susceptibility

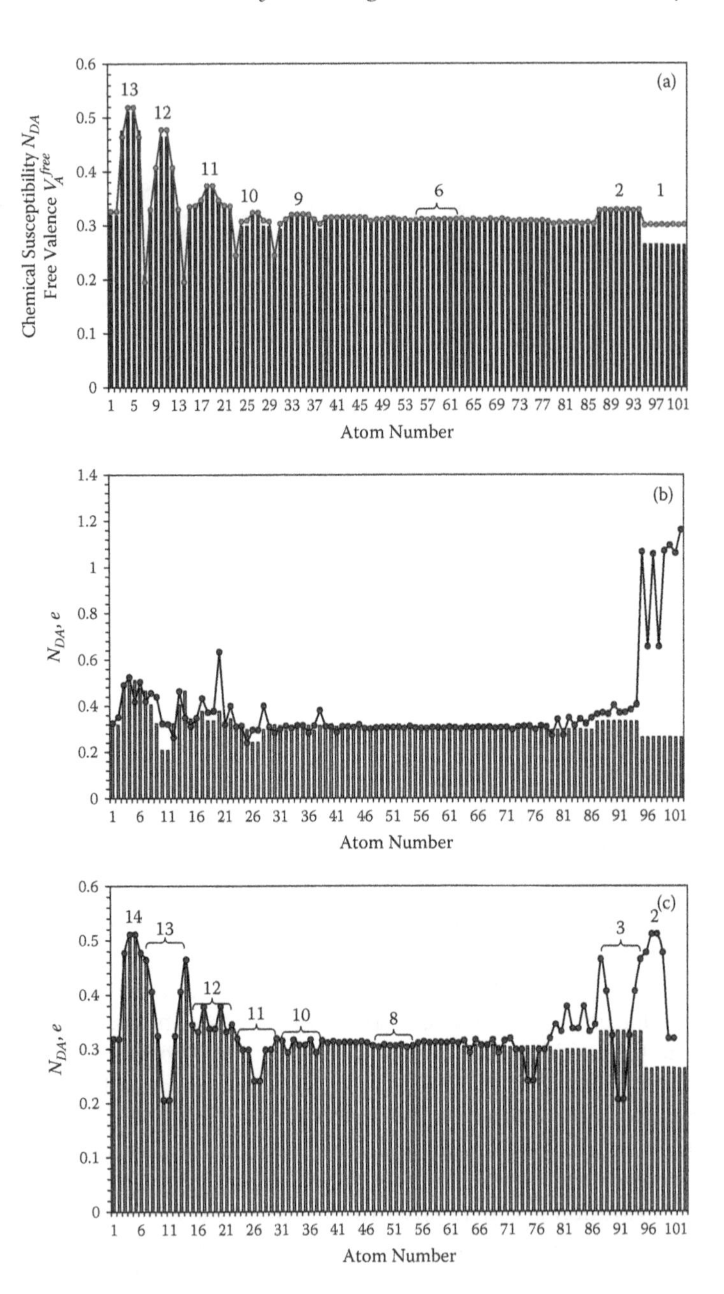

Figure 11.3 Atomic chemical susceptibility N_{DA} maps. (a) NT1 fragment; histogram exhibits N_{DA} distribution; curve with dots represents free valence. (b) NT1 (bars) and NT2 (curve with dots) fragments. (c) NT1 (bars) and NT3 (curve with dots) fragments. Numerals match atom rows. (From Sheka, E. F., and Chernozatonskii, L. A., *Int. J. Quant. Chem.*, 110:1466–80, 2010.)

expressed by N_{DA}. According to the latter, the tube cap is the most reactive part, while the tube sidewall is more passive, with ill-pronounced selectivity along the tube.

Removing hydrogen atoms at the tube's open end, one obtains the N_{DA} map of the NT2 fragment shown in Figure 11.3(b). A tremendous contribution of the end atoms obviously dominates the map. This is due to the fact that the ethylene-like C-C bonds (see discussion regarding Figure 2.1 in Section 2.3) are replaced by the acetylene-like bonds at the tube end, where each atom has not one but two odd electrons. The transformation naturally results in increasing the total number of effectively unpaired electrons, N_D, from 32.38 to 39.59. The injection of additional effectively unpaired electrons disturbs the N_{DA} map of the hydrogen-terminated tube (shown in the figure by bars) quite considerably. It is important to note that the changes occur not only in the vicinity of the open end within rows 2 and 3, which is quite reasonable, but also in the opposite cap end (rows 14 to 11). Practically no changes occur along the tube sidewall, which seems to serve as a peculiar resonator for the electron conjugation. Addressing the chemical activity of the tube, dominant activity of the empty-end atoms is evident.

When the other end is capped, fragment NT3 (N_D = 33.35) becomes highly symmetrical (D_{2h}), which is reflected in its N_{DA} map, shown in Figure 11.3(c). The N_{DA} distribution of the fragment is specularly symmetrical with respect to the atoms of row *8* in the middle. As seen in the figure, besides the second cap region, the addition of the second cap results in no changes in the distribution, which is characteristic of the one-cap fragment NT1.

The main alterations caused by the defect introduction in the NT4 fragment concern the atoms of rows 5 to 7 that are involved in the defect structure (Figure 11.4). The remainder of the sidewall region, as well as the cap- and open-end regions of the tube, is much less affected. The significance of the open-end termination can be traced through charge characteristics of the fragments as well. In series NT1, NT2, and NT3 dipole moments of the fragments are 9.93, 0.44, and 0.03 D, respectively. A drastic drop of the value when terminating hydrogen atoms are removed is evidently connected with the charge redistribution over the tube atoms. Figure 11.5 presents the corresponding plot for the fragments. Bars in the figure show the charge distribution along the NT1 fragment. The atom enumeration coincides with that of the N_{DA} map in Figure 11.3(a). As seen in the figure, attached hydrogens as well as adjacent carbon atoms provide the main contribution to the distribution. The influence of terminators is still sensed at the other two atomic layers toward the tube cap. The remainder of the tube sidewall is practically uncharged, and a weak charging of atoms appears, again, only at the end cap. The acquired total negative charge is fully compensated by the positively charged hydrogen

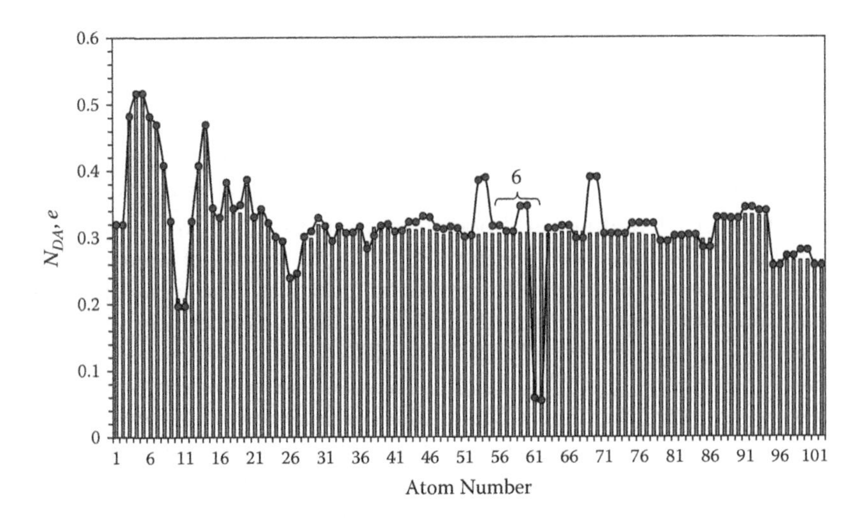

Figure 11.4 Atomic chemical susceptibility N_{DA} maps of NT1 (bars) and NT4 (curve with dots) fragments. (From Sheka, E. F., and Chernozatonskii, L. A., *Int. J. Quant. Chem.*, 110:1466–80, 2010.)

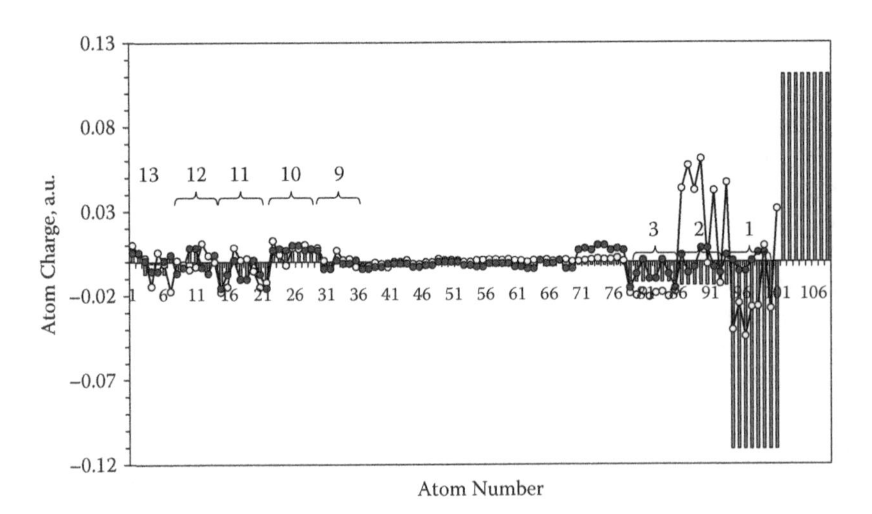

Figure 11.5 Charge distribution over atoms of fragments of group 1. Bars, and curve with light and dark dots plot data for NT1, NT2, and NT3, respectively. Numerals match atom rows of fragments NT1 and NT2. (From Sheka, E. F., and Chernozatonskii, L. A., *Int. J. Quant. Chem.*, 110:1466–80, 2010.)

atoms. However, the significantly charged tail of the fragment provides a high value of the dipole moment.

When hydrogen atoms are removed, the charge map is reconstructed. First, it concerns end carbon atoms whose charge greatly decreases but the charging area is still large, covering rows 1 to 3. Besides, the charge redistribution is seen in the cap region as well while along the tube sidewall changes are small. And again, as in the case of the N_{DA} distribution, one can speak about a resonant character of the disturbance transfer along the tube sidewall. One might suggest that this very charge transfer along the tube results in a significant decrease of the dipole moment value in spite of a still considerable charge on the tube open end.

Replacing the open end by the second cap does not violate the charge distribution in the region of the first cap and simply duplicates it in the region of the second cap just making the total tube N_{DA} map specularly symmetrical with respect to the middle row of atoms. Highly symmetric and having low value, the charge distribution produces an exclusively small dipole moment.

11.2.1.2 *Fragments of group 2*

The N_{DA} maps given for fragments NT5 and NT6 in Figure 11.6 show that similarly to members of group 1, considered in the previous subsection, the N_{DA} map of the fragments consists of three regions related to the cap end on the left, open end on the right, and an extended tube sidewall. As for both end regions, the data for the elongated tube fully reproduce those for the shorter one, irrespective of the open end being either hydrogen terminated (Figure 11.6(a)) or empty (Figure 11.6(b)). Similar behavior of the map should be expected if the open end is capped. The only difference in the distributions related to shorter and longer fragments consists in expending the distributions' fraction related to the tube sidewalls. Therefore, the N_{DA} map may be presented as consisting of three fragments, MapI, MapII, and MapIII. In the case of the two cap ends, MapIII should be replaced by MapI.

11.2.1.3 *Fragments of group 3*

Group 3 covers fragments with both open ends. Similarity in the tube length of fragments from group 2 and group 3 allows revealing the changes caused by replacing one or two capped ends by open ends. Figure 11.7(a) presents a comparative view of N_{DA} maps of fragments NT5 and NT7, which have similar right open ends, terminated by hydrogen, and different left ends, presented by the cap in the case of NT5 and by the hydrogen-terminated open end in NT7. The comparison shows that if the three-region pattern described above is characteristic of the capped NT5 fragment, this cannot be said about NT7. The N_{DA} map in this case is

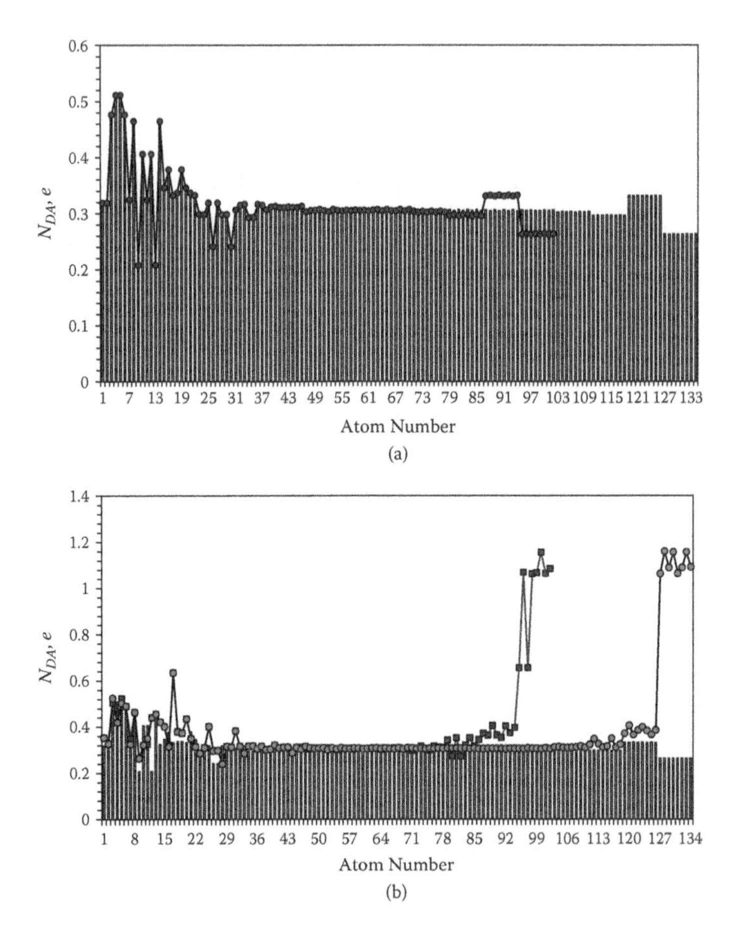

Figure 11.6 Atomic chemical susceptibility N_{DA} maps. (a) NT5 (bars) and NT1 (curve with dots) fragments. (b) Bars, and curves with light and dark dots plot data for NT5, NT6, and NT2 fragments, respectively. (From Sheka, E. F., and Chernozatonskii, L. A., *Int. J. Quant. Chem.*, 110:1466–80, 2010.)

specularly symmetrical with respect to the middle row, but is quite inhomogeneous, with a peculiar step-like character reaching minimum at the tube center. It should nevertheless be noted that the step character related to the last three rows on the right end is the same as that for the NT5 open end.

Removing hydrogen atoms on the right end of NT7 brings the N_{DA} map of fragment NT8 to the three-region form described in the previous subsection (see Figure 11.7(b)).

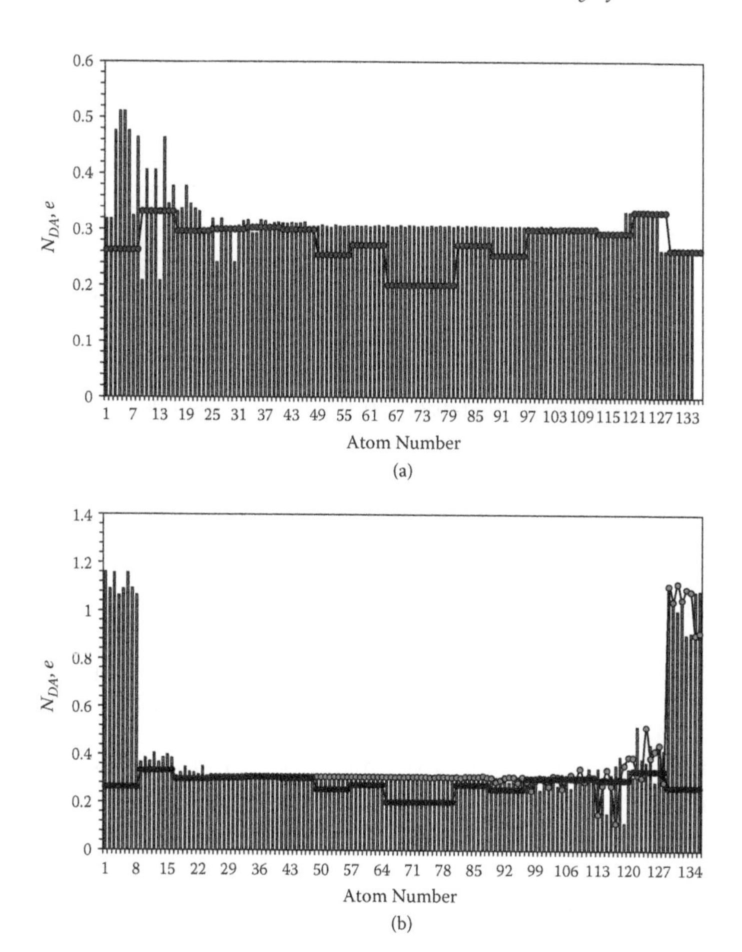

Figure 11.7 Atomic chemical susceptibility N_{DA} maps. (a) NT5 (bars) and NT7 (curve with dots) fragments. (b) Bars, and curves with light and dark dots plot data for NT9, NT8, and NT7 fragments, respectively. (From Sheka, E. F., and Chernozatonskii, L. A., *Int. J. Quant. Chem.*, 110:1466–80, 2010.)

11.2.2 (n,n) and (m,0) Single-walled carbon nanotubes

11.2.2.1 (n,n) Single-walled carbon nanotubes

Calculations show that the three-region pattern of the N_{DA} maps of (4,4) fragments discussed above is generally supported for fragments of any (n,n) composition. Thus, the graphical views of the N_{DA} maps for (5,5), (6,6), (7,7), and (8,8) members are quite similar to those of the (4,4) set, while being naturally different numerically. The corresponding N_{DA} values are summarized in Table 11.1. When the scattering of data is big,

Table 11.1 Atomic chemical susceptibility of single-walled carbon nanotubes

	Diameter, Å	N_{DA},[a] e			
				Open ends	
		Cap	Sidewall	Empty	H terminated
		SWCNTs (n,n)[b]			
4,4					
NT5, NT5*	5.76	0.51–0.32	0.31	1.16; 1.06	0.26
5,5					
NT10, NT10*	7.06	0.17–0.03	0.30–0.26	1.16; 0.66	0.26
6,6					
NT11, NT11*	8.34	0.29–0.06	0.25	1.16; 0.67	0.26
7,7					
NT12, NT12*	9.79	0.34–0.06	0.26	1.07; 0.67	0.25
8,8					
NT13, NT13*	10.91	0.34–0.06	0.24	1.16–1.06; 0.67–0.68	0.25
9,9					
NT14, NT14*	12.64	0.26–0.04	0.23; 0.03	1.12–0.67	0.25; 0.09
10,10					
NT15, NT15*	14.00	0.26–0.05	0.23; 0.19–0.17; 0.10–0.06	1.16–0.97; 0.68	0.24–0.22; 0.19–0.17; 0.06–0.02
		SWCNTs (m,0)			
8,0					
NT16, NT16*	6.20	0.45–0.11	0.29–0.28	1.80–1.11	0.49–0.24
10,0					
NT17, NT17*	7.97	0.35–0.19	0.26	1.36; 1.14	0.45
12,0					
NT18, NT18*	9.61	0.37–0.04	0.26–0.21	1.36; 1.15	0.47–0.11

Source: Sheka, E. F., and Chernozatonskii, L. A., *Int. J. Quant. Chem.*, 110:1466–80, 2010.

[a] Within each cell the data are presented by single numbers or a set of single numbers when their scattering around the corresponding number is less than 0.01 *e*. Otherwise, the data are presented by intervals or groups of intervals.

[b] NT*N* nominates a capped-end/H-terminated open-end fragment while NT*N** does the same after removing hydrogen atoms from the open end (see NT5 and NT5* (that is, NT6) in Figure 11.2). Fragments from NT14 to NT18 are shown in Figure 11.8.

the related interval of the value is shown. The analysis of the data given in the table makes allowance for exhibiting the dependence of the ACS on the SWCNT diameter. As seen from the table, the dependence is different in different regions. Thus, ACS of the H-terminated ends does not show any dependence at all. Carbon atoms of the empty ends form two groups, which are characterized by different N_{DA} values. The bigger value of 1.16 does not change when the diameter increases, while the lower one drops from 1.06 to 0.67 when passing from (4,4) SWCNT to (5,5), and then remains unchanged. The sidewall ACS decreases gradually when the diameter increases and approaches a constant value when the diameter grows further. Qualitatively, the ACS behavior in this region looks like that expected on the basis of the Haddon approach. As for the cap region, large scattering of the ACS values as well as their peculiar change reflect a severe reconstruction of the C-C bond net in the region, adapting it to the minimal stress of the tube body as a whole and its cap end, in particular. From this viewpoint, small N_{DA} values at the cap of the (5,5) tube are obvious. The tube symmetry is C_{5v} and the cap structure is comfortably centered around the pentagon. The C-C bonds are the least stressed and lengthened, which causes the smallest N_{DA} values.

As seen from Table 11.1, opposite to the previous case, scattering of the N_{DA} values of tubes (9,9) and (10,10) is significant. Particularly, small N_{DA} values in the sidewall and open H-terminated regions should be mentioned. This feature was attributed to the great stress of the tube bodies, which is clearly seen in Figure 11.8. As follows, in this case one cannot expect regular C-C bond lengths' distribution in the cap region, or on the sidewall and end atoms as well. The observed decrease of the N_{DA} values is connected with the C-C bond shortening that reflects the body stress. At the same time, the largest values in the sidewall and H-terminated end region are similar to those of the previous group and clearly show their saturation under the tube diameter growing. Similar characteristics might be expected for tubes of larger diameter as well.

In spite of large scattering of the N_{DA} data in the sidewall region, their distribution retains a regular view. The matter is that the scattering takes place within each individual row of the sidewall while being repeated regularly over rows. The N_{DA} distribution within one row of the studied tubes is shown in Figure 11.9. As seen in the figure, it looks quite similar for both tubes as well for other rows of the two tubes.

11.2.2.2 (m,0) Single-walled carbon nanotubes

While (n,n) SWCNTs are all semiconductive, the (m,0) SWCNT family consists of both semiconductive and metallic (really, semimetallic) tubes, depending on whether m is divisible by 3. Therefore, a series of (8,0), (10,0), and (12,0) fragments covers two semiconductive and one metallic tube and offers two new aspects for the comparative study. The latter concerns

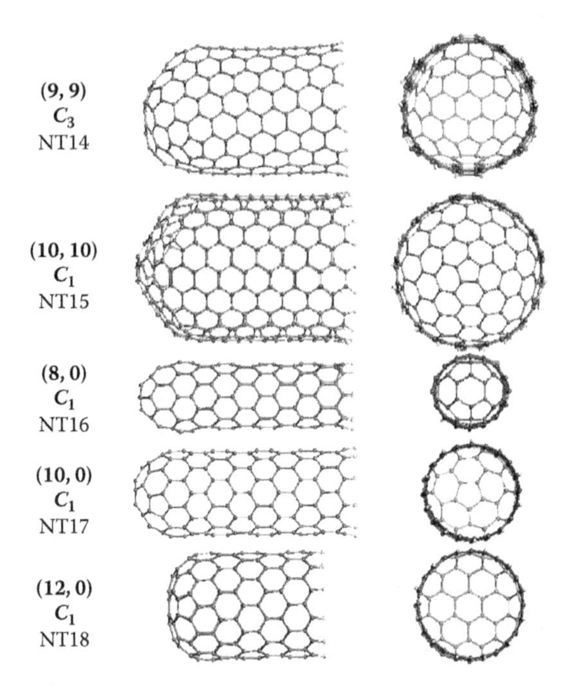

Figure 11.8 Equilibrium structures of armchair (9,9) (NT14) and (10,10) (NT15) fragments and zigzag (8,0) (NT16), (10,0) (NT17), and (12,0) (NT18) SWCNT fragments in two perpendicular projections. (From Sheka, E. F., and Chernozatonskii, L. A., *Int. J. Quant. Chem.*, 110:1466–80, 2010.)

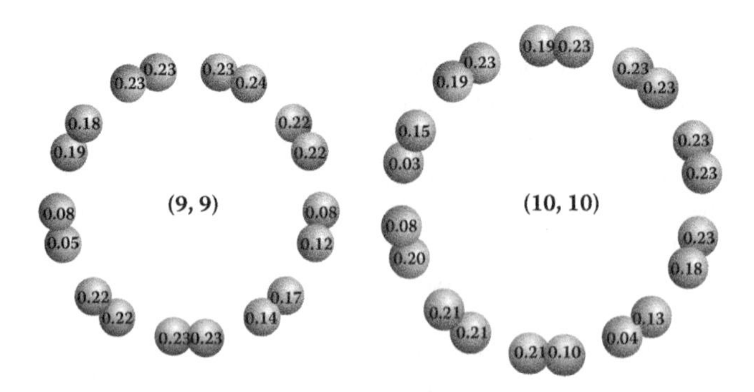

Figure 11.9 Atomic chemical susceptibility N_{DA} maps of the cross-sections of (9,9) (NT14) and (10,10) (NT15) SWCNTs. (From Sheka, E. F., and Chernozatonskii, L. A., *Int. J. Quant. Chem.*, 110:1466–80, 2010.)

changes of the N_{DA} maps caused by (1) replacing armchair open ends of (n,n) tubes by zigzag ones for the (m,0) family and (2) changing the tube conductivity within the (m,0) family.

In response to the first inquiry, the calculations performed in Sheka and Chernozatonskii[28] have revealed a large similarity in the behavior of (n,n) and (m,0) fragments. A comparative view on the N_{DA} maps of (4,4) and (8,0) fragments, which are both semiconductive and close in diameter, is presented in Figure 11.10(a) and (b). As seen from the figure, N_{DA} maps of the two tubes are both qualitatively and quantitatively similar. First, the maps consist of three parts related to variable N_{DA} values at capped and open ends, and to rather homogeneous quantities along the tube sidewalls. In both cases the open empty ends are the most chemically reactive. Following these places in activity are the end caps and sidewall. Second, not only the shape of the N_{DA} maps, but numerical plottings related to the end caps and sidewalls (Table 11.1) are practically the same. The only difference concerns zigzag open ends, both H terminated and empty, that exhibit about a 15% increase of the reactivity in comparison with the one of the armchair ends. These regularities remain to a great extent for SWCNTs of bigger diameter: compare the data in Table 11.1 for (6,6) and (10,0) and for (7,7) and (12,0) tubes, as well for graphene, as will be shown in Section 12.2.

A comparison of (7,7) and (12,0) tubes is of particular interest since (7,7) SWCNT is semiconductive while (12,0) is metallic. A more detailed comparison between the maps of the tubes of different conductivity is visualized in Figure 11.10(c) for (10,0) and (12,0) SWCNTs belonging to the same family. As seen in the figure, plottings for both tubes are well similar in both shape and numbers and do not exhibit any remarkable characteristic difference that might be related to a change in the tubes' conductivity. Therefore, the chemical reactivity of atoms of both semiconductive and metallic SWCNTs is comparable, which should not cause different behavior of the tubes with respect to similar chemical reactions. This is consistent with the reality one faces when using sidewall covalent chemistry for tube separation.[29] Partial luck in achieving the goals of distinguishing semiconductive and metallic tubes by using chemical modification seems to be connected with the peculiarities of intermolecular interaction between the tube and corresponding additives caused by the donor–acceptor interaction.

11.3 General view on single-walled carbon nanotubes' chemical reactivity

Exhibited peculiarities of the obtained SWCNT ACS (N_{DA}) maps allowed making the following conclusions concerning addition reactions to be expected:[28]

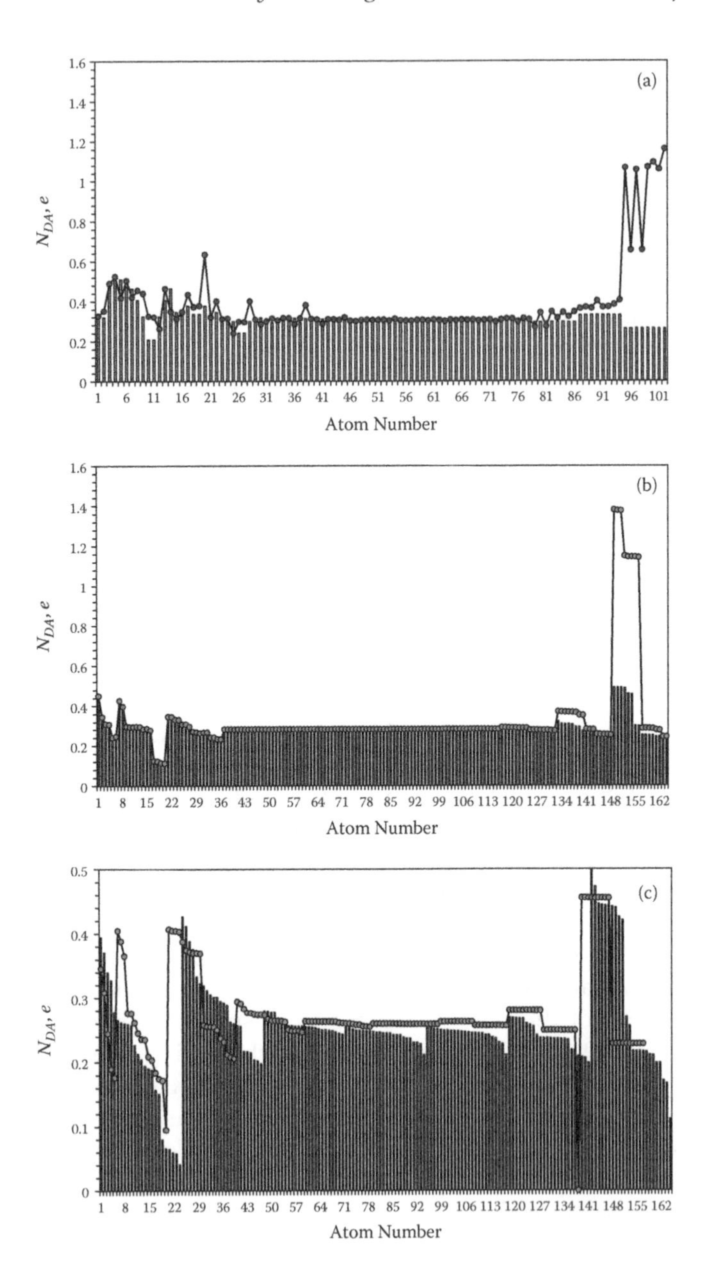

Figure 11.10 Atomic chemical susceptibility N_{DA} maps. (a) (4,4,) SWCNT. Histogram and curve with dots are related to fragments NT5 and NT6, respectively. (b) (8,0) single-walled carbon nanotube. Histogram is related to NT16 fragment, while curve plots N_{DA} quantities for NT16* when hydrogen atoms are removed from the open end of NT16. (c) NT17 (histogram) and NT18 (curve) fragments. (From Sheka, E. F., and Chernozatonskii, L. A., *Int. J. Quant. Chem.*, 110:1466–80, 2010.)

1. The space of chemical reactivity of any SWCNT coincides with its coordinate space, while remaining different for particular structure elements. This simultaneously both complicates and facilitates chemical reactions involving the tubes, depending on a particular reaction goal. A summarized view on chemical reactivity of the tubes, presented in detail by the N_{DA} maps in Figures 11.3, 11.4, 11.6, 11.7, and 11.10, is given in Table 11.1. Shifting the activity to either cap- or empty-end regions, depending on the tube shape and composition, may be done to look for practical ways for tube chemical covalent modification.
2. Local additions of short-length addends (involving individual atoms, simple radical, and so forth) to any SWCNT are the most favorable at open empty ends, both armchair and zigzag ones, and the latter somewhat more effective. Following these places in activity are end caps, defects in the tube sidewall, and the sidewall itself.
3. Chemical reactivity of the open but chemically terminated ends of SWCNTs exceeds that of the sidewall only for tubes with zigzag ends.
4. Single local addition of long-length addends (polymers) will follow similar rules. However, wrapping along the SWCNT sidewall will be more favorable due to a large number of local contacts on the way.
5. Chemical contacts of SWCNTs with spatially extended partners (substrates of different kind, thin films, graphene sheets, etc.) form an interface between the bodies, the configuration of which depends on whether the tube is oriented normally to or parallel over the partner. In the latter case, a particular situation occurs when mono-atomic-layered graphene may act additionally as a cutting blade.[30]
6. Addition reactions with participation of multiwalled carbon nanotubes will proceed depending on the target atoms involved. If the tubes' empty open ends are the main targets, the reaction will proceed as if one were dealing with an ensemble of individual SWCNTs. If the sidewall becomes the main target, the reaction output will depend on the accessibility of the inner tubes' sidewall, in additional to the outer one.

11.4 Comparison with experiment

Lack of experimental data for individual single-walled carbon nanotubes complicates the exact verification of the obtained data. However, there are some generalized observations that can be discussed in view of the performed computations. Actually, we will consider only a restricted number of examples from a practically endless ocean of experimental evidence now available. These examples are nevertheless characteristic enough to present an empirical picture of the modern chemistry of CNTs.

1. Studies performed by Mawhinney et al.[31,32] can be attributed to revealing empty open ends interacting with primarily esters and quinones via investigation of oxidatively cut SWCNT caps of ~1 nm in diameter. Such oxidized SWCNTs have been assembled on a number of surfaces, including silver,[33] highly oriented pyrolytic graphite,[34] and silicon.[35] High coverage densities and orientation normal to the surface have been shown; the latter are suggestive of higher degrees of functionalization at the nanotube ends. These findings correlate well with the calculated high reactivity of the SWCNT empty open ends.[28]
2. Frequently, intrinsic defects on SWCNTs are supplemented by oxidative damage to the nanotubes' framework by strong acids, which leave holes functionalized with oxygenated groups such as carboxylic acid, ketone, alcohol, and the ester group.[36] This and other studies (see reviews[6,9,37]) point to high chemical reactivity of the defects, which is in good agreement with the calculated data.[28]
3. Attributing end caps of SWCNTs to the chemically active regions is widely accepted.[5,38] However, as shown by calculations, this region is not always the most active since its activity can be overcome by a particular composition of open ends.
4. Some time ago there was skepticism concerning the sidewall activity, which was considered much smaller with respect to that of fullerenes.[4–15] However, producing small-diameter SWCNTs has removed the doubts, opening the way to large-scale hydrogenation,[39] fluorination,[40] amination,[40] and a great number of other addition reactions[10,41] involving SWCNT sidewalls.

Therefore, available chemical data correlate well with predictions made on the basis of the computations described in Section 11.2. Particularly, the available data on the experimental study of hydrogenated SWCNTs[39] should be noted. As shown, a significant perturbation of SWCNTs' structure under hydrogenation occurs. This finding seems to be understood on the basis of data[28] concerning the tube with both H-terminated open ends (see Figure 11.7). The computations clearly highlight that attaching hydrogen atoms to the most active places of the tube causes a significant disturbance of its N_{DA} map and, consequently, the tube structure that is spread over a large region. Obviously, the more atoms that are attached, the more drastic are the changes that occur and are observed experimentally.

11.5 Electronic characteristics of single-walled carbon nanotubes

A quite extended set of data, obtained for (4,4) SWCNT fragments, makes allowance for presenting a summarized view on electronic properties of the tubes listed in Table 11.2:

Table 11.2 Basic electronic characteristics of fragments of (4,4) single-walled carbon nanotube[a]

	UBS HF singlet state					
Fragments	Symmetry	ΔH, kcal/mol	D, D	I, eV	ε, eV	N_D
NT1	C_{2v}	1,182.60 1,342.84[b]	9.928	9.19	2.19	32.38
NT2	C_2	1,648.34	0.440	9.71	3.16	39.59
NT3	D_{2h}	1,447.39	0.029	9.81	2.45	33.35
NT4	C_S	1,271.77	9.954	9.13	2.21	32.60
NT5	C_{2v}	1,476.6	10.390	9.16	2.26	42.17
NT6	C_2	1,962.91	0.428	9.71	3.14	50.47
NT7	C_{4h}	1,271.42	0.003	8.20	2.38	37.96
NT8	C_{4h}	1,689.58	10.990	9.19	2.92	47.36
NT9	C_{2h}	2,136.90	0.225	9.66	3.04	54.90

Source: Sheka, E. F., and Chernozatonskii, L. A., *Int. J. Quant. Chem.*, 110:1466–80, 2010.

[a] ΔH, D, I, and ε represent the heat of formation, dipole moment, ionization potential, and electron affinity, respectively.

[b] Restricted Hartree-Fock data. A similar ~14% increase in the energy value has been observed for all listed fragments when going from UBS HF to the restricted Hartree-Fock calculation scheme.

1. According to the fragment energy, end-cap/H-terminated open-end and H-terminated open-end/H-terminated open-end tubes are the most energetically stable among unilength fragments.
2. Removing eight hydrogen terminators from end-cap/H-terminated open-end tubes requires 465.7 and 486.3 kcal/mol in the case of shorter and longer tubes, respectively. Supposing this energy is needed for the disruption of C-H bonds, one obtains 58.2 and 60.8 kcal/mol (or 2.5 and 2.6 eV) per one bond, which is consistent with both experimental[39] and calculated data[14,15] estimating the energy of ~2.5 eV for the formation of one C-H bond.
3. Substitution of eight terminators at the open end by the second cap requires 264.8 kcal/mol, which is essentially less than that needed for emptying the end.
4. Emptying one open end of the H-terminated open-end/H-terminated open-end fragment costs 418.2 kcal/mol (52.3 kcal/mol per one C-H bond), while additionally removing eight hydrogen terminators from the other end adds 447.3 kcal/mol (55.9 kcal/mol) to the energy cost. As seen, the obtained data for the end-cap/open-end tube and open-end/open-end one are well consistent.
5. All three end-cap/H-terminated open-end tubes are polarized with a large dipole moment of ~10 D, while the end-cap/empty open-end and

end-cap/end-cap tubes have a very low dipole moment, if any. Practically, this circumstance is very important when applying SWCNTs either to improve the characteristics of nonlinear optical devices based on liquid crystalline media[42,43] or to design new hybrid materials based on SWCNTs and electron donor–acceptor nanocomposites.[10]

6. Among open-end tubes, the H-terminated open-end/empty open-end tube is the most polarized, while H-terminated open-end/H-terminated open-end and empty open-end/empty open-end tubes have small dipole moments.
7. It is important to note the high level of donor–acceptor characteristics of the tubes. All tubes are characterized by not very high ionization potentials and high electron affinity. The characteristic value I_D–ε_A is in the interval from 7.4 to 5.9 eV, and it meets the requirements of the formation of tightly coupled adducts involving two or more SWCNTs, leading to their donor–acceptor-stimulated adhesion, similarly to the oligomerization of fullerene C_{60} considered in Chapter 8. At the same time, this explains the high acceptor ability of SWCNTs in numerous donor–acceptor nanocomposites.[10]

Once related to single-walled carbon nanotubes of the (4,4) family, the obtained regularities can be spread over other (n,n) and (m,0) tubes due to the close similarity in their behavior, exhibited in the discussions presented in the chapter.

References

1. Tasis, T., Tagmatarchis, N., Bianco, N., and Prato, M. 2006. Chemistry of carbon nanotubes. *Chem. Rev.* 106:1105–36.
2. Meyer, R. R., Sloan, J., Dunin-Borkowski, R. E., Kirkland, A. I., Novotny, M. C., Baily, S. R., Hutchison, J. L., and Green, M. L. H. 2000. Discrete atom imaging of one-dimensional crystals formed within single-walled carbon nanotubes. *Science* 289:1324–26.
3. Mickelson, E. T., Huffman, C. B., Rinzler, A. G., Smalley, R. E., Hauge, R. H., and Margrave, J. L. 1998. Fluorination of single-wall carbon nanotubes. *Chem. Phys. Lett.* 296:188–94.
4. Hirsh, A. 1999. Principles of fullerene reactivity. *Top. Curr. Chem.* 199:1–65.
5. Niyogi, S., Hamon, M. A., Hu, H., Zhao, B., Bhomik, P., Sen, R., Itkis, M. E., and Haddon, R. C. 2002. Chemistry of single-walled carbon nanotubes. *Acc. Chem. Res.* 35:1105–13.
6. Bahr, J. L., and Tour, J. M. 2002. Covalent chemistry of single-wall carbon nanotubes. *J. Mater. Chem.* 12:1952–58.
7. Strano, M. S., Dyke, C. A., Usrey, M. I., Barone, P. W., Allen, M. J., Shan, H., Kittrell, C., Hauge, R. H., Tour, J. M., and Smalley, R. E. 2003. Electronic structure control of single-walled carbon nanotube functionalization. *Science* 301:1519–22.

8. Dyke, C. A., and Tour, J. M. 2004. Covalent functionalization of single-walled carbon nanotubes for materials applications. *J. Phys. Chem. A* 108:11151–59.
9. Banerjee, S., Hemray-Benny, T., and Wong, S. S. 2005. Covalent surface chemistry of single-walled carbon nanotubes. *Adv. Mater.* 17:17–29.
10. Guldi, D. M., Rahman, G. M. A., Zerbetto, F., and Prato, M. 2005. Carbon nanotubes in electron donor-acceptor nanocomposites. *Acc. Chem. Res.* 38:871–78.
11. Herrero, A., and Prato, M. 2008. Recent advances in the covalent functionalization of carbon nanotubes. *Mol. Cryst. Liq. Cryst.* 483:21–32.
12. Chen, Z. F. Thiel, W., and Hirsch, A. 2003. Reactivity of the convex and concave surfaces of single-walled carbon nanotubes (SWCNTs) towards addition reactions: Dependence on the carbon-atom pyramidalization. *Chemphyschem* 4:93–97.
13. Chen, Z. F., Nagase, S., Hirsch, A., Haddon, R. C., and Thiel, W. 2004. Sidewall opening of single-walled carbon nanotubes (SWCNTs) by chemical modification: A critical theoretical study. *Angew. Chem. Int. Ed.* 43:1552–54.
14. Yildirim, T., Gülseren, O., and Ciraci, S. 2001. Exohydrogenated single-wall carbon nanotubes. *Phys. Rev. B* 64:075404.
15. Gülseren, O., Yildirim, T., and Ciraci, S. 2002. Effects of hydrogen adsorption on single-wall carbon nanotubes: Metallic hydrogen decoration. *Phys. Rev. B* 66:121401.
16. Choi, W. I., Park, S., Kim, T.-E., Park, N., Lee, K.-R., Ihm, J., and Han, S. 2006. Band-gap sensitive adsorption of fluorine molecules on sidewalls of carbon nanotubes: An *ab initio* study. *Nanotechnology* 17:5862–65.
17. Mercury, F., Sgamelotti A., Valentini, L., Armentano, I., and Kenny, J. M. 2005. Vacancy-induced chemisorption of NO_2 on carbon nanotubes: A combined theoretical and experimental study. *J. Phys. Chem. B* 109:13175–79.
18. Mercury, F., and Sgamelotti, A. 2006. Functionalization of carbon nanotubes with Vaska's complex: A theoretical approach. *J. Phys. Chem. B* 110:15291–94.
19. Mercury, F., and Sgamelotti, A. 2007. Theoretical investigations on the functionalization of carbon nanotubes. *Inorg. Chim. Acta* 360:785–93.
20. Lee, Y.-S., and Marzari, N. 2006. Cycloaddition functionalizations to preserve or control the conductance of carbon nanotubes. *Phys. Rev. Lett.* 97:116801.
21. Lee, Y.-S., and Marzari, N. 2008. Cycloadditions to control bond breaking in naphthalenes, fullerenes, and carbon nanotubes: A first-principles study. *J. Phys. Chem. C* 112:4480–85.
22. Zhao, J.-X., and Ding, Y.-H. 2008. Chemical functionalization of single-walled carbon nanotubes (SWNTs) by aryl groups: A density functional theory study. *J. Phys. Chem. C* 112:13141–49.
23. Haddon, R. C. 1993. Chemistry of the fullerenes: The manifestation of strain in a class of continuous aromatic molecules. *Science* 261:1545–50.
24. Chen, Y., Haddon, R. C., Fang, S., Rao, A. M., Eklund, P. C., Lee, W. H., Dickey, E. C., Grulkem, E. A., Pendergrass, J. C., Chavan, A., Haley, B. E., and Smalley, R. E. 1998. Chemical attachment of organic functional groups to a single-walled carbon nanotube material. *J. Mater. Res.* 13:2423–31.
25. Hammon, M. A., Itkis, M. E., Niyougi, S., Alvarez, T., Kuper, C., Menon, M., and Haddon, R. C. 2001. Effect of rehybridization on the electronic structure of single-walled carbon nanotubes. *J. Am. Chem. Soc.* 123:11292–93.

26. Sheka, E. F., and Chernozatonskii, L. A. 2007. Bond length effect on odd electrons behavior in single-walled carbon nanotubes. *J. Phys. Chem. C* 111:10771–79.
27. Chen, Z., Jiang, D., Lu, X., Bettinger, H. F., Dai, S., Schleyer, P. R., and Houk, K. N. 2007. Open-shell singlet character of cyclacene and short zigzag nanotubes. *Org. Lett.* 9:5449–52.
28. Sheka, E. F., and Chernozatonskii, L. A. 2010. Broken symmetry approach and chemical susceptibility of carbon nanotubes. *Int. J. Quant. Chem.* 110:1466–80.
29. Hersam, M. C. 2008. Progress towards monodisperse single-walled carbon nanotubes. *Nature Nanotechnol.* 3:387–94.
30. Sheka, E. F., and Chernozatonskii, L. A. 2010. Graphene-carbon nanotube composites. *J. Compt. Theor. Nanosci.*, 7:1814–24.
31. Mawhinney, D. B., Naumenko, V., Kuznetsova, A., and Yates, Jr., J. T. 2000. Infrared spectral evidence for the etching of carbon nanotubes: Ozone oxidation at 298 K. *J. Am. Chem. Soc.* 122:2383–84.
32. Mawhinney, D. B., Naumenko, V., Kuznetsova, A., Yates, Jr., J. T., Liu, J., and Smalley, R. E. 2000. Surface defect site density on single walled carbon nanotubes by titration. *Chem. Phys. Lett.* 324:213–16.
33. Wu, B., Zhang, J., Wei, Z., Cai, S., and Liu, Z. 2001. Chemical alignment of oxidatively shortened single-walled carbon nanotubes on silver surface. *J. Phys. Chem. B* 105:5075–78.
34. Yanagi, H., Sawada, E., Manivannan, A., and Nagahara, L. A. 2001. Self-orientation of short single-walled carbon nanotubes deposited on graphite. *Appl. Phys. Lett.* 78:1355.
35. Chattopadhyay, D., Galeska, I., and Papadimitrakopoulos, F. 2001. Metal-assisted organization of shortened carbon nanotubes in monolayer and multilayer forest assemblies. *J. Am. Chem. Soc.* 123:9451–52.
36. Chen, J., Hamon, M. A., Hu, H., Chen, Y., Rao, A. M., Eklund, P. C., and Haddon, R. C. 1998. Solution properties of single-walled carbon nanotubes. *Science* 282:95–98.
37. Khabashesku, V. N., Bilups, W. E., and Margrave, J. L. 2002. Fluorination of single-walled carbon nanotubes and subsequent deviation reactions. *Acc. Chem. Res.* 35:1087–95.
38. Rakov, E. 2006. Chemistry of carbon nanotubes. In *Nanomaterial handbook*, ed. Y. Gogotsi, 105–76. Boca Raton, FL: Taylor & Francis.
39. Nikitin, A., Ogasawara, H., Mann, D., Denecke, R., Zhang, Z., Dai, H., Cho, K., and Nilsson, A. 2005. Hydrogenation of single-walled carbon nanotubes. *Phys. Rev. Lett.* 95:225507.
40. Tasis, D., Tagmatarchis, N., Georgakilas, V., and Prato, M. 2003. Soluble carbon nanotubes. *Chem.-Eur. J.* 9:4000–8.
41. Dyke, C. A., and Tour, J. M. 2004. Overcoming the insolubility of carbon nanotubes through high degrees of sidewall functionalization. *Chem-Eur. J.* 10:812–17.
42. Vivien, L., Riehl, D., Hashe, F., and Anglaret, E. 2000. Nonlinear scattering origin in carbon nanotube suspensions. *J. Nonlinear Opt. Phys. Mater.* 9:297–307.
43. Shulev, V. A., Filippov, A. K., and Kamanina, N. V. 2006. Laser-induced processes in the IR range in nanocomposites with fullerenes and carbon nanotubes. *Technol. Phys. Lett.* 32:694–97.

chapter twelve

Chemical reactivity of graphene

12.1 Introduction

Graphene is currently one of the most actively studied objects of modern nanoscience. Countless theoretical and experimental studies have already been performed, targeting electronic, magnetic, thermal, optical, structural, and vibrational properties (see Peres and Ribeiro[1] and other FOCUS papers in the September 2009 issue of *New Journal of Physics*). Studies that modify pristine graphene, aiming at finding new physics and possible new applications, take a particular place. These include patterning nanoribbons and quantum dots, exposing graphene's surface to different chemical species, studying multilayer systems, and inducing strain and curvature. Among the latter, chemical modification or chemical decoration of graphene sheets is considered the most promising and predicted manner to tune the graphene properties in the intended direction. This explains a great number of theoretical and experimental efforts directed to working out reliable algorithms of the chemical treatments of graphene. And nevertheless, there is an empty niche in the studies that is connected with disclosing the grounds of the graphene chemical reactivity.

A conventional way to tackle the problem from the computational viewpoint, now widely practiced, consists of the following. It is accepted that the chemical treatment implies mainly adsorption of different chemical units on the graphene surface. To describe the adsorption event computationally, one constructs a periodic supercell of graphene containing as many atoms as is necessary to comfortably accommodate the unit within the cell. Due to a large number of supercell atoms, an efficient computational scheme, predominantly based on the density functional theory (DFT), is then applied. Usually, DFT schemes are limited by restricted approaches. Real progress is that not only restricted but also unrestricted DFT schemes are in operation. The latter are mainly aimed at considering magnetic properties of graphene ribbons, while adsorption still remains the object for restricted calculation schemes (look at two of the recent publications on the topic[2,3]).

The adsorption capable of modifying the graphene properties is obviously connected with chemically attached species in due course of addition chemical reactions occurring at a graphene sheet. The first unrestricted broken symmetry (UBS) DFT examination of such chemical reaction between a hydrogen-terminated graphene ribbon and common radicals[4] disclosed *unpaired π electrons* (authors' nomination) distributed over zigzag edges in 0.14 *e* on each atom (N_{DA} in terms of the current book). The finding permitted the authors to make a conclusion about the open-shell character of the graphene singlet ground state of the ribbon, as well as of special chemical reactivity of the atoms, which leads to a partial radicalization of the species.

This finding, which seems quite natural from the viewpoint disclosed throughout the book, is just an independent confirmation of peculiarities of the odd electron coupling in the sp^2 electron system of graphene due to enlarging the C-C bond length in comparison with benzene. Actually, as shown in Figure 2.1, the graphene bonds are remarkably longer than those in benzene, which must cause a noticeable radicalization of the species, thus enhancing its chemical reactivity. Once we are coherent with the discussions presented in the previous papers, we can look at the problem from the UBS Hartree-Fock (HF) point of view.

12.2 Broken symmetry Hartree–Fock approach to chemical reactivity of graphene

Low and homogeneous chemical reactivity of atoms throughout a graphene sheet is usually expected by the predominant majority of scientists dealing with graphene chemistry. However, the first UBS HF calculations[5,6] showed that this is not the case, since the length of equilibrium C-C bonds of graphene exceeds the critical value of 1.395 Å, over which odd electrons become partially unpaired (see Section 2.3). The calculated results for hydrogen-terminated graphene sheets of different size (nanographenes (NGrs)) are listed in Table 12.1. Rectangular NGrs are nominated as (n_a·n_z) structures following Gao et al.[7] Here n_a and n_z match the

Table 12.1 Atomic chemical susceptibility of H-terminated nanographenes

Nanographenes (n_a·n_z)[a]	N_{DA}, *e*		
	Armchair edge	Central part	Zigzag edge
(12,15)	0.28–0.14	0.25–0.06	0.52–0.28
(12,15)[b]	1.18–0.75	0.25–0.08	1.56–0.93
(7,7)	0.27–0.18	0.24–0.12	0.41–0.28
(5,6)	0.27–0.16	0.23–0.08	0.51–0.21

Source: Sheka, E. F., and Chernozatonskii, L. A., *J. Exp. Theor. Phys.*, 110, 121–32, 2010.

[a] Following Gao et al.,[7] n_a and n_z match the numbers of benzenoid units on the armchair and zigzag ends of the sheets, respectively.

[b] After removing hydrogen terminators.

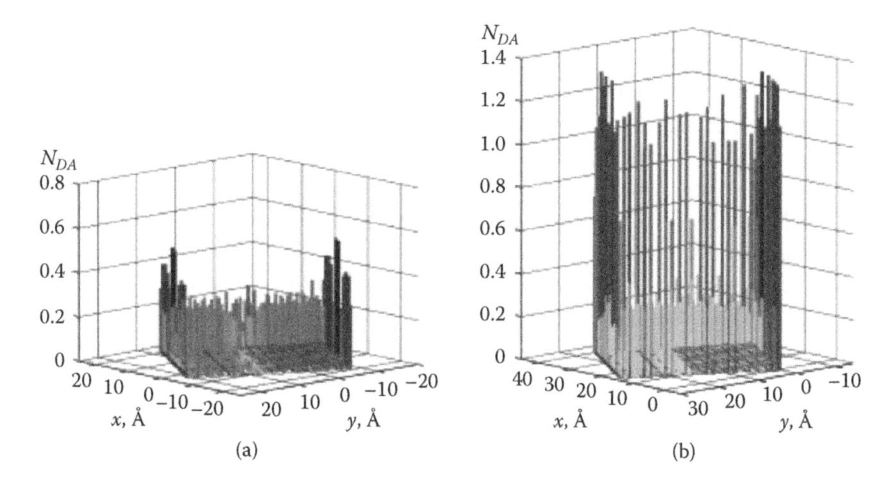

Figure 12.1 Distribution of atomic chemical susceptibility over atom of rectangular nanographene NGr (12,15) with hydrogen terminated (a) and empty (b) edges. UBS HF solution. Singlet state. (From Sheka, E. F., and Chernozatonskii, L. A., *J. Exp. Theor. Phys.*, 110, 121–32, 2010.)

number of benzenoid units on the armchair and zigzag edges of the sheets, respectively. The atomic chemical susceptibility (ACS) (N_{DA}) profile for NGr (12,15) with hydrogen-terminated edges, presented in Figure 12.1(a), demonstrates a rather significant variation of the quantity over atoms due to a noticeable dispersion of the C-C bond lengths. Shown by computations, the equilibrium sheet structure is characterized by a remarkable varying of the C-C bond lengths, while the structure optimization in all cases starts with practically constant bond lengths throughout the sheet, with a C-C bond length of 1.43 to 1.42 Å. It should be noted that this variation is a direct consequence of taking effectively unpaired electrons into account. When the spin peculiarities are omitted and calculations are performed in the closed-shell restricted Hartree-Fock and DFT approximations, the bond length variation is absent.

As seen in Figure 12.1, the highest atomic chemical susceptibilities are characteristic of carbon atoms at the zigzag edges, while those of the armchair edges are similar to the ACS values of the sheet inner atoms and are comparable with those of fullerenes (ca. Figure 3.8) and single-walled carbon nanotube (SWCNT) sidewalls (Figure 11.3). When hydrogen terminators are removed, the ACS profile over the sheet remains unchanged, while N_{DA} values on both zigzag and armchair edges grow significantly (Figure 12.1(b)), still conserving bigger values for zigzag edges.

The obtained results made allowance for the following conclusions concerning chemical reactivity of nanographenes:[5,6]

1. Any chemical addend will first be attached to the graphene zigzag edges, both hydrogen terminated and empty.
2. Slightly different by activity nonterminated armchair edges compete with zigzag ones.
3. Chemical reactivity of inner atoms does not depend on the edge termination and is comparable with those of SWCNT sidewalls and fullerenes, thus providing a large range of addition reactions at the graphene surface.
4. The disclosed chemical reactivity of both edges and inner atoms of graphene causes a particular two-mode pattern of the nanographene attaching to any spatially extended molecular object, such as either a carbon nanotube or substrate surface, namely, a normal mode and a tangent or parallel one.

In view of these data, let us look at the results of the UBS DFT studies of nanographenes that led to the foundation of today's presentations by the scientific community concerning chemical reactivity of graphene. First notification about peculiar edge states of graphene ribbons appeared as early as 1987,[8] but further extended study started about ten years later.[9,10] Since that time, three main directions of the peculiarity investigation have shaped up, focusing on (1) edge states within the band structure of graphene, (2) chemical reactivity, and (3) magnetism of graphene's ribbon zigzag edges. The first topic lies mainly within the framework of the solid-state theory concerning the formation of localized states caused by the breakage of translational symmetry in a certain direction that occurs when a graphene sheet is cut into graphene ribbons. This fundamental property was well disclosed computationally and independently of the technique used[10,11] and was approved experimentally.[12,13] Two other topics are intimately connected with UBS DFT[14–16] itself and demonstrate a spin-contaminated character of the obtained solutions.

As already mentioned, the first UBS DFT examination of the chemical reaction[4] disclosed *unpaired π electrons* (authors' nomination) distributed over zigzag edges in $N_{DA} = 0.14\,e$ on each atom. The next authors' conclusion concerned nonedge ribbon carbon atoms, armchair atoms, and carbon nanotube (presumably sidewall) atoms that show little or no radical character. Therefore, both UBS DFT and UBS HF approaches disclosed the open-shell character of the ground singlet state of graphene and established the availability of effectively unpaired electrons. However, both the total numbers of effectively unpaired electrons, N_D, and their distribution over atoms, N_{DA}, differ by an order of magnitude that limits the UBS DFT discussion of the chemical reactivity of graphene to zigzag edge atoms only. The fixation of the open-shell character of the graphene ground singlet state by both UBS techniques was obvious due to the single-determinant character of the wave functions in the two cases. The feature was revealed due to considerable weakening of the odd electron interaction in graphene caused by rather large C-C bond lengths. As for the magnitude of the unpaired odd (π) electrons numbers, N_{DA}, its decreasing by one order of magnitude, compared to the UBS HF data, might indicate a pressed-by-functional character of the UBS DFT calculations.[17] The functional-dependent character of the UBS DFT solutions was thoroughly analyzed just recently.[18,19] At any rate, the results clearly exhibited much lower sensitivity of the UBS DFT approach to chemical reactivity of atoms, which can be imagined as lifting a zero reading level up to 0.2 to 0.3 *e* in Figure 12.1(a), and up to 1.1 *e* in Figure 12.1(b); thereafter, the fixation of values below the level becomes impossible.

The close-to-zero chemical reactivity of graphene inner atoms predicted by the UBS DFT calculations strongly contradicts the active chemical adsorption of individual hydrogen and carbon atoms on graphene's surface, recently disclosed experimentally.[20] To the most extent, the chemical reactivity of inner atoms has been proven by the formation of a chemically bound interface between a graphene layer and silicon dioxide over the course of the graphene sheet,[21] as well as by producing a new particular one-atom-thick CH species named graphane.[22] At the same time, the empirical observations are well consistent with the UBS HF data obtained in Sheka and Chernozatonskii.[5,6] That is why one can accept the obtained data on chemical reactivity of graphene as quite reliable and use ACS values as quantified pointers for predicting chemical reactions or modifications to which graphene can be subjected. Thus, the disclosed reactivity of both graphene edge and inner atoms, as well as a possible two-mode pattern of an approaching nanographene sheet to a carbon nanotube, suggests a number of peculiar graphene-nanotube composites whose appearance might be expected in the near future.

12.3 Carbon nanotube–graphene composites

Intermolecular interaction in CNT-graphene dyads slightly, if at all, differs from that in fullerene-fullerene (Section 8.3) and CNT-fullerene (Section 8.5) dyads, due to the similarity of both ionization potentials (IPs) and electron affinities (EAs) of all constituents. Therefore, as in the latter two case, the potential energy surfaces (intermolecular interaction (IMI) terms), $E_{int}(r,R)$, for the ground state of dyads CNT + grph ($I + II$) (I and II label pair partners) are very similar and possess a characteristic two-well structure as a function of intermolecular coordinates R. The formation of a stationary product AB at the R^{+-} minimum is accompanied by the creation of intermolecular chemical bonds between A and B partners. In contrast, widely spaced neutral moieties form a charge transfer (CT) complex $A + B$ in the vicinity of the R^{00} minimum. In ensembles of the above partners, the availability of a pairwise interaction at the R^{00} minimum leads to the clusterization of the dopants in solutions of the *sol-sol* $[A]_n$ and *sol-solv* $[A]_k[B]_l$ types; this actually takes place in the case of fullerenes (Chapter 10). Thus, as recently observed, the transformation of hydrothermally cut graphene sheets into quantum dots of 5 to 13 nm is accompanied with a blue luminescent,[23] which is characteristic for transfer charge complex clusters (see Section 10.8) and might be considered a confirmation of the formation of graphene + graphene charge transfer complex clusters. Electronically, the formed nanosize clusters provide good conditions for electron property space confining that might be responsible for the enhancement of the dopant properties previously mentioned.[24–28] Leaving a detailed consideration of some of them to Chapter 10, we will concentrate on the formation of covalently bound CNT-NGr composites, $(I)_k$ $(II)_l$, in the vicinity of the R^{+-} minimum.

The formation of mixed composites has just recently become known.[29] The finding has disclosed that the formation of CNT-graphene constructions is possible. This makes the expectation of CNT-NGr compositions in, say, diluted solutions quite promising. Nothing is known about the shape and properties of such composites. Based on atomic chemical susceptibility (ACS) of both CNTs and graphene, we will suggest a set of $(I)_k$ $(II)_l$ ($k = 1, 2; l = 1, 2$) composites exhibiting the most probable structures to be experimentally synthesized.[30]

12.3.1 *Grounds for the computational synthesis of $(I)_k$ $(II)_l$ composites*

To simulate the formation of the $(I)_k$ $(II)_l$ product, one has to build a starting configuration consisting of components *I* and *II*. As has already been discussed, the intermolecular C-C distances at the reactive spot should be shorter than R^{00} but longer than R^{+-}. Thoroughly analyzed by calculating the dimerization barrier for fullerene C_{60} (see Section 8.2), the distance should be 1.6 to 1.9 Å. Since in the composites under study intermolecular chemical bonds are formed by carbon atoms, conclusions obtained for fullerenes should be valid for the current case as well.

A choice of the atomically matched reactive spot is the most complicated part of the simulation. Following the procedure applied to fullerenes and CNTs, let us use the ACS maps discussed in detail for SWCNTs in Section 11.2, and for NGrs in the previous section. Combining the peculiarities of the two species, two basic groups of composites, below referred to as *hammer* and *cutting-blade* structures, can be suggested. The former follows from the fact that empty ends of SWCNTs are the most chemically active, so that the tubes might be willingly attached to any NGr sheet, forming a "hammer handle." In turn, the sheet surface is reactive enough to provide chemical bonding with the tube. As for the second type of structure, this is a result of exclusive chemical reactivity of both zigzag and armchair edges of nonterminated NGr, so that the latter can tangentially touch a SWCNT sidewall that is chemically active as well.

12.3.2 *Hammer* $(I)_{1,2}$ $(II)_{1,2}$ *composites*

The selected composites presented in Figure 12.2 well exhibit general tendencies that govern the formation of the SWCNT-NGr interface. To

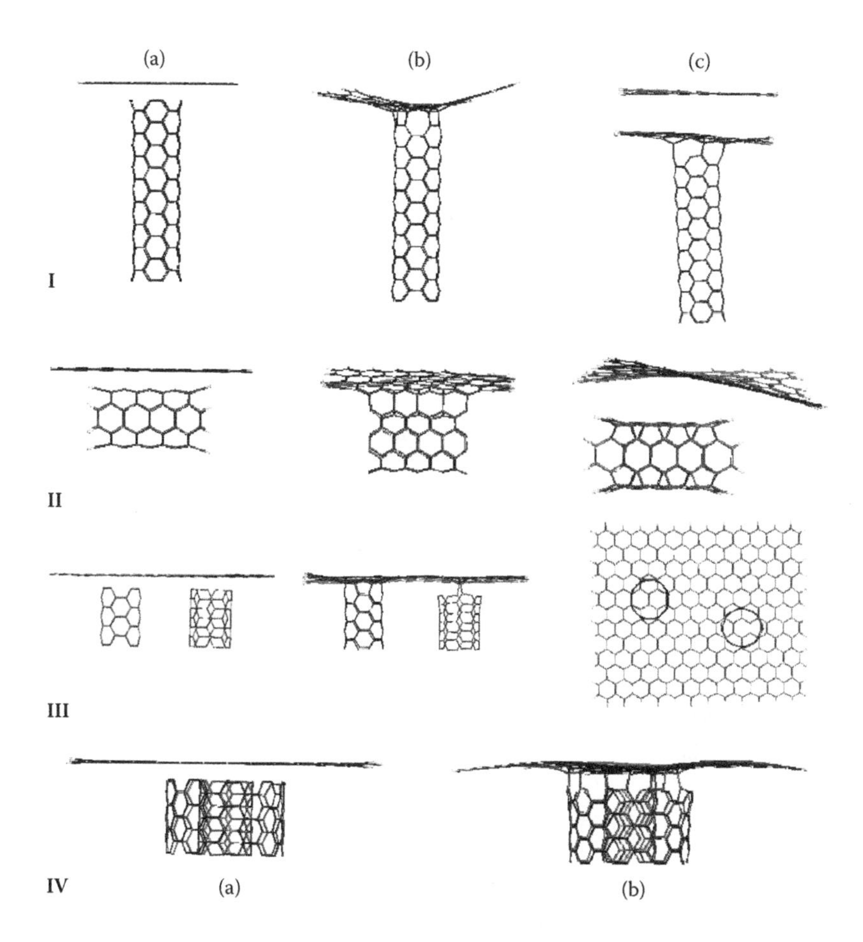

Figure 12.2 Hammer CNT-graphene composites. Starting (a) and equilibrated (b and c) compositions. I: (4,4) SWCNT with one (b) and two (c) NGrs (7,7). II: Ngr (7,7) and (4,4) SWCNT with both empty (b) and hydrogenated (c) ends. III: Two (4,4) SWCNTs and NGr (11,12); side (a, b) and top (c) views. IV: Ngr (11,12) and (9,9+4,4) DWCNT. (From Sheka, E. F., and Chernozatonskii, L. A., *J. Compt. Theor. Nanosci.* 7, 1814–24, 2010.)

simplify their description, each structure is nominated by a combined notation, like Ia, Ib, and so forth. The obtained composites are characterized by the coupling energy, E_{cpl}, per one intermolecular C-C bond formed at the interface, determined as

$$E_{cpl} = \left(\Delta H_{cps} - k\Delta H_{SWCNT} - i\Delta H_{Ngr}\right) / n_{C\text{-}C} \tag{12.1}$$

where ΔH_{cps}, ΔH_{SWCNT}, and ΔH_{Ngr} present the heats of formation of the composite, nanotube, and NGr, respectively, and $n_{C\text{-}C}$ is the number of intermolecular C-C bonds formed. The obtained data are listed in Table 12.2.

Table 12.2 Coupling energy at the interface of CNT + Ngr composites per one intermolecular C-C bond formed, kcal/mol

Nomination[a]	Composites	E_{cpl}
	Hammer $(I)_1$ $(II)_{1,2}$ structures	
Ib	(4.4) SWCNT + (7,7) Ngr (7)[b]	–32.24
Ic	(4.4) SWCNT + 2*(7,7) Ngr (7)	–28.13
IIb	(4,4) SWCNT + (7,7) Ngr (7)	–11.30
IIc	(4,4) SWCNT + (7,7) Ngr (0)	+4.11[c]
IIIb	2* (4,4) SWCNTs + (11,12) Ngr (8+2)	–42.64
IV	(4,4 + 9,9) DWCNT + (11,12) Ngr (8)	–10.00
	Cutting-blade $(I)_1$ $(II)_{1,2}$ structures	
Va	(8,0) SWCNT + (7,7) Ngr z-(8)[d]	–52.49
Vb	(8,0) SWCNT + (7,7) Ngr a-(8)	–66,28
VIIa	(8,0) SWCNT + 2*(7,7) Ngr zz-(15)	–54.58
VIIb	(8,0) SWCNT + 2*(7,7) Ngr aa-(16)	–54.07
VIII	(8,0) SWCNT + 2*(7,7) Ngr aa-(16)	–62.43
IXa	(4,4) SWCNT + (7,7) Ngr z-(7)	–70.75
IXb	(4,4) SWCNT + (7,7) Ngr a-(4)	–77.84
XIa	(4,4) SWCNT + 2*(7,7) Ngr zz-(14)	–118.97
XIb	(4,4) SWCNT + 2*(7,7) Ngr zz-(16)	–83.50
XII	2*(4,4) SWCNT + (7,7) Ngr zz-(13)	–69.86

Source: Sheka, E. F., and Chernozatonskii, L. A., *J. Compt. Theor. Nanosci.* 7, 1814–24, 2010.)

[a] See corresponding structures in Figures 12.2 to 12.4.
[b] The figure in parentheses indicates the number of intermolecular C-C bonds formed.
[c] The total interaction energy (see text).
[d] z(a) — (*n*) and zz(aa) – (*n*) indicate zigzag (armchair) single or double Ngr attachment to the tubes sidewall, respectively, accompanied by the formation of *n* intermolecular C-C bonds.

12.3.2.1 Composites I

A fragment of (4,4) SWCNT with both open empty ends is normally oriented to the NGr (7,7) sheet at a starting distance of 1.8 Å (Ia). Seven intermolecular C-C bonds of 1.51 Å in length are formed joining the tube with the sheet after the structure optimization (Ib). As seen in Table 12.2, the corresponding coupling energy is quite large, pointing to a strong connection at the interface. In addition to the energy of the bond formation, the coupling energy involves the deformation energy of both SWCNT and NGr, caused by the reconstruction of the electron configuration of the interface carbon atoms from sp^2 to sp^3. However, when this reconstruction touches mainly upon the structure of the tube end, the NGr is significantly deformed as a whole, which leads to the transformation of a flat "roof" of the starting composition in Ia to a "Chinese pagoda" pattern when the structure is equilibrated (Ib). However, when the second NGr sheet is added to the first one of the starting composition in Figure 12.2, Ia at 3.35 Å above, the intermolecular interaction between two sheets evidently smoothes the deformation of the first one, while the second is only slightly deformed (Ic). At the same time, the coupling energy of the tube to the sheet pair remains big and decreases rather slightly. Evidently, coupling a three-layer graphene stack with the nanotubes is strong as well, while the deformation of the upper sheets is practically negligible. As might be expected, the obtained results are in line with Viena Abinitio Simulation Package (VASP) DFT calculations of the connection of the graphene sheet stack with the SiC substrate,[31] promoting the growth of epitaxial graphene.[32]

12.3.2.2 Composites II

When the tube is placed parallel to the sheet plane (IIa) at the same distance as in case I, the equilibrium structure depends on whether the tube open ends are either empty or terminated (by hydrogens in our case). In the first case, the tube and the sheet attract each other willingly, and seven newly formed intermolecular C-C bonds provide the tight connection between the partners (IIb). When tracing subsequent steps of the joining (optimization), one can see that the coupling starts at the tube ends by the formation of a single bond at first, and then a pair of C-C bonds at each end. Afterwards, these bonds play the role of the strops of gymnastic rings, which pull the tube body to the sheet. In contrast, when the tube ends are hydrogen terminated, no intermolecular C-C bonds are formed (IIc) and the total coupling energy becomes repulsive (see Table 12.2). However, even under this repulsive interaction the sheet is deformed, showing a tendency to wrap up the tube.

12.3.2.3 Composites III

Attachment of two SWCNTs to a single NGr (IIIa) shows that the final result depends on the topological coherence between the tube projection and the benzenoid structure of the sheet (IIIc). The left joining in IIIb evidently occurs under better coherence than the right one. This causes the different number of formed intermolecular C-C bonds, namely, eight in the first case and two in the second. Both attachments are accompanied with the sheet deformation that causes a remarkable roughening of the latter. The coupling energy is comparable with that for a single SWCNT attachment.

12.3.2.4 Composites IV

The fragment of a double-walled CNT (DWCNT) in IVa consists of fragments of (4,4) and (9,9) SWCNTs with the same number of benzoid units along the vertical axis and open empty ends on both sides. However, since the periodicity of the Kekule-incomplete Clar-complete networks[33] is slightly different in the two tubes, the fragment lengths do not coincide exactly. As a result, the attachment of the joint fragment to the graphene sheet starts from the formation of intermolecular C-C bonds, with either the inner or outer tube, depending on which is closer to the sheet. When opposite free ends of the tube are not fixed, the remaining fragment slides outward, transforming the composition into a peculiar telescope system. When the free ends are fixed, both inner and outer fragments are joined to the sheet (IVb). The coupling energy given in Table 12.2 cannot be directly compared with that of SWCNT since it is affected by both the sheet deformation and the free-end fixation. Nevertheless, it is large enough to provide a strong coupling between the graphene sheet and the DWCNT that explains a high stability of recently synthesized MWCNT-graphene composite under conditions when one end of each MWCNT was fixed.[29]

Summarizing, it is possible to conclude the following:

1. The normal attachment of an empty-end SWCNT to a graphene sheet is energetically favorable.
2. The horizontal attachment of the tube is also possible, but much weaker.
3. H termination of the tube ends renders the horizontal attachment impossible and severely weakens the normal one.
4. Both multiple normal attaching of SWCNTs and a single and multiple attaching of a DWCNT are energetically favorable, and graphene sheets can be easily fixed over tubes in case their open ends are empty. This conclusion is in a perfect agreement with experimental observations presented in Kondo et al.[29]
5. Graphene sheets are extremely structure flexible, and even a weak intermolecular interaction causes a loss of the sheet flatness.

12.3.3 Cutting-blade $(I)_{1,2}$ $(II)_{1,2}$ composites

Typical compositions to be formed in this case are presented in Figures 12.3 and 12.4 by two SWCNT fragments, namely, (8,0) and (4,4), as well as NGr

(7,7). At the start each time, the NGr edge was oriented parallel to the cylinder axe in the vicinity of SWCNT along a line of sidewall atoms in such a way to maximize the number of expected intermolecular C-C bonds. Since ACS distribution over cross-section atoms of both tubes is well homogeneous (see Figures 11.3 and 11.10), there is no azimuthal selectivity of the line position in this case. As for NGrs, zigzag and armchair edges of NGr with empty edges are comparable while somewhat different. Due to this, two NGr orientations with respect to the tube sidewall were examined.

12.3.3.1 Composites V

The formation of either zigzag (Va) or armchair (Vb) attached monocomposites (*monoderivatives*) in Figure 12.3 is followed by the creation of eight intermolecular C-C bonds at the interface in both cases. However, the two composites differ by the coupling energy (Table 12.2); thus, Vb is more energetically profitable in spite of the zigzag edges, and ACS is slightly higher. The difference is a consequence of the interface different structure that causes different deformation energies (positive by sign), which is obviously bigger for composite Va.

12.3.3.2 Cross-sections VI

To look for a proper spot for the attachment of the second NGr, let us analyze the ACS profile over the cross-section of the tube body of both composites. The relevant distributions VIa and VIb were quite similar, keeping the same view in all cross-sections along the tube in both cases. It should be noted that there is a vivid transformation of a circular cross-section of a free tube into a drop-like one with the apex on atom A. It is obviously caused by the $sp^2 \rightarrow sp^3$ transformation at the point. In both cases, atom A, matched by small black circles in Va and Vb, is involved in the line through which the first NGr is attached to the tube. As shown in Figure 11.10, ACS is distributed over (8,0) SWCNT cross-section atoms quite homogeneously, at an almost constant level of $N_{DA} \sim 0.29\ e$. When one NGr is attached, atom A is involved in the formation of an intermolecular C-C bond, and consequently, its N_{DA} falls to zero. The remainder value of 0.01 can point to the reliability of the N_{DA} value's determination. Consequently, a redistribution of ACS over the cross-section atom occurs; this is clearly seen in the VIa and VIb diagrams. According to the basic concept of the computational synthesis of odd electron systems with unpaired electrons, the next attachment will be the most profitable at atoms with the highest N_{DA} values. Looking at diagram VIa, we can conclude that such atoms are located on the right and on the left from atom A. However, attaching NGr along the corresponding lines of atoms may meet sterical difficulties. The next suitable atom, matched as B, is located three atoms from atom A. Atom B′ is symmetrically located with similar characteristic. Between atoms B and B′ there are seven more atoms with the same ACS. As a result, the nearest

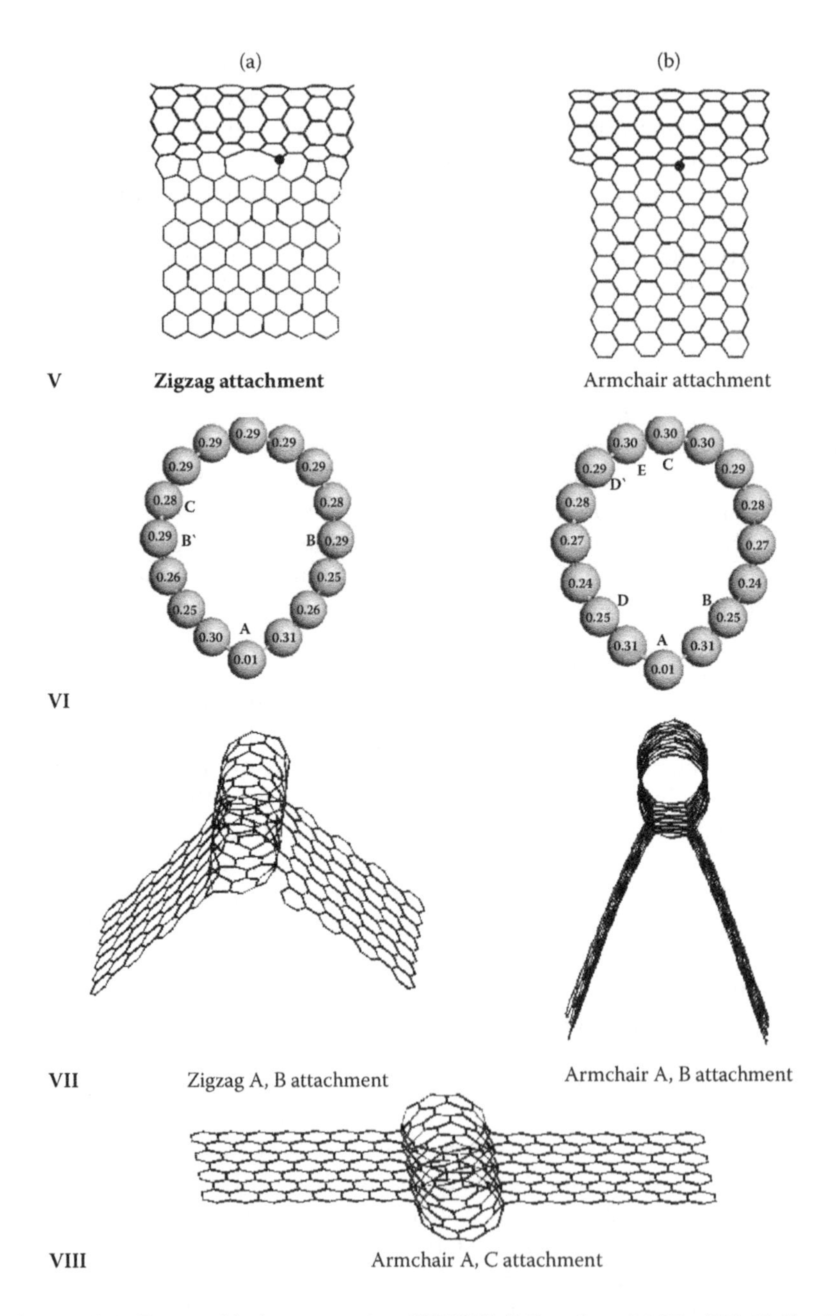

Figure 12.3 Cutting-blade composites SWCNT (8,0) + 1 or 2 NGr (7,7). V: Equilibrium structures of monocomposites with the interface along the line of atoms A at the tube sidewall; zigzag (a) and armchair (b) attachments. VI: Distribution of the ACS N_{DA} values over the tube cross-section atoms in *(continued on next page)*

attachment of the second NGr should occur along atom lines involving either the B or B′ atom, or any of seven atoms between them. The similarly analyzed VIb diagram allows concluding that the preferable place of the second NGr attachment to the tube is the atom line involving atom C, while the line with atom B is the least preferable.

12.3.3.3 Composites VII and VIII

Realizing the first of the above conclusions, dicomposite (*diderivative*) VIIa was obtained. The creation of seven new intermolecular C-C bonds accompanies the composite formation with the coupling energy averaged over fifteen C-C bonds at both interfaces presented in Table 12.2. Comparing composites Va and VIIa, one can see that the coupling energy remains practically the same, with a slight increase in the latter case.

Composite VIIb is a result of attaching the second NGr along the line with atom B on diagram VIb, while composite VIII is formed when the second NGr is attached along the line with atom C. Eight new C-C bonds are formed in both cases, with the coupling energy averaged over all formed C-C bonds for both composites shown in Table 12.2. According to the data, the two composites are characterized by large coupling energy, among which the glider-like composite VIII is the most energetically favorable. The energy difference for VIIb and VIII composites could have been considered as evidence of the preference of the addition reaction in place C against place B. However, one has to take into account the deformation energy, which seems to be larger in case B.

The series of $(I)_1$ $(II)_i$ composites is not restricted to i = 1 and 2. To proceed further, we have to determine the place of the next attachment, for which we have to look at the ACS profile over the cross-section of the tube body after the second reaction. Naturally, the related diagrams differ from VIa and VIb. As follows from the calculations performed, in the case of composite VIIa, atom C (see VIa) has the highest N_{DA} value, pointing to the place of the third attachment. In the case of composites VIIb and VIII, the corresponding places are marked by E and D (D′) (see VIb), respectively. Therefore, when the NGr attaches the (8,0) SWCNT via zigzag edges, a series of multiadditions will look like A(1) → B(2) → C(3) →, and so on (see mapping in VIa). In the case of attaching via armchair edge, composite VIII is energetically preferable, so that the multiaddition will

Figure 12.3 (continued) Va (a) and Vb (b) composites, respectively. The NGr contacts the tube along the line involving atom A (A attachment). VII: Equilibrium structures of dicomposites related to A, B (VIa) zigzag (a) and A, B (VIb) armchair (b) attachments. VIII: The same as in VII but for A, C (VIb) armchair attachment (see text). (From Sheka, E. F., and Chernozatonskii, L. A., *J. Compt. Theor. Nanosci.*, 7, 1814–24, 2010.)

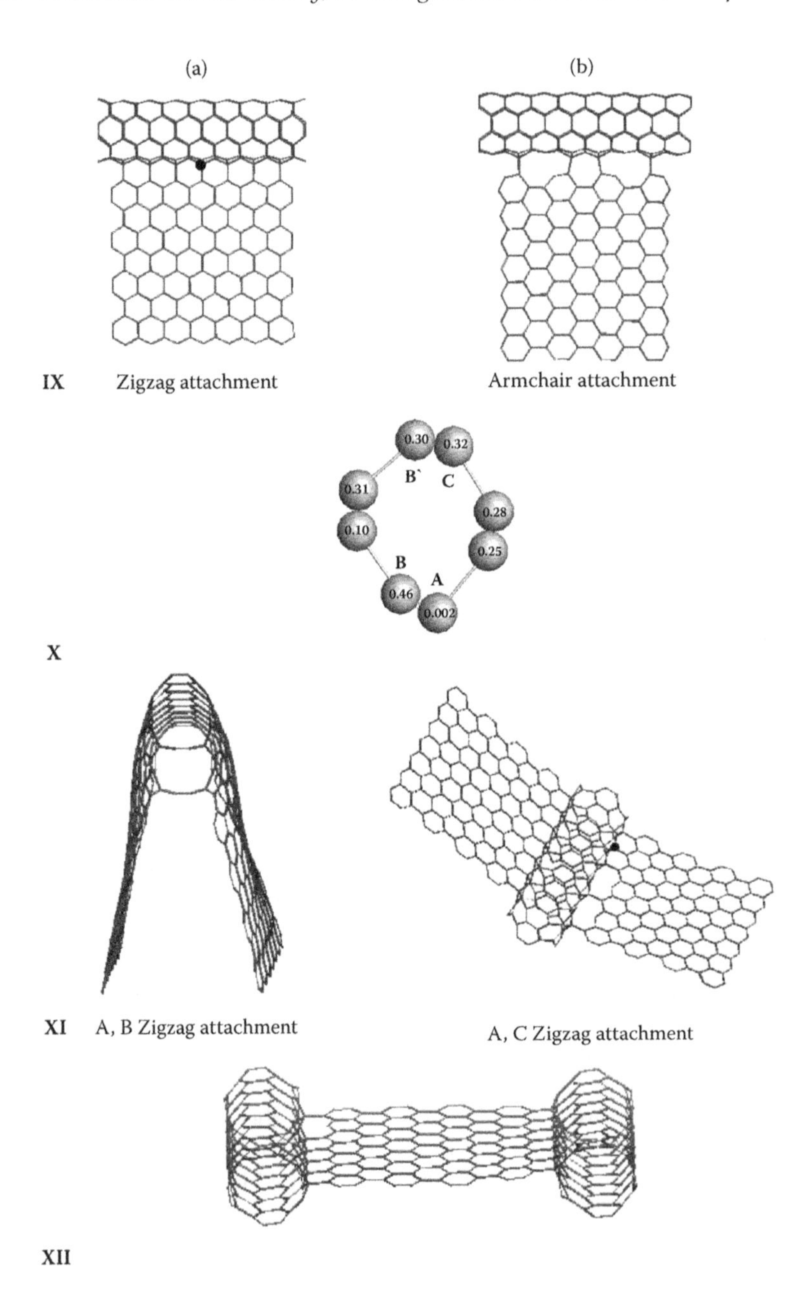

Figure 12.4 Cutting-blade composites SWCNT (4,4) + 1 or 2 Ngr (7,7). IX: Equilibrium structures of monocomposites with the interface along the line of atoms A at tube sidewall; zigzag (a) and armchair (b) attachments. X: Distribution of the ACS N_{DA} values over the tube cross-section atoms in IXa composite. The NGr contacts the tube along the line involving atom A. *(continued on next page)*

follow the scheme A(1) → C(2) → D(3) or D'(3) →, and so on (VIb), until a six- to seven-tooth NGr gear is formed on the basis of the tube.

We have considered the simplest cases. In reality, addition reactions might be more complex since a mixed attachment cannot be excluded, when one NGr attaches a tube via the armchair edge, while the other via a zigzag one. The number of intermolecular C-C bonds formed at the interface will play an important role as well, since this determines the total coupling energy. However, the presented scheme shows a definite way how any individual addition reaction can be traced.

Equilibrated structures of composites based on (4,4) SWCNT are shown in Figure 12.4.

12.3.3.4 Composites IX

In contrast to the composites based on (8,0) SWCNT, the results of mono-addition, IXa and IXb, show a clear preference of the zigzag attachment. Although the coupling energy seems to favor the second one, a great number of C-C bonds formed, as well as much weaker deformation at the interface, evidently support the above preference.

12.3.3.5 Cross-section X

The NGr is attached to the tube along the atom line involving atom A matched by a small black circle in IXa. Similarly to the previous case, a drastic falling of the reactivity of atom A causes a severe distortion of the circular cross-section of the tube that demonstrates the ACS profile X passing through atom A. The cross-section of the ACS distribution exhibits atoms B and C with the highest N_{DA}. Earlier we mentioned that placing the second NGr in the exact neighborhood of the fixed atom may cause distortion of the structure due to sterical constraint.

12.3.3.6 Composites XI

As expected, A-B attachment causes a severe distortion at the interface, fully destroying the tube structure (composite XIa). Importantly, the composite is characterized by a large coupling energy as well as by a big number of intermolecular C-C bonds formed. In contrast to this case, the formation of composite XIb does not cause any destroying of the tube body, and dicomposite XIb is quite similar to VIII considered above for (8,0) SWCNT. The coupling energy is large and the number of the formed C-C bonds is big, which favor such composites formation in the diluted solutions. Nevertheless, composite XIa is obviously more

---

Figure 12.4 (continued) XI: Equilibrium structures of dicomposites; A,B (a) and A,C (b) zigzag attachments. XII: Equilibrium structure of a crandle dicomposite; A—zigzag attachment of both tubes. (From Sheka, E. F., and Chernozatonskii, L. A., *J. Compt. Theor. Nanosci.*, 7, 1814–24, 2010.)

energetically favorable, so that the real situation in a laboratory flask might be quite complex.

Analyzing the ACS profile over a cross-section of the composite XIb tube passing through an atom marked by a small black circle, one finds atoms B and B′ (see X), with practically equal N_{DA} values (B = 0.52 *e* and B′ = 0.49 *e*) that are more than twice the related values for other atoms. Under these conditions, the third attachment should occur along lines passing through these atoms. But those are in a direct neighborhood of atoms A and C, so that one may expect a severe destroying of the tube body structure, similar to that of composite XIa. It cannot be excluded that this destroying is characteristic for small-diameter tubes only, and for large-diameter tubes a sequential addition of a number of NGr will result in the formation of a multitooth gear, as was the case on (8,0) SWCNT. Both questions require further detailed investigations.

12.3.3.7 Composite XII

Particular attention should be paid to this cradle-like composite. As shown in Table 12.2, it is energetically stable. On the other hand, the calculations show one of several possible ways of individual graphene sheet fixation under conditions of the least perturbation of the sheet. Obviously, not (4,4) SWCNT, but much larger tubes should be taken as supporters. Since ACS of SWCNTs does not drastically depend on the tube diameter, as we saw in Chapter 11, the cradle composite formation can be provided by any tubes, even if they different in diameter within the pair.

In spite of the doubtless exemplary of considered composites, analysis of their properties makes allowances for the following general conclusions:

1. The formation of the hammer and cutting-blade $(I)_k(II)_l$ composites is energetically favorable, not only as a monoaddition of NGr to the tube body and vice versa, but also as a multiaddend attachment.
2. A strong contact between the tube and NGr is provided by the formation of an extended set of the intermolecular C-C bonds, the number of which is comparable with the number of either the tube end or NGr edge atoms.
3. The contact strength is determined by both the energy of the newly formed C-C bonds and their number. Optimization of the latter dictates a clear preference toward zigzag or armchair edges of the attaching NGr, depending on the tube configuration. Thus, (8,0) SWCNT (as all other members of the (m,0) family) prefers armchair contacts that maximize the number of point contacts. In its turn, (4,4) SWCNT (as well as other members of the (n,n) family) favors zigzag contacts for the same reasons.

4. The total coupling energy between the NGr addend and the tube involves both the energy of the C-C bond formed and the energy of deformation caused by the reconstruction of the sp^2 configuration for the carbon atom valence electrons into an sp^3 one.
5. In general, the coupling energy of cutting-blade composites is much more than that of hammer ones, which is important for a practical realization of the $(I)_k(II)_i$ composites production.
6. The final product will depend on whether both components of the composition are freely accessible or one of them is rigidly fixed. Thus, in diluted solutions where the first requirement is met, one can expect the formation of cutting-blade composites due to significant preference in the coupling energy. In contrast, in gas reactors, where often either CNTs or graphene sheets are fixed on some substrates, the hammer composites will be formed as has recently been shown.[29]

12.4 Concluding remarks

Recently started, the manufacturing of nanocarbon-based composite materials pursues well-defined goals to provide the best conditions for the exhibition and practical utilization of extraordinary thermal, mechanic, electronic, and chemical properties of nanocarbons. Obvious success in the case of carbon nanotubes (CNTs)[23–27] and graphene[28,34,35] dissolved in different polymers points to great perspectives of a new class of composites and their use in a variety of applications. These nanocarbons (dopants) homogeneously dispersed in polymer matrices at low concentration exhibit electrical and mechanical properties at the level of meeting the requirements of their exploitation at practice. Simultaneously, not only the preservation of the properties of individual tubes or graphene sheets, but also a considerable enhancement of the latter has been observed. Since low-concentration solutions of individual CNTs and graphene sheets can be obtained, one can ask: What can we expect when both CNTs and graphene are dissolved simultaneously?

SWCNT + graphene composites have a few basic problems due to the extreme specificity of both components. Thus, they are both good donors and acceptors of electrons, and this significantly complicates the intermolecular interaction, leading to a two-well shape of the ground-state energy term. This provides the formation of two modes of composites, one of which consists of weakly interacting components located at a comparatively large distance, and the other of which is formed in the range of short interatomic distances and corresponds to strongly coupled composition. Both composites, expected to be drastically different by properties, may coexist and should be differentiated.

The next problem is related to the odd electron character of both components that is revealed via an enhanced chemical reactivity of the species. This makes it possible to consider graphene–carbon nanotube composites as mono- or multiderivative structures when both tubes and graphene either serve as main bodies or are attached additives. So far, the composites have been synthesized computationally in the framework of the unrestricted broken spin-symmetry approach implemented via the Hartree-Fock approximation. Computed profiles of the atomic chemical susceptibility (ACS) along the tube and across their body, as well as over graphene sheets, served as quantified pointers that allowed localizing the most active contact zones of interacting partners.

Due to the fact that the space of chemical reactivity of both CNTs and graphene coincides with the coordinate space of their structures, addition reactions that lead to the composite formation are not local but largely extended in the space. This greatly complicates the construction of starting diads, triads, and more complex configurations of components, making their number practically endless. However, a thorough analysis of the ACS profiles of both components made it possible to select two main groups of the composites, conditionally called hammer and cutting-blade structures. The former follows from the fact that empty ends of SWCNTs are the most chemically active, so that the tubes might be willingly attached to any NGr, forming a hammer handle. The latter is a consequence of exclusive chemical reactivity of both zigzag and armchair edges of nonterminated NGr, so that NGr can touch a SWCNT sidewall tangentially as a blade. As occurred, the coupling energy of the cutting-blade composites exceeds that of the hammer ones, which is important for a practical realization of the $(I)_k(II)_l$ composite production. The final product will depend on whether both components of the composition are freely accessible or one of them is fixed. Thus, in diluted solutions where the first requirement is met, one can expect the formation of the multiaddend cutting-blade composites due to significant preference in the coupling energy. Among the latter, a particular "cradle" composite is suggested for an individual graphene sheet to be fixed by a pair of nanotubes. In contrast, in gas reactors, where often either CNTs or graphene sheets are fixed on some substrate, the hammer composites will be formed, as has recently been shown.[29] It is evident that these new materials exhibit an extremely wide range of electronic and mechanical properties, so that they can be of great demand for various physics-chemical applications, such as super light conducting and plastic material, material for gas storage, CNT-NGr (graphene nanoribbon) junction elements for nanoelectronic schemes, etc.

12.5 Synopsis of features concerned with chemical reactivity of nanocarbons

12.5.1 Fullerenes

From the UBS HF viewpoint, the C_{60} and C_{70} molecules are characterized by a partial exclusion of the odd electrons from the covalent bonding that results in the appearance of effectively unpaired electrons constituting a ~15 to 20% odd electron fraction and exhibiting the radicalization of the molecules. The distribution of effectively unpaired electrons over the molecules atoms (N_{DA} distribution) completes the structural description of the C_{60} and C_{70} molecules by their chemical portraits.

According to the C_{60} chemical portrait in Figure 3.6, the initial step of any addition chemical reaction involves atoms of the highest N_{DA} of group 1. There are twelve identical atoms that form six short C-C bonds belonging to six identical naphthalene-core fragments. One may choose any of the pairs to start the reaction of attaching any addend to the fullerene cage. When the first adduct, $C_{60}R_1$, is formed, the reaction proceeds around its cage atoms, with the biggest N_{DA} values resulting in the formation of adduct $C_{60}R_2$. A new N_{DA} map reveals the sites for the next addition step, and so on. The reaction stops when all N_{DA} values are fully exhausted. Following this methodology, a long list of various fullerene C_{60} derivatives can be computationally synthesized. Among the latter are C_{60} fluorinates and hydrogenates, different fulleroamines, fullerosterenes, trannulenes, and others.

In the case of the C_{70} fullerene, addition reactions will start in the belt region (see Figure 3.9) and the polyderivatives formation will be subordinated to high-rank N_{DA} data that should be obtained after each step of addition.

12.5.2 Single-walled nanotubes

A thorough analysis of the peculiarities of the SWCNT N_{DA} distributions allows for making the following conclusions concerning addition reactions to be expected:

1. The space of chemical reactivity of SWCNTs coincides with the coordinate space of their structures, while different for particular structure elements. This both complicates and facilitates chemical reactions involving the tubes, depending on a particular reaction goal.
2. Local additions of short-length addends (involving individual atoms, simple radical, and so forth) to any SWCNT are the most favorable at open empty ends, both armchair and zigzag ones, with the latter being

more effective. Following these places in activity are cap ends, defects in the tube sidewall, and the sidewall itself. The reactivity of the latter is comparable with the highest reactivity of fullerene atoms.

3. Chemical contacts of SWCNTs with spatially extended reagents (graphene sheets) can occur in three ways: when the tube is oriented either normally or parallel to the surface and when graphene acts as a "cutting blade."
4. Addition reactions with the participation of multiwalled CHTs will proceed depending on the target atoms involved. If empty open ends of the tubes are the main targets, the reaction will occur as if one were dealing with an ensemble of individual SWCNTs. If the sidewall becomes the main target of the reaction, the output will depend on the accessibility of the inner tubes, in addition to the outer one.

12.5.3 *Graphene*

The chemical reactivity of nanographenes is subordinated to the following rules:

1. Any chemical addend will first be attached to the nanographene zigzag edges, both hydrogen terminated and empty.
2. Slightly different by activity nonterminated armchair edges compete with zigzag ones.
3. Chemical reactivity of inner atoms does not depend on the edge termination and is comparable with that of SWCNT sidewalls and fullerenes, thus providing a large range of addition reactions at the nanographene surface.
4. The disclosed chemical reactivity of both edges and the main body of nanographenes causes a particular two-mode pattern of their attaching to any spatially extended molecular object, such as either a carbon nanotube or substrate surface, namely, a normal mode and a tangent or parallel one.

The above conclusions, related to possible chemical reactions that involve fullerenes, carbon nanotubes, and graphene, can be illustrated by composites presented in Figures 8.9, 8.10, 12.2 to 12.4, and obtained so far only computationally.

References

1. Peres, N. M. R., and Ribeiro, R. M. 2009. Focus on graphene. *New J. Phys.* 11:095002.
2. Boukhalov, D. W., and Katsnelson, M. I. 2008. Tuning the gap in bilayer graphene using chemical functionalization: Density functional calculations. *Phys. Rev. B.* 78:085413.

3. Boukhalov, D. W., Katsnelson, M. I., and Lichtenstein, A. I. 2008. Hydrogen on graphene: Electronic structure, total energy, structural distortions and magnetism from first-principles calculations. *Phys. Rev. B* 77:035427.
4. Jiang, D., Sumper, B. G., and Dai, S. 2007. Unique chemical reactivity of a graphene nanoribbon's zigzag edge. *J. Chem. Phys.* 126:134701.
5. Sheka, E. F., and Chernozatonskii, L. A. 2009. Chemical reactivity and magnetism of graphene. arXiv: 0901.3757v1 [cond-mat.mes-hall].
6. Sheka, E. F., and Chernozatonskii, L. A. 2010. Broken spin symmetry approach to chemical reactivity and magnetism of graphenium species. *J. Exp. Theor. Phys.* 110:121–32.
7. Gao, X., Zhou, Z., Zhao, Y., Nagase, S., Zhang, S. B., and Chen, Z. 2008. Comparative study of carbon and BN nanographenes: Ground electronic states and energy gap engineering. *J. Phys. Chem. A* 112:12677–82.
8. Stein, S. E., and Brown, R. L. 1987. pi-Electron properties of large condensed polyaromatic hydrocarbons. *J. Am. Chem. Soc.* 109:3721–29.
9. Fujita, M., Wakabayashi, K., Nakada, K., and Kusakabe, K. 1996. Peculiar localized state at zigzag graphite edge. *Phys. Soc. Jpn.* 65:1920–23.
10. Nakada, K., Fujita, M., Dresselhaus, G., and Dresselhaus, M. S. 1996. Edge state in graphene ribbons: Nanometer size effect and edge shape dependence. *Phys. Rev. B* 54:17954–61.
11. Miyamoto, Y., Nakada, K., and Fujita, M. 1999. First-principles study of edge states of H-terminated graphitic ribbons. *Phys. Rev. B* 59:9858–61.
12. Kobayashi, Y., Fukui, K., Enoki, T., Kusakabe, K., and Kaburagi Y. 2005. Observation of zigzag and armchair edges of graphite using scanning tunneling microscopy and spectroscopy. *Phys. Rev. B* 71:193406.
13. Niimi, Y., Matsui, T., Kambara, H., Tagami, K., Tsukada, M., and Fukuyama, H. 2006. Scanning tunneling microscopy and spectroscopy of the electronic local density of states of graphite surfaces near monoatomic step edges. *Phys. Rev. B* 73:085421.
14. Chen, Z., Jiang, D., Lu, X., Bettinger, H. F., Dai, S., Schleyer, P. R., and Houk, K. N. 2007. Open-shell singlet character of cyclacene and short zigzag nanotubes. *Org. Lett.* 9:5449–52.
15. Lee, H., Son, Y.-W., Park, N., Han, S., and Yu, J. 2005. Magnetic ordering of graphitic fragments: Magnetic tail interaction between the edge-localized states. *Phys. Rev. B* 72:174431.
16. Jiang, D., Sumper, B. G., and Dai, S. 2007. First principles study of magnetism in nanographenes. *J. Chem. Phys.* 127:124703.
17. Adamo, C., Barone, V., Bencini, A., Broer, R., Filatov, M., Harrison, N. M., Illas, F., Malrieu, J. P., and Moreira, I. de P. R. 2006. Comment on "About the calculation of exchange coupling constants using density-functional theory: The role of the self-interaction error" [*J. Chem. Phys.* 123, 164110 (2005)]. *J. Chem.Phys.* 124:107101.
18. Hod, O., Barone, V., and Scuseria, G. E. 2008. Half-metallic graphene nanodots: A comprehensive first-principles theoretical study. *Phys. Rev. B* 77:035411.
19. Dutta, S., Lakshmi, S., and Pati, S. K. 2008. Electron-electron interaction on the edge states of graphene: A many-body configuration interaction study. *Phys. Rev. B* 77:073412.
20. Meyer, J. C., Girit, C. O., Crommie, N. F., and Zettl, A. 2008. Imaging and dynamics of light atoms and molecules on graphene. *Nature* 454:319–22.

21. Shemella, Ph., and Nayak, S. K. 2009. Electronic structure and band-gap modulation of graphene via substrate surface chemistry. *Appl. Phys. Lett.* 94:032101.
22. Elias, D. C., Nair, R. R., Mohiuddin, T. M. G., Morozov, S. V., Blake, P., Halsall, M. P., Ferrari, A. C., Boukhvalov, D. W., Katsnelson, M. I., Geim, A. K., and Novoselov, K. S. 2009. Control of graphene's properties by reversible hydrogenation: Evidence for graphane. *Science* 323:610–13.
23. Pan, B. D., Zhang, J., Li, Z., and Wu, M. 2009. Hydrothermal route for cutting graphene sheets into blue-luminescent graphene quantum dots. *Adv. Mater.* 21:1–5.
24. Liu, H. 2005. Homogeneous carbon nanotube/polymer composites for electrical applications. *Appl. Phys. Lett.* 83:2928–30.
25. Grossiord, N., Loos, J., and Koning, C. E. 2005. Strategies for dispersing carbon nanotubes in highly viscous polymers. *J. Mater. Chem.* 15:2349–52.
26. McLachian, D. S., Chiteme, C., Park, C., Wise, K. E., Lowther, S. E., Lillehei, P. T., Siochi, E., and Harrison, J. S. 2005. AC and DC percolative conductivity of single wall carbon nanotube polymer composites. *J. Polym. Sci. B* 43:3273–87.
27. Guldi, D. M., Rahman, G. M. A., Zerbetto, M., and Prato, M. 2005. Carbon nanotubes in electron donor–acceptor nanocomposites. *Acc. Chem. Res.* 38:871–78.
28. Stankovich, S., Dikin, D. A., Dommett, G. H. B., Kohlhaas, K. M., Zimney, E. J., Stach, E. A., Piner, R. D., Nguyen, S. T., and Ruoff, R. S. 2006. Graphene-based composite materials. *Nature* 442:282–86.
29. Kondo, D., Sato, S., and Awano, Y. 2008. Self-organization of novel carbon composite structure: Graphene multi-layers combined perpendicularly with aligned carbon nanotubes. *Appl. Phys. Express* 1:074003.
30. Sheka, E. F., and Chernozatonskii, L. A. 2010. Graphene-carbon nanotube composites. *J. Compt. Theor. Nanosci.* 7:1814–24.
31. Varchon, F., Feng, R., Hass, J., Li, X., Ngoc Nguyen, B., Naud, C., Mallet, P., Veuillen, J.-Y., Berger, C., Conrad, E. H., and Magaud, L. 2007. Electronic structure of epitaxial graphene layers on SiC: Effect of the substrate. *Phys. Rev. Lett.* 99:126805.
32. Berger, C., Song, Z., Li, T., Li, X., Orgbazghi, A. Y., Feng, R., Dai, Zh., Marchenkov, A. N., Conrad, E. H., First, P. N., and de Heer, W. A. 2004. Ultrathin epitaxial graphite: 2D electron gas properties and a route toward graphene-based nanoelectronics. *J. Phys. Chem. B* 108:19912–16.
33. Matsuo, Y., Tahara, K., and Nakamura, E. 2003. Theoretical studies on structures and aromaticity of finite-length armchair carbon nanotubes. *Org. Lett.* 5:3181–84.
34. Eda, G., and Chhowalla, M. 2009. Graphene-based composite thin films for electronics. *Nano Lett.* 9:814–18.
35. Park, S., An, J., Jung, I., Piner, R. D., An, S. J., Li, X., Velamakanni, A. and Ruoff, R. S. 2009. Colloidal suspensions of highly reduced graphene oxide in a wide variety of organic solvents. *Nano Lett.* 9:1593–97.

chapter thirteen

Magnetism of fullerenes and graphene

13.1 Introduction

Since the discovery of magnetism in all-carbon crystals consisting of polymeric layers of covalently bound C_{60} fullerene molecules,[1] the still unclear nature of this phenomenon offers an intriguing issue for researchers. The number of publications devoted to this problem is in the dozens, which naturally led to a recent monograph[2] systematizing the main concepts and results achieved in this valley. The main issue is still far from being completely clear, and we are encouraged to undertake one more attempt to shed light on this phenomenon. The proposed approach is connected with the main idea concerning unpairing of odd electrons disclosed in the book, and is based on a traditional notion that the magnetism of any crystalline substance must be related to features of the electron structure of the main structural block that supplies unpaired (magnetic) electrons (spins). It is also well known that, in addition to the primary electron structure, the package of these blocks is the second operation factor.

As for the structural elements of polymeric fullerene crystals, these are C_{60} molecules (monomers). It is conventionally assumed that the singlet C_{60} molecule has no unpaired electrons. This assumption is quite well substantiated in experiment: no significant electron paramagnetic resonance (EPR) signals from fullerene samples have been observed in either the gas or solid state. Consequently, vanishingly small diamagnetism of the molecule was observed.[3] Once not subjected to question, this fact underlies all quantum chemical calculations of the singlet ground state of both C_{60} molecules and the crystals built of them, which have been performed until recently without allowance for the electron spin, within the framework of the closed-shell approximation. In this approximation, a thorough analysis of the possibility that magnetic electrons appear as a result of polymerization led to a negative result,[4–6] from which it has been concluded that the perfect crystals of C_{60} polymer must be nonmagnetic. Since this statement was at variance with experimental data, several theoretical models have been proposed, which explain the appearance of unpaired electrons by radical[7–9] or topological[10,11] defects. However, the presence of such defects was not experimentally confirmed, the more so that, in order to provide for the magnetic susceptibility in bulk crystals, such defects must either be arranged in a uniform ordered manner or form rather dense structures ensuring percolation spin ordering.

Recent experiments revealed some new features of the phenomenon. It was established that the observed magnetism is not a stable internal property of perfect crystals, but its observation is directly related to certain structural imperfections of crystals.[12–15] These observations stimulated us to revise formulation of the problem. We have attempted to answer two related questions: (1) Is it possible to obtain significant magnetic susceptibility in defect-free fullerene polymers based on C_{60} molecules in the singlet state, and (2) how can the magnetic behavior of such systems be related to the real structure of crystalline samples?

The singlet character of the ground state of the C_{60} molecule is caused by the antiferromagnetic exchange interaction of its electrons. As is known,[16] the magnetic susceptibility of the electron system in such cases can be increased only by the admixture of states with higher multiplicities, after which it is important to know the energy gap between these states and the singlet ground level. A quite correct estimation of the gaps is provided by the well-known relation

$$E^{PS}(S) = E^{PS}(0) - S(S+1)J \tag{13.1}$$

where E^{PS} is the energy of the pure spin state, J is the exchange integral, and S is the total spin of the electron system. Obviously, the exchange integral J is the measure of an energetic spacing between pure spin states; that is why it is called the exchange coupling or magnetic coupling constant.[17] Apparently, the effect of admixture increases with decreasing exchange interaction.

It is commonly accepted that this interaction in the C_{60} molecule is intrinsically strong, which ensures, on one hand, complete covalent pairing of its odd electrons and, on the other hand, results in a negligibly small magnetic susceptibility. However, as we already know, the molecule possesses effectively unpaired electrons, which is indicative of a relatively weak electron-electron interaction. These electrons seem to be a real candidate for providing enough of a spin excess for a noticeable magnetic susceptibility. Let us look at the problem of the C_{60} crystal from this viewpoint.

13.2 *Why are C_{60} and C_{70} molecules nonmagnetic?*

To determine the energy of pure spin states given by Equation 13.1, it is necessary to know the energy of the pure-spin singlet state, $E^{PS}(0)$, and the exchange integral J. Both values present the main problem in the molecular magnetism theory due to weakness of the electron interaction, which forces us to address complicated schemes of calculations involving configurational interaction (see a brief description of the problem in Adamo et al.[17]). As was discussed in Chapter 2 in detail, the unrestricted broken symmetry (UBS) approach assists in the problem solution to a great extent. Within the framework of the UBS Hartree-Fock (HF) approach, the coupling constant J and energy $E^{PS}(0)$ can be straightforwardly determined using Equations 2.2 and 2.3, thus giving the possibility to evaluate the ability of C_{60} to respond upon application of the magnetic field.

The energies of the pure-spin singlet state, $E^{PS}(0)$, and the total numbers of effectively unpaired electrons N_D of C_{60} and C_{70} molecules are given in Table 3.2. The corresponding J values constitute –1.86 and –1.64 kcal/mol, respectively. The negative sign of the constants shows that the Coulomb contribution prevails in the exchange interaction between odd electrons in both molecules. At the same time, the large magnitude of this interaction hinders effective mixing of the singlet state with the states of higher multiplicity, which accounts for the vanishingly small magnetic susceptibility of free molecules. The same should be expected from the pristine cubic C_{60} crystal, whose properties are described by the unit cell containing four weakly bound molecules. But what happens to the exchange integral J in the course of polymerization? In order to elucidate this question, a set of calculations has been performed related to C_{60} and C_{70} oligomers, presented in Figure 8.8.[18]

13.3 *Effectively unpaired electrons in monomer molecules of oligomers*

The properties of effectively unpaired electrons and their influence on the magnetic susceptibility of crystals are considered first for a monomer molecule, which is the main structural block in polymeric fullerene crystals. Let us assume that all the central monomer units (discriminated by colors in Figure 8.8) are in positions closely analogous to those in the relevant perfect crystals. The properties of thus selected monomer molecules include the values of exchange integrals (J), total numbers of effectively unpaired electrons (N_D), and their ACS distribution over atoms (N_{DA} maps). While the latter two quantities can be readily extracted from data (output files) related to the corresponding oligomers, the evaluation of exchange integrals encounters certain difficulties. It might seem that, extracting the corresponding monomer molecules from the complete structure and calculating their energies in the singlet state and in the state with maximum spin ($S_{max} = 30$) at a fixed geometry, one could determine the J value using Equation 2.2. However, the actual rupture of the C-C bond of cyclobutane

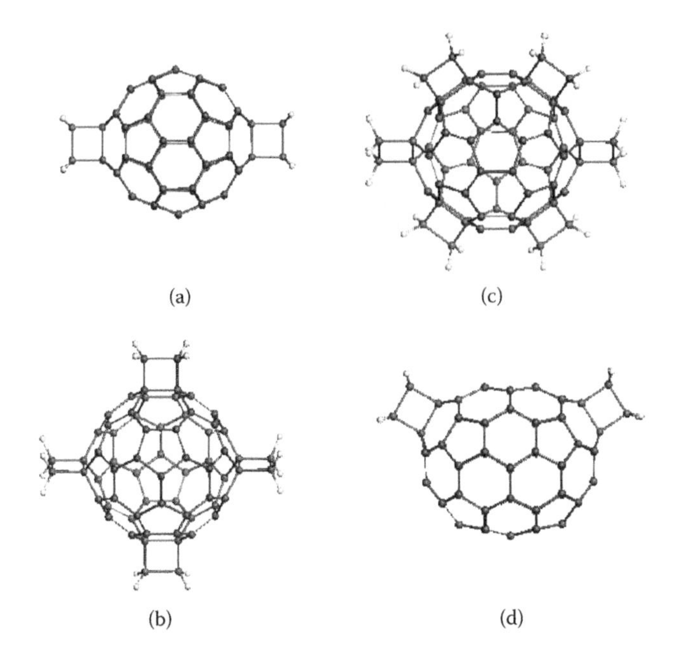

Figure 13.1 Terminated monomer molecules corresponding to darkened molecules in (a) Figure 8.8a, (b) 8.8b, (c) 8.8c, and (d) 8.8e. UBS HF AM1 singlet state. (From Sheka, E. F., et al., *J. Exp. Theor. Phys.*, 103, 728–39, 2006.)

rings that join neighboring molecules renders the electron structure of a monomer molecule perturbed, and hence the integrals determined in this case can somewhat differ from real values.

A definite way out of this difficult situation can be provided by extraction of monomers from the optimized oligomer structures, together with the adjacent cyclobutane rings. The corresponding structures are presented in Figure 13.1. In order to avoid the introduction of additional perturbations to the electron system and to restrict the region of existence of the odd electrons to the fullerene core, the ruptured bonds at each external carbon atom of the cyclobutane ring were terminated with a couple of hydrogen atoms. The calculations for such furnished monomer molecules were performed for all carbon atoms fixed in positions corresponding to those in oligomers, with complete optimization of the positions of terminal hydrogen atoms.[18] The values of N_D and J determined in this way are presented in Table 13.1.

13.3.1 Peculiarities in the odd electrons behavior

The numbers of effectively unpaired electrons for the monomer molecules of $2C_{60}$ (L), $3C_{60}$ (L), and $4C_{60}$ (L) oligomers presented in Table 13.1 (n^*C_{60}

Table 13.1 Number of unpaired electrons (per monomer), and exchange integral in C_{60} and C_{70} oligomers (UBS HF AM1 singlet)

	N_D		J, kcal/mol	
Oligomers	n^*C_{60}[a]	MM[b]	n^*C_{60}[a]	MM[b]
C_{60}	9.84	—	−1.86	−1.86
2^*C_{60}	10.96	—	−1.01	—
3^*C_{60}	11.26	—	−0.67	—
4^*C_{60}	11.45	—	−0.50	—
5^*C_{60}	12.02	12.04	−0.38	−2.10
9^*C_{60} (*Tg*)	13.41	11.50	−0.23	−2.10
9^*C_{60} (*Hg*)	9.23	9.41	−0.21	−2.17
10^*C_{60} (*Hg*)	10.43	—	−0.18	−2.16
C_{70}	14.39	—	—	−1.64
5^*C_{70}	16.02	13.95	−0.39	−1.84

Source: Sheka, E. F., et al., *J. Eksp. Theor. Phys.*, 103, 728–39, 2006.

[a] Data for oligomers shown in Figure 8.8.
[b] Data for terminated monomers shown in Figure 13.1.

column) represent the average values determined according to Equation 2.18 as $N_D = 2\langle S^2_{olig}\rangle/n$, where $\langle S^2_{olig}\rangle$ is the total spin square of oligomers in the singlet state. For the monomers incorporated into $5C_{60}$ (*L*), $9C_{60}$ (*Tg*), $9C_{60}$ (*Hg*), and $5C_{70}$ (*L*) oligomers, the N_D values were determined as $N_D = \Sigma_A N_{DA}$, that is, by separating data for atoms of the specified molecules from the entire N_{DA} list in the output files related to the oligomer. The next column of the table contains N_D values for terminated monomer molecules shown in Figure 13.1, which were determined using Equation 2.23.

According to Table 13.1, polymerization leads to significant changes in the electron structure of monomers. The numbers of effectively unpaired electrons significantly increase in all oligomers except $9C_{60}$ (*Hg*), for which N_D exhibits a small decrease. The growth of N_D is indicative, on the one hand, of a decrease in the extent of covalent pairing in monomer molecules and, on the other hand, of a generally enhanced chemical activity of these molecules as compared to the pristine one. The sequence of oligomer structures arranged in the order of decreasing N_D of the monomers is as follows: (*L*) > (*Tg*) > (*Hg*), and in terms of the corresponding crystal structures: (*O*) > (*T*) > (*R*).

Changes in the total number of effectively unpaired electrons are accompanied by a considerable variation of ACS N_{DA} maps of the monomers. Figure 13.2 shows the N_{DA} maps of the pristine molecules (dotted curves), in decreasing order of value. Since, as we know, the N_{DA} plotting provides a chemical portrait of a molecule, which elucidates many

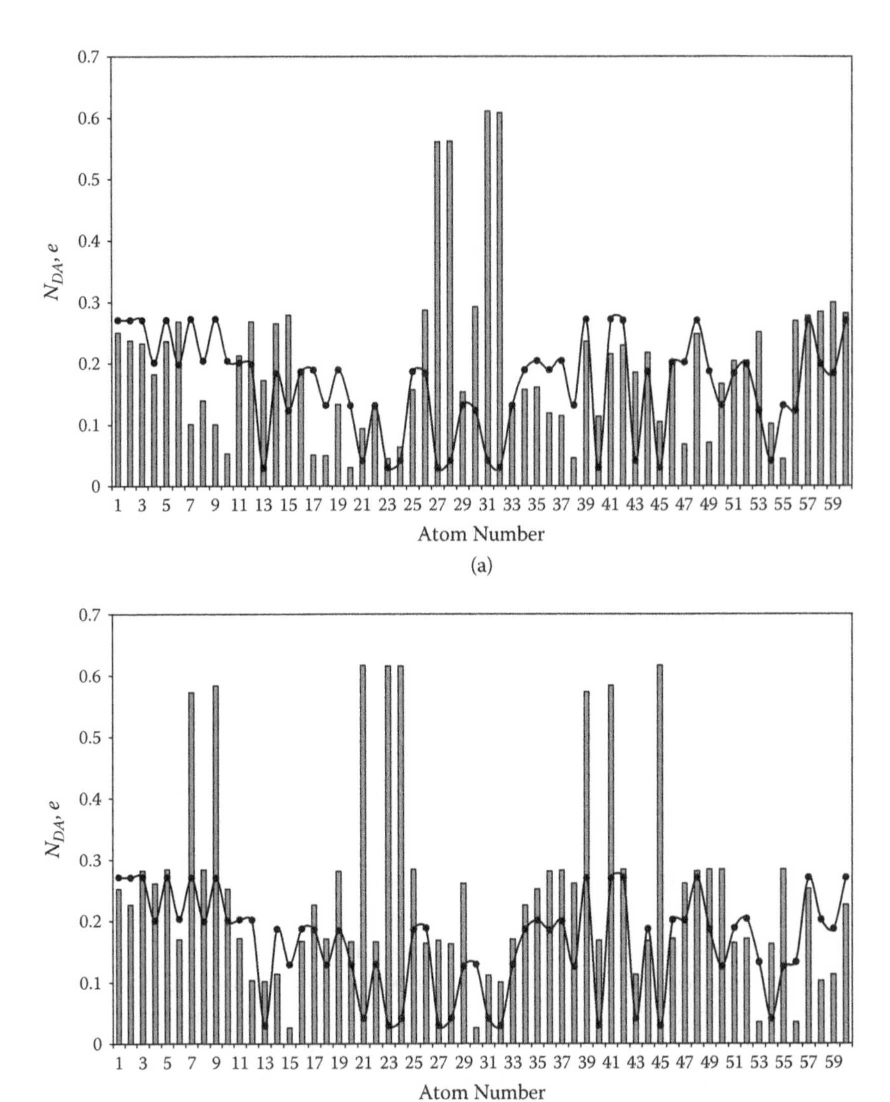

Figure 13.2 Atomic chemical susceptibility N_{DA} maps for monomer molecules: (a) $5C_{60}$ (*L*), (b) $9C_{60}$ (*Tg*), (c) $9C_{60}$ (*Hg*), (d) $5C_{70}$ (*L*). Dotted curves show N_{DA} maps of the pristine molecule. (From Sheka, E. F., et al., *J. Eksp. Theor. Phys.*, 103, 728–39, 2006.)

structural features concerning its core, the N_{DA} maps of monomers show substantial changes in the pristine chemical portrait. While molecular chemical susceptibility, N_D, is responsible for the general potential reactivity of the species, the efficiency of these reactions is determined by high-rank N_{DA}. In the case of polymerized crystal, the efficiency of the

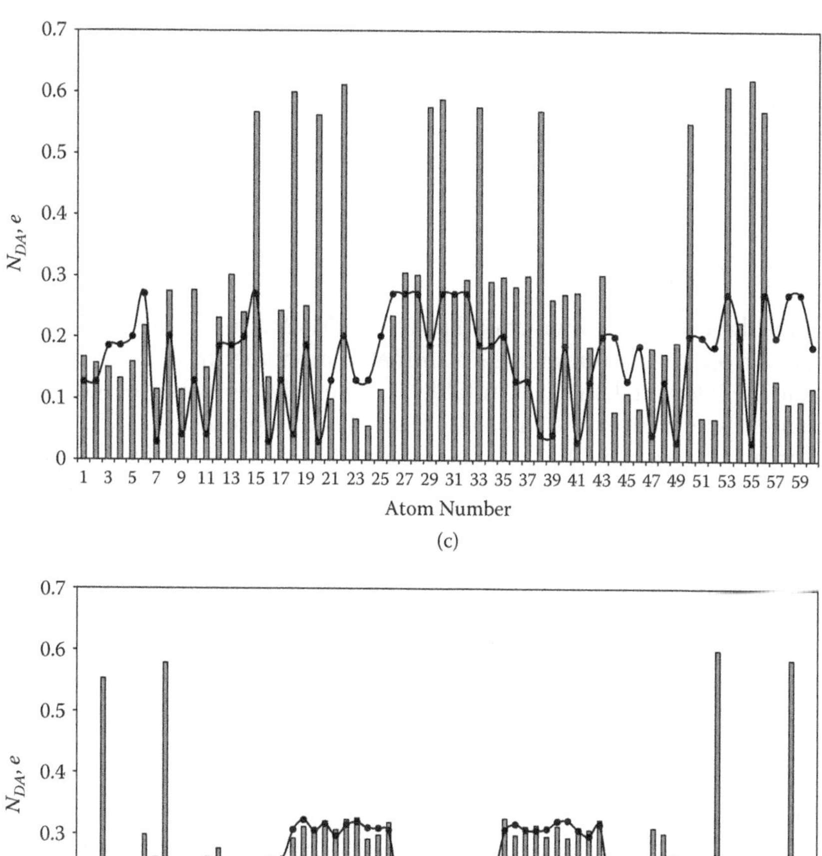

Figure 13.2 (continued).

relaxation of stresses related to changes in the molecular structure as a result of the new adduct formation, as well as the corresponding energy consumption, should be taken into account. From the standpoint of high-rank N_{DA} values presented in Figure 13.2, the oligomers are ordered as follows: (*Hg*) > (*Tg*) > (*L*). The same sequence is valid for crystalline modifications in view of the relaxation of stresses, which proceeds more effectively in less thermodynamically stable systems. Thus, taking into account both factors, the efficiency of chemical reactions (at least at the initial stages) in polymeric C_{60} crystals decreases in the following order: (*R*) > (*T*) > (*O*). In complete agreement with the above considerations, this very sequence was experimentally observed for the reactions of fluorination of polymerized C_{60} crystals.[19]

13.3.2 Exchange integral *J*

As for the exchange interaction between odd electrons, this can only be judged from the *J* values for the terminated monomer molecules. According to the relevant data listed in Table 13.1, the polymerization of molecules apparently enhances the interaction, so that this process as such cannot favor an increase in the magnetic susceptibility of samples. However, an analysis of the characteristics of the whole oligomer shows that the *J* value significantly changes—approximately in proportion to $1/n$, where n is the number of monomer units in the oligomer. This decrease in the exchange integral with increasing n indicates that the interaction of odd electrons decreases in the oligomers of finite size, which favors growth in the magnetic susceptibility. Decreasing in magnitude, the integral (initially negative due to the predominating Coulomb interaction) can change the sign—for example, as a result of interaction between various oligomers—thus altering the multiplicity of the ground state. Among mechanisms describing the induction of a magnetic order in systems with the singlet ground state,[16] one promising factor is the van Vleck polarization induced by the applied magnetic field.[20] The magnetic susceptibility of an electron system in this case can increase due to the admixture of states with higher multiplicity to the singlet state, which makes important the energy gap between the singlet and high-spin states. In particular, in regards to a triplet admixture, a decrease in *J* on the passage, for example, from the pristine C_{60} molecule to the $9C_{60}$ (*Hg*) carpet oligomer, implies a ninefold decrease in the singlet-triplet gap ($\Delta E = 2J$). A decrease in the gap to 0.018 eV makes possible the formation of a spin-mixed ground state upon magnetic field application. This particular size-dependent effect appears to be very important in regards to the magnetism of polymerized fullerenes.

13.4 Nanostructures and magnetism in polymeric C_{60} crystals

A decrease in magnitude of the exchange integral in finite-size oligomers suggests that only nanostructural samples can be magnetic in all-carbon materials—and there is numerous experimental evidence for the validity of this hypothesis. For example, it was demonstrated that the photo oligomerization of C_{60} molecules in the pristine crystal leads to the development of a significant magnetic susceptibility in the case of formation of linear oligomers with an average chain length of $n = 20$.[21] We believe that nanostructurization is also responsible for the magnetism of carpet polymers. Indeed, the magnetism in crystals of the *R* modification is observed for the samples prepared in the range of temperatures and pressures close to critical with respect to the stability of crystals.[13] Under these conditions, even very small changes in the technological parameters can lead to the formation of magnetic carbon phases with various contributions of the ferromagnetic component. Magnetic samples of the *R* modification exhibit high mosaicity[14] and inhomogeneity of the magnetic structure, which accounts for no more than 30% of the total sample volume.[12] Based on these results, we can propose a scale-like model of the *R* type magnetism. Apparently, the graphite-like polymer structure in such crystals favors the formation of nanosize scaly oligomer nanoclusters under rigid conditions. These clusters contain dozens of molecules and are characterized by small exchange integrals. The number and size of scales evidently depend on the sample preparation technology. However, such structural features are not present in tetragonal crystals, which are well known to not exhibit magnetism. The hypothesis concerning the nanostructural nature of magnetic carbons is consistent with the observations of magnetic properties of nanostructural graphite formed under proton beam irradiation,[22] local ferromagnetism in microporous graphite possessing a zeolite-Y-like structural ordering with nanosize pores,[23] and the magnetism of a nanostructural carbon foam.[24] An increase in the diamagnetism of shungite rocks with a structure containing bent graphene carpets also exhibits a clear nanostructural character.[25]

13.5 *Magnetism of zigzag edge nanographenes*

The magnetic phenomenon, predicted and studied computationally for graphene nanoribbons, is one of the hottest issues of graphene science. At the heart of the graphene magnetism are localized states whose flat bands are located in the vicinity of the Fermi level and whose peculiarities were attributed to zigzag edges.[26–30] In numerous UBS density functional theory (DFT) studies, this fact was connected with the spin density on edge atoms. Computations were carried out in presumably Ψ-contaminated UBS DFT approximations following such a logical scheme: taking into account spins of edge atoms at the level of wave function; considering so-called antiferromagnetic (AFM) and ferromagnetic (FM) spin configurations with spin alignment up on one edge and down (up) on the other edge, respectively, or a nonmagnetic configuration when up-down spin pairs are located at each edge; and performing calculations for these spin configurations. The obtained results have shown that (1) the antiferromagnetic configuration corresponds to the open-shell singlet ground state and is followed in stability by ferromagnetic and then nonmagnetic states, and (2) the calculated spin density on edge atoms corresponds to the input spin configurations in all cases. It should be added that numerical results obtained in different studies differ from each other when different functionals are used under calculations.

However, the UBS DFT antiferromagnetic (singlet) state is as spin contaminated as the UBS HF one, and the availability of the spin density is just a strong confirmation of the contamination. Nevertheless, the presence of spin density at zigzag edge atoms was accepted as a decisive point in heralding the magnetism of graphene ribbons, after which the phenomenon was considered to be confirmed, giving rise to a big optimism toward the expectation of a number of exciting applications of the material, in spintronics, for example.[31]

Since spin density is direct evidence of the solution spin contamination, particularly for the singlet state, it is worth comparing spin-density data computed at the UBS DFT and UBS HF levels of the theory. The UBS HF spin-density distribution over NGr (12,15) atoms with hydrogen-terminated and empty edges[32,33] is demonstrated in Figure 13.3. As seen in the figure, the spin density is available at all atoms of the graphene sheet. In both cases, its summation over all atoms gives zero since the ground state is singlet. The spin density at zigzag edge atoms is the highest, even absolutely dominating when the edges are emptied. In contrast, UBS DFT data are related to zigzag edge atoms only and spin-density absolute values vary from 0.26 to 0.47 when the local density functional is replaced by

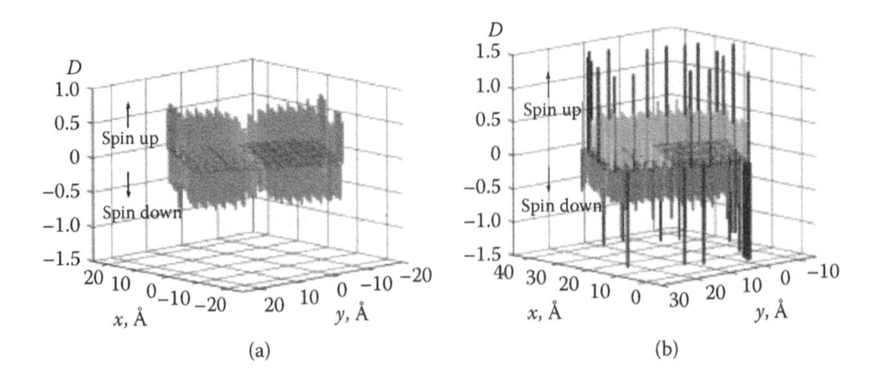

Figure 13.3 Distribution of spin density over atom of rectangular nanographene NGr (12,15) with hydrogen terminated (a) and empty (b) edges. UBS HF AM1 singlet state. (From Sheka, E. F., and Chernozatonskii, L. A., *J. Exp. Theor. Phys.*, 110, 121–32, 2010.)

the screen exchange hybrid density functional.[29] To see only these atoms in Figure 13.3 means lifting the zero reading level up (down) to about ±0.4 in the first case and ±1.3 in the second one, which, in other words, means lowering the sensitivity in recording the density values. This is the same situation caused by the pressed-by-functional character of the UBS DFT solution that was discussed for the atomic chemical susceptibility profiles in Section 12.2.

It should be noted that the UBS HF spin density on the zigzag edge is distributed quite peculiarly, not following the above-mentioned up- and down-edge antiferromagnetic configuration assumed for the ground state by UBS DFT. Remembering that the spin-density value is sensitive to the C-C bond length, it becomes clear why varying the latter produces variation in the density distribution as well. Therefore, the UBS HF data differ from those of UBS DFT both qualitatively and quantitatively, not supporting a ranged configuration of spins on zigzag edge atoms only. At the same time, the UBS HF data well correlate with, so far, the only many-body configuration interaction calculations of the edge states of graphene.[30] The latter infer that although the electrons have the tendency to accumulate at the edges, the spin distribution is quite irregular, so that a net spin polarization of the edges is highly improbable. Therefore, as in the case of the chemical reactivity of graphene discussed in Section 12.2, many-body configurational interaction calculations support the ability of UBS HF to highlight main physical features of weakly interacting electrons.

Coming back to the magnetism of graphene ribbons, let us start from the fact that the real ground state of the object is a pure-spin singlet. This means that the real spin density at each atom is zero. We can nevertheless discuss the possibility of the magnetic behavior of the object, not from the spin-density viewpoint, but addressing the energy difference between states of different spin multiplicity, as was discussed in Section 13.3.2.

An attempt to go outside the spin-density concept at the UBS DFT level was made in Hod et al.[29] This time the main attention was concentrated on the difference in position of singlet and high-spin (mainly triplet) states of graphene nanoribbons, thus implicitly appealing to the J value. However, as said in Section 13.1, the magnetic coupling constant, J, should be attributed to the difference of pure-spin-state energies, while the UBS DFT states under discussion are spin-mixed so that their energies do not correspond to those of pure spin states that makes the authors' conclusions quite uncertain.

In contrast to UBS DFT, UBS HF offers a straightforward way to determine pure spin states. Computed accordingly to Equation 2.3, $E^{PS}(0)$, $E_B(0)$, and J values related to the number of nanographenes are listed in Table 13.2. As seen from the table, the ground state of all species is singlet, so that a question arises if the magnetization of the singlet-ground-state

Table 13.2 Electronic characteristics of nanographenes[a] in kcal/mol

Nanographenes[b]	The number of "magnetic" (odd) electrons	$E_B(0)$	J	$E^{PS}(0)$	Singlet-triplet gap[c]
(12,15)	400	1,426.14	−0.42	1,342.14	0.84
(7,7)	120	508.69	−1.35	427.69	2.70
(5,6)	78	341.01	−2.01	262.72	4.02

Source: Sheka, E. F., and Chernozatonskii, L. A., *J. Exp. Theor. Phys.*, 110, 121–32, 2010.

[a] Tabulated energies $E^{PS}(0)$ and $E_B(0)$ correspond to the heats of formation of the relevant states.

[b] Following Gao et al., n_a and n_z match the numbers of benzenoid units on the armchair and zigzag ends of the sheets, respectively.

[c] For pure spin states the singlet-triplet gap $E^{PS}(1) - E^{PS}(0) = -2J$.

object is possible. As discussed earlier, the phenomenon may occur as a consequence of mixing the state with those of high multiplicity following, say, van Fleck mixing promoted by applied magnetic field.[20] Since the effect appears in the first-order perturbation theory, it depends on J, which determines the energy differences in denominators. Consequently, J should be small to provide noticeable magnetization. Obviously, singlet-triplet mixing is the most influential. As follows from Table 13.2, the energy gap to the nearest triplet state for the studied nanographenes constitutes 1 to 4 kcal/mol. The value is large to provide a noticeable magnetization of these molecular magnets. However, the value gradually decreases when the number of odd electrons increases. The behavior is similar to that obtained for fullerene oligomers discussed in the previous section, which led to the suggestion of a scaly mechanism of nanostructured solid-state magnetism of the polymerized fullerene C_{60}.

In view of this idea, it was possible to estimate how large the nanographene should be to provide a noticeable magnetization. As known,[35] molecular magnetism can be fixed at a J value of 10^{-2} to 10^{-3} kcal/mol or less. Based on the data presented in Table 13.2 and supposing the quantity to be inversely proportional to the number of odd electrons, N should be ~10^5. In rectangular nanographenes N coincides with the number of carbon atoms, which is determined as[34]

$$N = 2(n_\alpha n_z + n_\alpha + n_z) \tag{13.2}$$

To fit the needed N value, the indices n_α and n_z should be in the hundreds, which leads to linear sizes of nanographenes of a few nanometers. The estimation is rather approximate, but it nevertheless correlates well with experimental observations of the magnetization of activated carbon fibers consisting of nanographite domains of 3 to 5 nm in size.[36,37]

13.6 About size-dependent magnetism

The obtained results highlight another important aspect of the fullerene and graphene magnetism exhibiting the belonging of the phenomenon to particular nanosize-effect ones. The latter means that both the fullerene oligomers and graphene magnetization are observed for nanosize samples only and, moreover, for samples whose size is within a particular interval, while the phenomenon does not take place in either very small or macroscopically large samples. Such size-dependent phenomena are not rare, and photoluminescence of nanosize silicon crystals[38] and other semiconductive grains[39] is one of the best examples. Thus, in the case of graphene, an individual benzenoid unit (including benzene molecule) is nonmagnetic (only slightly diamagnetic). When the units are joined to form a graphene-like cluster, effectively unpaired electrons appear due to weakening the interaction between odd electrons. The weakening accelerates when the cluster size increases, which is followed with the magnetic constant J decreasing until the latter achieves a critical level that provides a noticeable mixing of the singlet ground state with high-level spin states for the cluster magnetization to be fixed. And until the enlargement of the cluster size does not violate the molecular (cluster)-like behavior of odd electrons, the cluster magnetization will grow. However, as soon as the electron behavior becomes spatially quantized, the molecular character of the magnetization will be broken and will be substituted by the one determined by the electron band structure that is based on the properties of a unit cell. A joint unit cell of graphene involves two atoms that form one C-C bond of the benzenoid unit, which is why one comes back to the case of large magnetic constant J when magnetization becomes nonobservable. A similar situation takes place in the case of polymerized C_{60} fullerene crystals. The crystal unit cells involve either one (tetragonal and orthorhombic) or two (hexagonal) diamagnetic molecules, so that the cell magnetic constant is either J or $J/2$, both large, which makes no allowance for the magnetization of a perfect crystal to be fixed. On the other hand, when the crystal is nanostructured by producing nanosize scales, the molecular-like behavior of odd electrons of the clusters provides a significant weakening of the interaction between them that gives rise to small J and cluster magnetization. In both cases, the critical cluster size is a few nanometers, which should be compared to the electron mean free path, l_{el}. Evidently, when the cluster size exceeds l_{el}, the spatial quantization quenches the cluster magnetization. An accurate determination of l_{el} for odd electrons either in polymerized fullerene crystal or in graphene is not known, but the analysis of a standard database for electron mean free paths in solids[40] shows that the quantity should be in the vicinity of 10 nm, which is supported by experimental data of a 3 to 7 nm electron free path in thin films of Cu-phthalocyanine.[41]

13.7 *Odd electrons as they are seen today*

This chapter has completed the consideration of the basic problem of weak interaction between odd electrons in carboneous species within the framework of the broken spin-symmetry single-determinant approach. The modern implementations of the approach in the form of either the unrestricted Hartree-Fock (UBS HF) scheme or spin-polarized DFT (UBS DFT) were discussed with particular attention to the applicability of spin-contaminated solutions of both techniques for the description of electronic properties of the species. As for carbon nanotubes and graphene, the UBS DFT applications generally reveal the open-shell character of the singlet state of the object and manifest an extra spin density concentrated on zigzag edge atoms. Similarly, the UBS HF approach not only supports these findings, but exhibits the common behavior of odd electrons in fullerenes, carbon nanotubes, and graphene. The understanding of the behavior peculiarities, which is suggested by the broken symmetry approach, permits us to quantitatively describe both enhanced chemical reactivity and magnetism. The former is presented in terms of a quantified atomic chemical susceptibility that is continuously distributed over all atoms with the value, which is similar for fullerenes, the carbon nanotube (CNT) sidewall, and graphene inner atoms, and which exceeds the latter at end atoms of carbon nanotubes as well as at edge atoms of graphene. The UBS HF calculations also highlighted that the magnetic response of polymerized fullerenes and graphene is provided by a collective action of all odd electrons and is size dependent. The relative magnetic coupling constant J decreases when the size of either fullerene oligomers or the graphene sheet increases and J approaches the limit value of 10^{-2} to 10^{-3} kcal/mol, needed for the object magnetization to be recorded, when the size is of a few nanometers, which is well consistent with experimental findings.

A common view of both chemical reactivity and magnetism of fullerenes, carbon nanotubes, and graphene, physically clear and transparent, witnesses the internal self-consistency of the UBS HF approach and exhibits its high ability to quantitatively describe practically important consequences of weak interaction between odd electrons. The statement is well supported by a deep coherency of the obtained UBS HF results with those found from the application of many-body configurational interaction calculation schemes to polyacenes and graphene.

References

1. Makarova, T. L., Sundqvist, B., Höhne, R., Esquinazi, P., Kopelevich, Ya., Scharff, P., Davydov, V. A., Kashevarova, L. S, and Rakhmanina, A. V. 2001. Magnetic carbon. *Nature* 413:716–19.
2. Makarova, T. L., and Palacio, F., eds. 2006. *Carbon based magnetism: An overview of the magnetism of metal free carbon-based compounds and materials.* Amsterdam: Elsevier.

3. Ruoff, R. S., Beach, D., Cuomo, J., McGuire, T., Whetten, R. L., and Diederich, F. 1991. Confirmation of a vanishingly small ring-current magnetic susceptibility of icosahedral buckminsterfullerene. *J. Phys. Chem.* 95:3457–59.
4. Okada, S., and Saito, S. 1999. Electronic structure and energetics of pressure-induced two-dimensional C_{60} polymers. *Phys. Rev. B* 59:1930–36.
5. Okada, S., and Oshiyama, A. 2003. Electronic structure of metallic rhombohedral C_{60} polymers. *Phys. Rev. B* 68:235402.
6. Miyake, T., and Saito, S. 2003. Geometry and electronic structure of rhombohedral C_{60} polymer. *Chem. Phys. Lett.* 380:589–94.
7. Andriotis, A. N., Menon, M., Sheetz, R. M., and Chernozatonski, L. 2003. Magnetic properties of C_{60} polymers. *Phys. Rev. Lett.* 90:026801.
8. Ribas-Ariño, J., and Novoa, J. J. 2004. Evaluation of the capability of C_{60}-fullerene to act as a magnetic coupling unit. *J. Phys. Chem. Sol.* 65:787–91.
9. Belavin, V. V., Bulusheva, L. G., Okotrub, A. V., and Makarova, T. L. 2004. Magnetic ordering in C_{60} polymers with partially broken intermolecular bonds. *Phys. Rev. B* 70:155402.
10. Park, N., Yoon, M., Berber, S., Ihm, J., Osawa, E., and Tomanek, D. 2003. Magnetism in all-carbon nanostructures with negative Gaussian curvature. *Phys. Rev. Lett.* 91:237204.
11. Kim, Y.-H., Choi, J., Chang, K. J., and Tomanek, D. 2003. Defective fullerenes and nanotubes as molecular magnets. An *ab initio* study. *Phys. Rev. B* 68:125420.
12. Han, K.-H., Spemann, D., Höhne, R., Setzer, A., Makarova, T., Esquinazi, P., and Butz, T. 2003. Observation of intrinsic magnetic domains in C_{60} polymer. *Carbon* 41:785–95.
13. Wood, R. A., Lewis, M. H., Lees, M. R., Bennington, S. M., Cain, M. G., and Kitamura, N. 2002. Ferromagnetic fullerene. *J. Phys. Cond. Matter* 14:L385–91.
14. Tokumoto, M., Narymbetov, B., Kobayashi, H., Makarova, T. L., Davydov, V. A., Rakhmanina, A. V., and Kashevarova, L. S. 2000. Physical properties of highly-oriented rhombohedral C_{60} polymer. In *Electronic properties of novel materials—Molecular nanostructures*, ed. H. Kuzmani, J. Fink, M. Mehrang, and S. Roth, 73–76. Melville, NY: AIP.
15. Narozhnyi, V. N., Müller, K.-H., Eckert, D., Teresiak, A., Dunsch, L., Davydov, V. A., Kashevarova, L. S., and Rakhmanina, A. V. 2003. Ferromagnetic carbon with enhanced Curie temperature. *Physica B* 329–33:1217–18.
16. Zvezdin, A. K., Matveev, V. M., Mukhin, A. A., and Petrov, Yu. 1985. *Rear earth ions in magnetically ordered crystals* [in Russian]. Moskva: Nauka.
17. Adamo, C., Barone, V., Bencini, A., Broer, R., Filatov, M., Harrison, N. M., Illas, F., Malrieu, J. P., and Moreira, I. de P. R. 2006. Comment on "About the calculation of exchange coupling constants using density-functional theory: The role of the self-interaction error" [*J. Chem. Phys.* 123, 164110 (2005)]. *J. Chem. Phys.* 124:107101.
18. Sheka, E. F., Zaets, V. A., and Ginzburg, I. Ya. 2006. Nanostructural magnetism of polymeric fullerene crystals. *J. Eksp. Theor. Phys.* 103:728–39.
19. Gu, Z., Khabashesku, V. N., Davydov, V. A., Rakhmanina, A. V., and Agafonov, V. 2006. Fluorination of crystalline polymerized phases of C_{60} fullerene. *Fullerenes Nanotubes Carbon Nanostr.* 14:303–6.
20. Van Fleck, J. H. 1932. *The theory of electric and magnetic susceptibilities*. Oxford: Clarendon Press.
21. Owens, F. J., Iqbal, Z., Belova, L., and Rao, K. V. 2004. Evidence for high-temperature ferromagnetism in photolyzed C_{60}. *Phys. Rev. B* 69:033403.

22. Esquinazi, P., Spemann, D., Höhne, R., Setzer, A., Han, K.-H., and Butz, T. 2003. Induced magnetic ordering by proton irradiation of graphite. *Phys. Rev. Lett.* 91:227201.
23. Kopelevich, Y., da Silva, R. R., Torres, J. H. S., Penicaud, A., and Kyotani, T. 2003. Local ferromagnetism in microporous carbon with the structural regularity of zeolite Y. *Phys.Rev. B* 68:092408.
24. Rode, A. V., Gamaly, E. G., Christy, A. G., Fitz Gerald, J. G., Hyde, S. T., Elliman, R. G., Luther-Davies, B., Veinger, A. I., Andrulakis, J., and Giapintzakis, J. 2004. Unconventional magnetism in all-carbon nanofoam. *Phys.Rev. B* 70:054407.
25. Kovalevski, V. V., Prikhodko, A. V., and Buseck, P. R. 2005. Diamagnetism of natural fullerene-like carbon. *Carbon* 43:401–5.
26. Fujita, M., Wakabayashi, K., Nakada, K., and Kusakabe, K. 1996. Peculiar localized state at zigzag graphite edge. *Phys. Soc. Jpn.* 65:1920–23.
27. Nakada, K., Fujita, M., Dresselhaus, G., and Dresselhaus, M. S. 1996. Edge state in graphene ribbons: Nanometer size effect and edge shape dependence. *Phys. Rev. B* 54:17954–61.
28. Lee, H., Son, Y.-W., Park, N., Han, S., and Yu, J. 2005. Magnetic ordering of graphitic fragments: Magnetic tail interaction between the edge-localized states. *Phys. Rev. B* 72:174431.
29. Hod, O., Barone, V., and Scuseria, G. E. 2008. Half-metallic graphene nanodots: A comprehensive first-principles theoretical study. *Phys. Rev. B* 77:035411.
30. Dutta, S., Lakshmi, S., and Pati, S. K. 2008. Electron-electron interaction on the edge states of graphene: A many-body configuration interaction study. *Phys. Rev. B* 77:073412.
31. Son, Y.-W., Cohen, M. L., and Louie, S. G. 2006. Half-metallic graphene nanoribbons. *Nature* 444:347–49.
32. Sheka, E. F., and Chernozatonskii, L. A. 2009. Chemical reactivity and magnetism of graphene. arXiv: 0901.3757v1 [cond-mat.mes-hall].
33. Sheka, E. F., and Chernozatonskii, L. A. 2010. Broken spin symmetry approach to chemical reactivity and magnetism of graphenium species. *J. Exp. Theor. Phys.* 110:121–32.
34. Gao, X., Zhou, Z., Zhao, Y., Nagase, S., Zhang, S. B., and Chen, Z. 2008. Comparative study of carbon and BN nanographenes: Ground electronic states and energy gap engineering. *J. Phys. Chem. A* 112:12677–82.
35. Kahn O. 1993. *Molecular magnetism*. New York: VCH.
36. Shibayama, Y., Sato, H., Enoki, T., and Endo, M. 2000. Disordered magnetism at the metal-insulator threshold in nano-graphite-based carbon materials. *Phys. Rev. Lett.* 84:1744–47.
37. Enoki, T., and Kobayashi, Y. 2005. Magnetic nanographite: An approach to molecular magnetism. *J. Mater. Chem.* 15:3999–4002.
38. Khryachtchev, L., Novikov, S., and Lahtinen, J. 2009. Thermal annealing of Si/SiO_2 materials: Modification of structural and photoluminescence emission properties. *J. Appl. Phys.* 92:5856–62.
39. Sroyuk, A. L., Kryukov, A. I., Kuchmij, S. Ya., and Pokhodenko, V. D. 2005. Quantum size effects in semiconductive photocatalysis [in Russian]. *Teoret. Eksperim. Khimia* 41:199–218.

40. Seach, M. P., and Dench, W. A. 2004. Quantitative electron spectroscopy of surfaces: A standard data base for electron inelastic mean free paths in solids. *Surf. Interf. Anal.* 1:2–11.
41. Komolov, S. A., Lazneva, E. F., and Komolov, A. S. 2003. Low-energy electron mean free path in thin films of copper phthalocyanine. *Technol. Phys. Lett.* 29:974–76.

chapter fourteen

Chemical and structural analogs of sp^2 nanocarbons

The sp^2 carbon masterpiece presented in this book will not be complete if we do not touch on the question regarding chemical and structural analogs of the carbons. First, it concerns siliceous counterparts, the discussion of which was a topic of large interest, since numerous attempts to get siliceous fullerene, nanotubes, and graphene were unsuccessful. Otherwise, there are fullerene-like, tube-like, and graphene-like structures of different halcogenides and boron nitride, while the atoms that form these structures are not analogous to carbon. We will briefly consider the basic grounds that lay the foundation of the species behavior and will try to explain their drastic difference from carboneous species.

14.1 Siliceous nanostructures

Silicon-based nanotechnology has attracted the hottest attention since the first steps of nanotechnology became a reality. Silicon-based nanoelectronics seemed an obvious extension of conventional silicon microtechnology. The first peak of interest was connected with the adventure of scanning tunneling microscopy (STM) "atom writing" on the silicon crystal surfaces. A controllable deposition or extraction of silicon atoms from the surfaces seemed to open a direct way to design electronic nanochips of any kind. However, the works, pioneered by Prof. M. Aono and his team in Japan, as well as by other groups throughout the world, met serious difficulties on the way that showed unexpected complications connected with the surfaces' properties. Thus, the most promising Si(111) (7 × 7) surface was metallic and magnetic, in contrast to semiconductive and nonmagnetic bulk silicon. Thus, Figure 14.1 presents the ACS N_{DA} distribution over the surface unit cell of the Si(111) (7 × 7) crystal, which demonstrates a peculiar distribution of 92.6 effectively unpaired electrons over the cell atoms in the singlet state and provides a peculiar magnetic and conductive behavior of the surface.[1] Consequently, silicon nanoelectronics did not meet with success at that time, although undertaken efforts stimulated a large realm of silicon surface science and the current extended materials' nanoarchitechtonics (see projects of the International Center for Materials Nanoarchitectonics at NIMS in Japan).

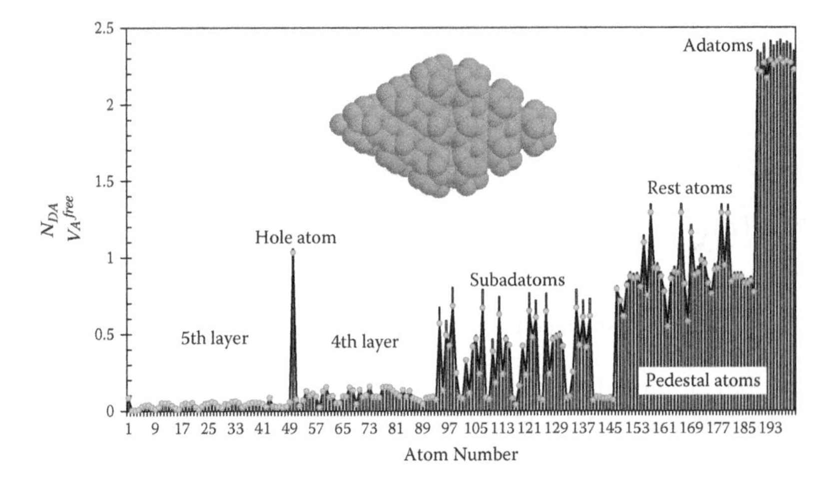

Figure 14.1 Atomic chemical susceptibility, N_{DA} (bars), and free valence, V_A^{free} (circles), distribution over five-layered Si(111) (7 × 7) surface unit cell (see insert). UBS HF AM1 singlet state. (From Sheka, E. F., *Int. J. Quant. Chem.*, 107, 2935–55, 2007.)

The next pulse of interest is being observed nowadays. It has been stimulated by extreme expectations connected with graphenium nanoprocessors. However, despite the reigning optimism about the devices, the graphene discoverers pointed out that the processors' appearance was unlikely for the next twenty years,[2] since replacement of the current silicon electronics technology is a tough hurdle. And again, compatibility of silicon-based nanoelectronics with the conventional one has focused attention on the question of whether carboneous graphene can be substituted by its siliceous counterpart, named silicene. Meeting the demands, the December 2008 Internet news brought information on "epitaxial growth of graphene-like silicon nanoribbons."[3] The report, based on the hexagon-patterned accommodation of silicon atoms adsorbed on the [110] Ag surface, heralded the silicene manifestation and is full of exciting applications to be expected.[4]

However, under detailed examination of the epitaxial grown silicon nanoribbons, the situation does not seem so transparent and promising. To clear the said, let us specify basic terms. First, we have to make clear what is implied by the term *silicene*. If any hexagon-packed structure of silicon atoms can be named silicene, then it has been known for a long time, for example, widely known silicon nanowires. However, four valence electrons of each silicon atom form the sp^3 configuration and participate in the formation of four chemical bonds in this case, so that nobody can pretend to look for a similarity between these species and carboneous graphene. Therefore, not hexagon packing itself, but a mono-atom-thick hexagon structure that dictates a *sp*^{2} configuration for atom valence electrons with the lack of one neighbor for each silicon atom meets requirements of comparison of silicene to graphene. Obviously, similar hexagon patterns should form the basis for silicon nanotubes. Only under these conditions can graphene and silicene, as well as carboneous and siliceous nanotubes, be considered on the same basis.

As for theoretical consideration, performed computations of silicene[5] and silicon nanotubes[6–8] meet the requirement completely. In contrast, experimental reports are full of appealing to silicon nanotubes (see the brief review in Perepichka and Rosei[9]) and silicene[3,4] (in the first announcement of the finding observed,[10] the latter was attributed to silicon nanowires) in spite of the evident *sp*^{3} configuration of silicon atoms in the structures observed. This fact was accepted by the experimentalists themselves. But the temptation to disclose silicon nanotubes and silicene seems to be so strong that the difference in the electron configuration is simply omitted. A detailed analysis of the available experimental data shows that silicon structures that can be compared to carbon nanotubes and graphene have not yet been observed. If we remember that fullerene Si_{60} has not been produced, we have to accept the availability of a serious reason that causes so drastic a difference between carboneous and siliceous analogs.

The problem is not new and is rooted deeply, so that "a comparison of the chemistry of tetravalent carbon and silicon reveals such gross differences that the pitfalls of casual analogies should be apparent"[11] (p. 505)—enough so to remind us that there is no silicoethylene or silicobenzene, as well as other aromatic-like molecules. A widely spread standard statement that "silicon does not like *sp*^{2} configuration" just postulates the fact, but does not explain the reason behind such behavior. A real reason was disclosed for the first time when answering the question of why fullerene Si_{60} does not exist.[12,13] The answer addressed changes in the electron interaction for C_{60} and Si_{60} species when their electron configurations are transformed from the *sp*^{3} to the *sp*^{2} type. The interaction of two odd electrons, which are formed under the *sp*^{3}-to-*sp*^{2} transformation of any interatomic bond, depends on the corresponding distance R_{int}, which is ~1.5 times larger for Si-Si chemical bonds than for C-C ones. As was shown in Section 2.3, generally, the distance R_{int} = 1.395 Å is critical for these electrons to be covalently coupled. Above that distance, the electrons become effectively unpaired, and the more so the larger the distance. In the case of graphene, distances between two odd electrons fill the interval 1.39 to 1.43 Å. Evidently, only parts of C-C bonds exceed the limit value that causes partial exclusion of odd electrons from the covalent coupling that makes the molecular species partially radicalized, as discussed in Chapter 12. The radicalization is rather weak since only ~10% of all odd electrons

Table 14.1 Energies[a] in kcal/mol and the number of effectively unpaired electrons in sp^2-configured siliceous species (see Figure 14.2)

Species	$N(N_2)$[b]	$E^R(0)$	$E_B(0)$	$E^{PS}(0)$	N_D
I	2	54,50	48,95	39,02	0.88
II	6	144,51	121,25	108,67	2.68
III	60	1295,99	1013,30	996,64	62.48
IVa	96 (24)	2530,19	1770,91	1749,56	128
IVb	96	1943,14	1527,77	1505,48	95,7
Va	100 (20)	2827,73	1973,67	1958,54	115,05
Vb	100	2119,60	1580,77	1559,64	100,12
VIa	60 (22)	1950,20	1359,44	1346,68	75,7
VIb	60	1253,39	1001,27	972,12	54,04

Source: Lee, B. X., et al., *Phys. Rev. B*, 61, 1685–87, 2000.

[a] Tabulated energies $E^R(0)$, $E_B(0)$ [or $E^U(0)$], and $E^{PS}(0)$ correspond to the heats of formation of the relevant states. For the energy nominations, see Chapter 1.

[b] Numbers N_2 in parentheses show how many two-neighbor edge atoms are involved in the total number N.

(equal to the number of atoms, N) are unpaired. In contrast to this case, R_{int} in siliceous species is 2.3 to 2.4 Å. This causes a complete unpairing of all odd electrons (see Figure 2.1) so that all siliceous species with the expected sp^2 configuration should be many-fold radicals.

The application of unrestricted broken symmetry Hartree-Fock (UBS HF) approach to the problem made these expectations evident.[13,14] Table 14.1 lists calculation results of the total number of unpaired electrons, N_D, as well as a set of energetic parameters for a number of siliceous sp^2-configured species shown in Figure 14.2. As seen from the table, there is a drastic lowering of the total energy of the species, constituting about 20 to 30% of the largest values, when a closed-shell restricted Hartree-Fock scheme is substituted by an open-shell UBS HF. Large numbers of effectively unpaired electrons, N_D, for all species indicate the highly spin-contaminated character of their singlet UBS HF state. The energy of the singlet pure spin states, $E^{PS}(0)$, was determined according to Equation 2.3. As should be expected, the energy is lower than both $E^R(0)$ and $E_B(0)$ values, while rather close to the latter. This shows a small value of exchange integral J, indicating that, if they existed, siliceous species would have been magnetic.

It should be emphasized that the numbers of effectively unpaired electrons, N_D, listed in the table coincide quite well with the total numbers of silicon atoms (N) in all cases when the edges of the considered siliceous species are terminated by hydrogen atoms and exceed N by

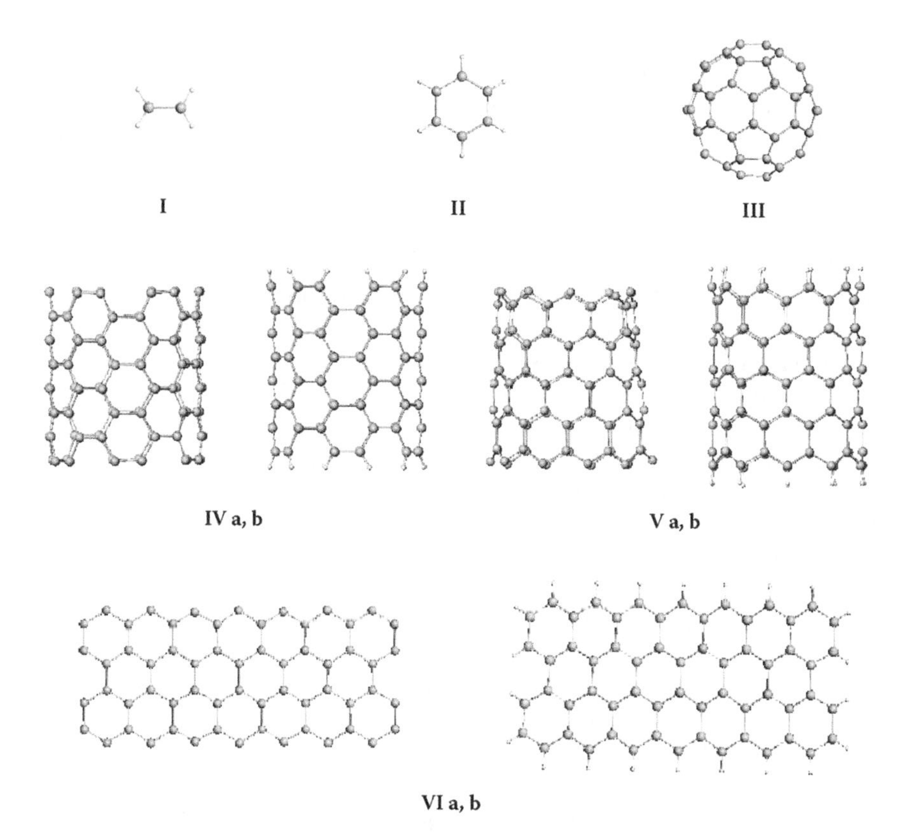

Figure 14.2 Equilibrium structures of sp^2-configured siliceous species: I, silicoethylene; II, silicobenzene; III, silicofullerene Si_{60}; IV, fragments of (6,6) SiNT with empty (a) and hydrogen-terminated (b) end atoms; V, the same but for (10,0) SiNT; VI, (7,3) silicene sheet with empty (a) and hydrogen-terminated (b) edges. UBS HF AM1 singlet state. (From Sheka, E. F., and Chernozatonskii, L. A., *J. Exp. Theor. Phys.*, 110, 121–32, 2010.)

the number of two-neighbor atoms (N_2) when hydrogen terminators are taken off of either tubes ends or silicene edges. This finding exhibits that silicon fullerene as well as silicon nanotubes and silicene are many-fold radicals and cannot exist under ambient conditions. Important to note is that no suitable passivation should be expected to provide the species stabilization since the passivation should be absolutely total, which will result in the transformation of all sp^2 silicon atoms to the sp^3 ones. That is why one observes sp^3 silicon nanowires instead of sp^2 SiNTs,[8] as well as sp^3-accommodated silicon atom adsorption layers on the (111) Ag surface instead of sp^2 silicene strips.[4,10]

Optimism expressed in theoretical papers where fullerene Si_{60},[15–18] silicon nanotubes,[6–8] and silicene[5] were considered is mainly due to the fact that the calculations were performed in the closed-shell approximation (similar to restricted Hartree-Fock (RHF)), so that the problem concerning weak interacting odd electrons was not taken into account.

14.2 *Boron nitride hexagon-packed species*

As known from molecular science, cyclic compounds are of high stability, among which carboneous benzenoid units are characterized as the most stable. However, this exclusive geometrical pattern is favorable not only in the atomic world but also in the macroscopic world, which was proven by Buckminster Fuller, who created his magnificent hexagon-based constructions. This means that the tendency to form hexagon-based structures is a common property of both the micro- and macroworld. Thus, the appearance of hexagon-based fullerene- and tube-like substances outside the organic world seemed surprising only because we were accustomed to their existence among carboneous species. The number of chemical entities forming such structures steadily increases, among which those consisting of boron and nitrogen atoms have attracted the most attention (see comprehensive reviews in Golberg et al.[19] and Pokropivny and Ivanovskii[20]). We will not go inside this interesting and intriguing topic, and will touch only the subject in comparison with carboneous species exemplified by a brief consideration of boron-nitride structural analogs.

Figure 14.3 presents structural hexagon-based (h) BN analogs to carboneous C_{60}, (4,4) and (8,8) single-walled carbon nanotubes, and (5,5) graphene based on computational studies.[21–23] All these structures are energetically stable, while not the best, as is the case of h-BN fulborene $B_{30}N_{30}$, which loses energy to other structures.[21] In appearance, they all look like their carboneous partners; even the spatial sizes of the structures are quite similar. Nevertheless, their behavior is absolutely different from that typical of carbons.[19,20] And, a high stability of the species against chemical treatments should be mentioned first. To those who have read this book attentively, the reason must be quite evident. Actually, in contrast to carboneous species, where a hexagon-based structure is the only frame that provides the best conditions for holding the odd electrons crowd, whose behavior is responsible for all peculiar properties of the substances, the hexagon frame of BN species is electron empty: there are no odd electrons since h-BN compounds are valence saturated due to the coincidence of numbers of valence electrons belonging to both boron and nitrogen atoms, and chemical bonds formed by each of them. That is why one should not expect from h-BN structures peculiar properties characteristic for carboneous sp^2 compositions, such as enhanced chemical activity or magnetism.

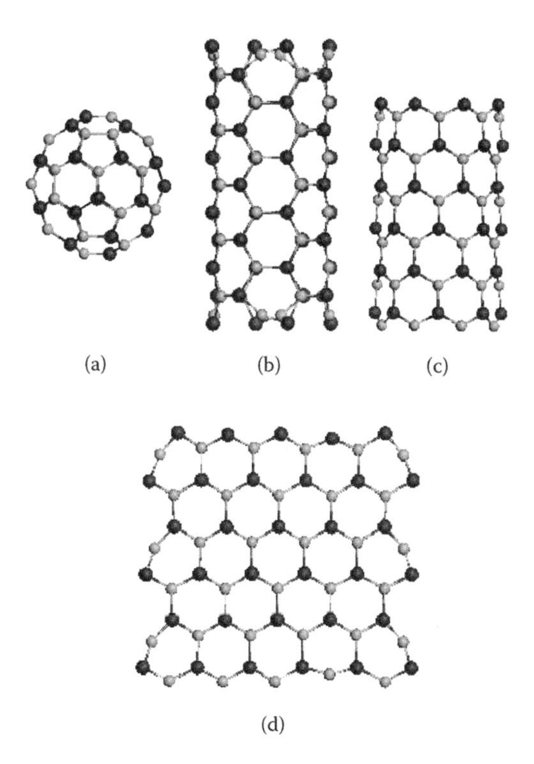

Figure 14.3 Equilibrium structures of hexagon-based boron-nitride structures. (a) h-Fulborene $B_{30}N_{30}$. Fragments of (4,4) (b) and (8,0) (c) single-walled h-BN nanotubes. (d) (5,5) h-BN nanographene. UBS HF AM1 singlet state.

References

1. Sheka, E. F. 2007. Odd electrons in molecular chemistry, surface science, and solid state magnetism. *Int. J. Quant. Chem.* 107:2935–55.
2. Geim, A. K., and Novoselov, K. S. 2007. The rise of graphene. *Nature Mater.* 6:183–91.
3. Kara A., Leandri C., Davila M. E., de Padova P., Ealet B., Ougaddou H., Aufray, B., and Le Lay, G. 2008. Physics of silicene stripes. arXiv:0811.2611v1.
4. Kara, A., Leandri, C., Davila, M. E., de Padova, P., Ealet, B., Ougaddou, H., Aufray, B., and Le Lay, G. 2009. Physics of silicene stripes. *J. Supercond. Novel Magn.* 22:259–63.
5. Guzman-Verri, G. G., and Lew Yan Voon, L. C. 2007. Electronic structure of silicon-based nanostructures. *Phys. Rev. B* 76:075131.
6. Fagan, S. B., Baierle, R. J., Mota, R., da Silva, A. J. R., and Fazzio, A. 2000. *Ab initio* calculations for a hypothetical material: Silicon nanotubes. *Phys. Rev. B* 61:9994–96.
7. Zhang, R. Q., Lee, S. T., Law, C.-K., Li, W.-K., and Teo, B. K. 2002. Silicon nanotubes: Why not? *Chem. Phys. Lett.* 364:251–58.

8. Yan, B., Zhou, G., Wu, J., Duan, W., and Gu, B.-L. 2006. Bonding modes and electronic properties of single-crystalline silicon nanotubes. *Phys. Rev. B* 73:155432.
9. Perepichka, D., and Rosei, F. 2006. Silicon nanotubes. *Small* 2:22–25.
10. de Padova, P., Quaresima, C., Perfetti, P., Olivieri, B., Leandri, M. E., Aufray, B., Vizzini, S., and Le Lay, G. 2008. Growth of straight, atomically perfect, highly metallic silicon nanowires with chiral asymmetry. *Nano Lett.* 8:271–75.
11. Gaspar, P., and Herold, B. J. 1971. Silicon, germanium, and tin structural analogues. In *Carbene chemistry*, ed. W. Kirsme, 504–38. 2nd ed. New York: Academic Press.
12. Sheka E. F. 2003. Violation of covalent bonding in fullerenes. In *Lecture notes in computer science, computational science—ICCS2003*, ed. P. M. A. Sloot, D. Abramson, A. V. Bogdanov, J. Dongarra, A. Y. Zomaya, and Y. E. Gorbachev, 386–98. Part II. Berlin: Springer.
13. Sheka, E. F. 2009. May silicone exist? arXiv:0901.3663v1 [cond-mat.mes-hall].
14. Sheka, E. F., and Chernozatonskii, L. A. 2010. Broken spin symmetry approach to chemical reactivity and magnetism of graphenium species. *J. Exp. Theor. Phys.* 110:121–32.
15. Lee, B. X., Cao, P. L., and Que, D. L. 2000. Distorted icosahedral cage structure of Si_{60} clusters. *Phys. Rev. B* 61:1685–87.
16. Slanina, Z., and Lee, S. L. 1994. A comparative study of C_{60}, Si_{60}, and Ge_{60}. *Fullerene Sci. Technol.* 2:459–69.
17. Lee, B. X., Jiang, M., and Cao, P. L. 1999. A full-potential linear-muffin-tin-orbital molecular-dynamics study on the distorted cage structures of Si_{60} and Ge_{60} clusters. *J. Phys. Condens. Matter* 11:8517–21.
18. Chen, Z., Jiao, H., Seifert, G., Horn, A. H. C., Y, D., Clark, T., Thiel, W., and v. R. Schleyer, P. 2003. The structure and stability of Si_{60} and Ge_{60} cages: A computational study. *J. Comput. Chem.* 24:948–53.
19. Golberg, D., Bando, Y., Tang, C., and Zhi, C. 2007. Boron nitride nanotubes. *Adv. Mater.* 19:2413–32.
20. Pokropivny, V. V., and Ivanovskii, A. L. 2008. New nanoforms of carbon and boron nitride. *Russ. Chem. Rev.* 77:837–73.
21. Szwacki, N. G. 2008. Boron fullerenes: A first-principles study. *Nanoscale Res Lett.* 3:49–54.
22. Rubio, A., Corkill, J. L., and Cohen, M. L. 1994. Theory of graphitic boron nitride nanotubes. *Phys. Rev. B* 49:5081–84.
23. Gao, X., Zhou, Z., Zhao, Y., Nagase, S., Zhang, S. B., and Chen, Z. 2008. Comparative study of carbon and BN nanographenes: Ground electronic states and energy gap engineering. *J. Phys. Chem. A* 112:12677–82.

chapter fifteen

Conclusion

This book, which is dedicated to the computational consideration of basic peculiarities that put fullerene nanoscience in a particular place, considers two basic problems, which are related not only to fullerenes but also to sp^2 nanocarbons. The former is connected with the odd electron character of their nature and addresses the weak interaction between the electrons. The latter originated with the strong donor–acceptor abilities of the species, which drastically influence the manner of the odd electron origin manifestation in due course of intermolecular interaction. The two features are tightly interconnected, which makes fullerene nanoscience so picturesque and attractive. The appearances of fullerene nanoscience peculiarities are quite extensive. They include nanochemistry, nanomedicine, nanophotonics, and nanomagnetism. They cover electronic and structural properties as well as a particular character of intermolecular interaction. An analysis of all this diversity can be done quantitatively, from the unified point of view, within the framework of the same scope of ideas and concepts.

Two concepts are used in an analysis of the peculiarities. The first one is the application of the broken spin-symmetry single-determinant Hartree-Fock approach. Providing a spin-contaminated solution of the odd electron problem, the approach manifests a high efficacy in determining pure spin states as well as in disclosing intimate characteristics of the species that are directly connected with the weakness of the electron interaction. The second appeals to the complicated structure of the terms of the potential energy related to intermolecular interaction caused by the contribution of the donor–acceptor interaction.

The basic initial information on the odd electron behavior that might be obtained in the UBS HF approach is contained in the extra spin density, D, with respect to that determined by the species spin multiplicity and exchange integral, J. The peculiarity of the situation lies in the fact that belonging to the spin-contaminated solution, these quantities retain their meaning and numerical values when passing to more sophisticated multideterminant solutions that include configurational interaction. The spin density is a measure of the odd electron unpairing and laid the foundation of the determination of the number of unpaired electrons, N_D. Within the framework of the UBS HF, the number is distributed over atoms so that the relevant fraction N_{DA} exhibits fractions of the unpaired electrons on each atom A. One of the main ideas of this book concerns the consideration of N_D and N_{DA} values as quantitative characteristics of molecular and atomic chemical susceptibilities, respectively. A practical realization of the idea leads to obtaining a chemical portrait of odd electron species that exhibits the targeting ability of each atom in regards to its ability to participate in an addition reaction, and thus to performing algorithmically clear computational synthesis of the reaction products. This idea led to the foundation of the book chapters related to nanochemistry and, partially, to nanomedicine. In contrast to spin density, revealing peculiarities in electron density distribution, the exchange integral J is related to the energy spectrum of the system of odd electrons. Once determined, it makes an allowance to fix the energy of pure spin states, and thus to show the way to quantitative characterization of the magnetic properties of the odd electron species. A practical realization of this idea has formed the basis for the consideration of the odd electron system magnetism presented in this book.

Another face of odd electron systems is connected with their high donor–acceptor abilities. The latter influences the intermolecular interaction of the systems quite considerably, leading to a multiwell structure of the potential energy surface in the space of intermolecular distances related to the ground state. The specialty of the structure concerns two regions of the well minima location, one of which is related to spacings characteristic for chemical bond lengths and attributed to the formation of new chemicals in due course of addition reactions, while the other is typically weakly bound to molecular species largely spaced from each other. In spite of the weakness, the interaction can be efficient enough to promote the species clusterization. Due to exclusive donor–acceptor (D–A) characteristics, a charge transfer occurs under the cluster photoexcitation. This specific clusterization led to the foundation of the nanophotonic features considered in this book. The latter involve photochemical reactions, photodynamic therapy, and enhancement of linear spectral and nonlinear optical properties.

Fullerene C_{60} has been the main object to which all the above concepts and ideas were applied, thus creating a conceptually unified computational vision of synergetic origin of the nanoscience of fullerenes. At the same time, addressing empirical reality has highlighted a few spots where the UBS HF vision turned out to be contradictory to experimental data. The discrepancy explanation had required some changing in the basic view on the molecule, thus forcing us to reject a conventional monoisomeric presentation of the molecule and take into account its extended isomerism, as well as to accept a continuous symmetry of its structure.

A general character of the odd electron concept has been illustrated by its application to carbon nanotubes and graphene and highlights a deep similarity in the behavior of all three species, just allowing speaking about a common computational nanoscience of *sp*^{2} nanocarbons. Addressing siliceous analogs of nanocarbons has confirmed a governing role of the distance between odd electrons that control the strength of the interaction between them, the extreme weakening of which leads to a complete radicalization of the *sp*^{2} nanosilicons, which makes them inaccessible in reality. However, the absence of odd electrons, which takes place in boron nitride analogs of nanocarbons, removes the latter of all peculiarities that are characteristic for carboneous species.

The author believes that the success in the interpretation and quantitative description of intimate properties of fullerenes as well as other *sp*^{2} nanocarbons in the framework of the UBS HF approach is quite convincing. It indicates that the basic assumption and models are correct, and that the theory behind it is efficient and heuristic. This allows enlarging the scope of the experiments that may be treated at the quantitative level. They have to cover phenomena that concern not only energy and wave functions of the ground state, but also those related to matrix elements that involve wave functions of both ground and excited states. Rather little is known about unrestricted approaches applied to the latter at the moment. However, there is a strong belief that the theory extension in this field will not only create a basis for a quantitative interpretation of a vast number of peculiar experimental data, but highlight spots pointing to a necessary revision of the presentations in use, as well as to bring to life new ideas, such as those concerning fullerene C_{60} isomerism and continuous symmetry of its structure, as in this book.

Index

F

I

T

Printed and bound by CPI Group (UK) Ltd, Croydon, CR0 4YY
21/10/2024
01777105-0005